The Physics of Stargates

Parallel Universes, Time Travel, and the Enigma of Wormhole Physics

Enrico Rodrigo

Eridanus Press

New York

Library of Congress Cataloging-in-Publication Data

Rodrigo, Enrico
The Physics of Stargates: Parallel Universes, Time Travel, and the Enigma of Wormhole Physics / Enrico Rodrigo

p. cm.

Includes bibliographical references and index.

ISBN-13 978-0-9841500-0-7 (pbk.)

1. General relativity (physics). 2. Time travel. 3. Quantum theory. I. Title
QC173.6.R637 2010
530.1'1—dc20

2010059893

Book design by Sharon Ross
Cover photographs © istockphoto.com

www.eridanuspress.com

Printed in the United States of America

10 9 8 7 6 5 4 3 2 1

Contents

Preface

Recently, I noticed an interesting television commercial. In it a young, overworked shipping clerk had hit it upon an idea. It would, he thought, solve his problem of shipping numerous packages quickly and safely to their various destinations. His idea, as you might have guessed, was to use wormholes as a delivery system. I found it interesting that the writers of the commercial were confident about a particular aspect of popular culture. They were sure that wormholes have become so familiar an idea that they need not be explained even to a general audience. Everyone has a rough sense of what they are. This permeation of wormholes into the general consciousness was likely accelerated by a fascinating phenomenon -- the broadcast within the last decade of no fewer than five popular television series thematically dependent on wormholes.

As the television commercial came to an end, the shipping clerk's enthusiasm for his brainstorm had diminished. He realized that he faced the daunting task of, as he put it glumly, figuring out "the whole spacetime continuum thing."

I wondered what this imaginary clerk might first read to introduce himself to wormhole physics in the least amount of time. I discovered that for this purpose the treatments found in extant books were either too advanced, or too thin, or too old. What he would need, it seemed to me, would be a collection of "bullet points" – facts about wormholes with as little interstitial material as required to render the facts intelligible. This is the sort of book that I have attempted to write.

These bullet points, which I elevated to the loft status of numbered "principles", require little or no mathematics to understand. Any mathematical equations that do appear in the text may be ignored with little penalty. One way to read this book is to read the principles first. The justifying verbiage may then be read for any particular principles of interest. Principles in the preliminary sections introducing general relativity and quantum theory are labeled "***PP***" (Preliminary Principle) to distinguish them from the wormhole-specific principles that are labeled simply "***P***".

The glossary is unusually comprehensive. In addition to defining terms used in this book, its purpose is to assist you in general readings in wormhole physics. Because I favor broad strokes to the point of using paint roller instead of a brush, the main text does not dwell on conceptual subtleties. The difference, for example, between causality violation and chronology violation appears only in the glossary. Those interested in seeing some of the equations caricatured in the text will find them in the appendix.

The principles are written in as plain a manner and with as little qualification as possible. Specialists will undoubtedly cringe at boldly asserted, insufficiently qualified statements. For our purposes, however, a simple statement that is usually true is better than a complicated one that is always true. The imprecision, then, of plain language and the absence of complicated qualification renders some of the principles "true" only in a generous sense. For tortuously rigorous versions of a principle, please consult an advanced text or the scientific literature.

The bulk of the chapters result from my attempt to distill the contents of seminal papers, a fifteen-year-old monograph, and a sampling of the hundreds of papers written more recently. Any errors that appear in descriptions of research are entirely mine.

Although I have chosen not to litter the text with reference numbers, the explicit mention of researchers and associated dates should quickly identify the relevant references in the list of sources.

Besides wildly ambitious shipping clerks, this book is also aimed at dentists, retail managers, engineers, musicians, students, business people, homemakers, military personnel, accountants, actors, executives, physicians, sales people, teachers, cooks, writers, politicians, science fiction fans or anyone else interested in a short, high-level summary of wormhole physics.

I would like to thank Dr. David Deutsch for countless hours of enlightening conversation over the years, the late Professor John Archibald Wheeler for allowing me to experience his understated brilliance first hand, and especially Tina Schifani for her constructive review of the first draft, encouragement, and guiding wisdom.

Enrico Rodrigo
2009

1. Introduction

Why *Wormhole* Physics?

Why would anyone who is not a science enthusiast of the most ardent order want to read a book about *wormhole* physics? Perhaps you came across a television show in which characters were stepping through a wormhole-based "stargate" and thereby traipsing throughout the galaxy. Or you encountered other televised versions of characters similarly moving quasi-instantaneously through space or time (*Farscape*, *Sliders*, *Star Trek*, *Time Tunnel*, *Doctor Who*, *Babylon 5*, *Lost*, *Primeval*, *The Event*). Or you might have seen a movie depicting this sort of thing (*Jumper*, *Déjà Vu*, *Timeline*, *Galaxy Quest*, *Event Horizon*, *Contact*, *Stargate*), or a documentary series (*Through the Wormhole with Morgan Freeman*), or read a novel (*Contact*, *Timeline*, *The Light of Other Days*, *Einstein's Bridge*, *Diaspora*). So you might have wondered, "How much of this is based in reality? Is wormhole travel possible, even theoretically?" If so, this book is for you.

If not, you might want to consider another reason for taking a look at it. Let me explain it with a few incidents from my own childhood.

When I was six, I was, like many children of that age, a firm believer in the existence of a "highest number". That such a number existed seemed perfectly obvious to me. Any language such as English consists of a limited number of words. Therefore, there could only be a limited number of words with which to name particular numbers. The sequence "million", "billion", "trillion", "quadrillion", etc., could not go on forever. The English language would run out of such names for large numbers. Hence, at any given time there is a "highest number" – the largest number for which there is a name. When I learned that I was wrong, that numbers can exist without having English names, that particular numbers can even exist without first having been held in anyone's mind, and that there is an unlimited quantity of such numbers, it changed my life. Ultimately, it switched my orientation from religion to science.

At about the same age I also believed that there were only two types of problems – those whose solution was obvious, and those that had no solution. This seemed reasonable to me, because whenever anyone asked me a question, I either knew the answer, or I didn't. When I soon learned about arithmetic and later algebra, I was amazed to realize that there exists another huge class of problems – those whose solution was not obvious *but could be worked out.* Understanding this widened my worldview -- drastically, permanently, and unforgettably.

My mother used to bring home large sheets of butcher paper for my siblings and me. We would spend hours scribbling crayon drawings on them. When I drew the inevitable scenes of the earth and other planets in outer space, the finiteness of the butcher paper posed a deep cosmological question that I would from time to time discuss with my brother. Does the universe, like the butcher paper, have an edge, or does it go on forever? It seemed clear to both of us that the universe could not have an edge for reasons of aesthetics if nothing else. It must be spatially infinite. Yet years later, upon entering high school, I learned another astounding fact: It is possible for the universe to be spatially *finite* and *not* have an edge. An aesthetically pleasing edgeless universe did not, I was amazed to learn, have to be spatially infinite.

These are three personal examples of what a 1960s hippy might pompously have called "mind expansion". Though a refugee from that era would use this term to refer to the experience of drug-induced hallucinations, I prefer to think of it as the sort of mental restructuring that occurs when fundamental assumptions long held to be obviously true are acknowledged instead to be false. Such restructuring occurs when you travel to a foreign country and observe the surprisingly efficacious operation of certain alien customs and modes of living. It likely occurs when you read a book whose author lived in a distant time or place. It also occurs when you read about fundamental aspects of reality about which you know little. It happens to me whenever I read pretty much anything outside of that tiny sliver of human knowledge that I know well.

Learning about an obscure subject -- be it Bactrian law, Aztec religious beliefs, or wormhole physics -- is of practical value. It enhances creativity, offers fresh perspectives on your problems, and thus leads to solutions that would otherwise not have occurred to you. Any book on unfamiliar physics, including this one, could furnish your mind with concepts that it does not yet possess. Some of these will conflict radically with your common sense. Your mind would require a bit of restructuring in order to accommodate them. It is not the mere awareness of these concepts that leads to significant restructuring, however. It is the understanding of how these concepts follow from known physical law. It is this understanding that results in the sober recognition that these concepts are not merely fanciful notions, but are instead likely representations of the actual nature of fundamental reality. Wormhole physics is particularly useful in this regard. Here the deepest problems of modern physics come home to roost. Wormholes not only force the consideration of time travel, but also that of travel between parallel universes. They demand a reevaluation of the fate of intelligent life in the universe of the distant future. They appear in attempts to unify the physics of the cosmos with that of the atom. They intrude on considerations of the truth of the principle of the indestructibility of information. To study wormholes, then, is to survey the great controversies of contemporary physics. It is, moreover, to confront startling implications for religion, ethics, and the future of humanity. Let us begin by stepping back to consider the implications for Western civilization of the existence of wormholes.

Wormholes and the West

Western culture now faces its greatest challenge. No, I don't mean the threat of Islamic jihad, or the decline of moral values, or global climate change, or any of the other commonly adduced hazards or ailments. I refer to something of far greater consequence. It's something more fundamental, the secular equivalent of what might be called a crisis of faith.

It Started with Thales

Let's begin at the beginning – Greek Ionia of the early 6th century B.C. Then it was universally believed that all physical phenomena were due to the whims of various gods. Thunderstorms were a manifestation of the ire of Zeus. A turbulent sea was a reflection of Poseidon's mood. In this intellectual environment a philosopher named Thales of Miletus hit upon a startling new idea. As he put it, "All things are full of gods." Rocks, grass, tables, water, each contained some sort of god. But these gods were different. They were not akin to the inscrutably temperamental personalities that occupied the Greek pantheon. The gods inhabiting rocks and grass were much simpler. So simple that with sufficient effort they could be fully understood. With this unique idea -- that "all things", the whole of reality, can be understood by the human mind -- Thales laid the cornerstone for Western culture.

Thales' radical idea has come to be known as "rationalism". Its prevalence in Western culture is the West's chief distinguishing characteristic. Understand that this sort of *philosophical* rationalism differs from *psychological* rationalism, the latter being tantamount to sanity. To say that rationalism is uniquely prevalent in the West is not, of course, to imply that the inhabitants of all other civilizations tend to be insane. Rather, it is to aver that only in the West was the mysticism that is at the heart of all cultures seriously challenged by a countervailing idea.

Creeping Naturalism and Thales "Lite"

The effect in Western culture of mysticism -- the Greek form of ancient times being later replaced by one of Jewish origin – has been to soften Thales' doctrine. This weakened version held that humanity can understand *much* of reality, but not *all* of it. That is to say, there exists a sector of reality that we cannot understand even *in principle* – irrespective of how many billions of man-years of effort we devote to attempting to do so. We call this unfathomable sector of reality "the supernatural". The weak version, then, of Thales' doctrine of rationalism leaves room for the supernatural to exist. The strong version asserts that it does not. Both versions agree, in contradistinction to mystical beliefs, that reality's underlying patterns are largely intelligible to the human mind.

The weak version of Thales' doctrine dominated Western thought throughout most of the West's history. Western intellectuals understood reality to have two distinct sectors: a natural one comprehensible to man and a supernatural one that is not. It was not until the 17th century that a serious problem with this dichotomy began to be noticed. Isaac Newton discovered that the same laws governing the interaction between apples and the earth governed that between celestial bodies. This meant that the behavior of the celestial sphere, long held to be part of the supernatural sector, was suddenly comprehensible. The supernatural sector shrunk, as the natural (comprehensible) sector expanded into the cosmos. The 19th century saw further reductions in the size of the supernatural sector. Increased understanding of the chemical basis for life destroyed the prevailing belief in vitalism, a supernatural life force. Moreover, the character of life on earth came to be understood to be a consequence of natural selection instead of one of divine deliberation.

It's not hard to see why the supernatural sector of reality has tended to shrink over time. It is because the natural and supernatural sectors come into contact. At the boundary between them, our understanding of the natural sector spills over into the supernatural sector. Once this occurs, the supernatural sector near the boundary ceases by definition to be supernatural, because it is now understood. If the effects of gravitation were confined to the surface of the earth, or if chemical laws did not operate within living cells, the natural and supernatural would not in these cases have come into contact. The cosmos and the essence of life would have remained within the province of the supernatural.

No one has found a way to stop this creeping naturalism and the corresponding shrinkage of the supernatural sector. This will not change as long as there exists any interface between the natural and the purportedly supernatural sectors of reality, i.e. as long as entities in the supernatural sector are believed to be able to affect us. This means that the weak form of Thales' doctrine of rationalism is untenable. We cannot reasonably assert that reality is only partially comprehensible. It is impossible to draw a stable boundary between what we can understand (the natural) and what we cannot (the supernatural).

Thales Undermines Himself

So we are left with the strong form of the doctrine: we can in principle comprehend the whole of reality[*]. Here we take "in principle" to mean that the laws of physics permit this comprehension. Astoundingly, the laws of physics -- as we have only recently come to understand them -- do *not* seem to permit this. This is the crisis of the West: Our own rationalism, which led to the development of physics, has in the end undermined itself.

Here is the problem. In order for the universe to be understood, some subsystem within it, such as a computer or a human society, must be able to contain that understanding. This understanding will take the form of a maximally concise yet faithful simulation of the universe running on the subsystem, which we shall henceforth take to be a future civilization. The better it understands the universe, the more faithfully and least expensively will this civilization be able to simulate it. As the complexity of the universe relentlessly increases, so too must the quantities of mass and energy that a civilization must devote to the task of understanding it, i.e. of concisely simulating it. The speed with which this civilization renders its simulation must, moreover, allow it to keep pace with the growth of complexity in the universe. Otherwise, this

[*] This ancient doctrine has found rigorous formulation in the form of the Turing Principle. See David Deutsch, *The Fabric of Reality* (Penguin 1997), p 131ff.

civilization's simulation would become progressively less complete. When the universe was believed to be destined to contract toward a "Big Crunch" – an infinitely dense fireball at the end of time – it was possible to accommodate the requirements of this future civilization. The increasing mass and energy densities of the universe would support the growing mass-energy needs of the simulating civilization. The ever decreasing size of the universe would induce a corresponding contraction in the physical dimensions of the civilization. This would ensure ever increasing processing speeds, as the maximum time required to communicate with any of the parts of the physically contracting civilization would continually fall. Its ever increasing processing speed would allow the civilization to perform an infinite number of processing steps, to think an infinite number of thoughts, before the Big Crunch arrives. Even when this final event is only a few seconds away according to conventional clocks, it will seem to members of this civilization with its ever accelerating speed of thought to be infinitely far in the future. They will have a subjectively infinite amount of time in which to understand the whole of reality. An example of a Big-Crunch-dependent cosmological model that assures that humanity's successors will thus enjoy unlimited understanding is the "Omega Point" scenario, which physicist Frank Tipler posited in the early 1980s. A Big Crunch, however, is no longer believed to be in our future.

The Shock of 1998

In 1998 two independent teams of astronomers jointly delivered a bombshell. Their observations showed that the expansion of the universe is not only failing to slow down, it is actually *accelerating*. Instead of evolving toward a Big Crunch the ultimate destiny of the universe now appears to be a "Big Freeze". This relentless expansion drives the energy density of the universe ever downward (at an accelerating rate), even as the complexity of the universe continues to rise. This means that a civilization attempting to simulate the universe would find perpetually decreasing quantities of mass and energy locally available for this purpose, despite its rising need for them. Once it has reached the limit imposed by the laws of physics on computational power per unit volume, the civilization's computational resources would necessarily spread. As the spatial volume occupied by the civilization expands, the rate of growth of its processing speed would decline, as the time required to communicate with its ever separating parts would rise. The civilization's simulation, its understanding, would eventually fall increasingly far behind the universe that it seeks to model. Moreover, vast quantities of the detectable universe will continually vanish, as the expansion takes these regions beyond the range of observation. The civilization would face strict time limits on its study of objects with these vanishing regions. The expansion of the universe thus prevents any civilization from simulating the universe arbitrarily well. It will in effect induce an ever worsening retardation in the growth of their collective mental capacity. It will only allow the civilization to perform a finite number of processing steps and study certain objects for a limited time before its ultimate demise. Unlike denizens of a Big Crunch universe, members of this civilization will see Doomsday as a fixed appointment looming balefully in their future. They will be unable to solve many of their problems, because they will simply run out of time. Put another way, the accelerating expansion of the universe invalidates the only tenable version of Thales' doctrine of rationalism. Humanity cannot understand the whole of reality after all, not even in principle.

Irrespective of any retarded growth in their collective mental capacity, the accelerating expansion of the universes, of the sort consistent with astronomical observations, dooms Humanity in a similar way. It will limit its supply of energy. Sources of energy beyond our cosmological horizon -- the radius at which the expansion-induced recession of an object from Earth exceeds the speed of light – are inaccessible. A limited supply of energy implies a limited quantity of computation – a limit on the total number of thoughts that Humanity can think. Again, indefinitely postponing Doomsday will not be an option.

Humanity might nevertheless attempt to postpone its demise by conserving its finite quantity of available thoughts – by reducing the rate at which it computes. Although this will postpone Doomsday according to a hypothetical external clock, it will have no effect whatever on Humanity's subjective impression of the proximity of its ultimate demise. This futile stratagem will moreover be thwarted by a lower limit below which Humanity, or whatever intelligence comes to dominate the future universe, cannot slow its thinking. It cannot think so slowly that its effective temperature sinks below the ambient temperature of the universe, lest it be unable to discard its waste heat. The latter temperature is the so-called "Hawking temperature", which turns out to be proportional to the rate at which the universe expands. Such an attempt to further slow its collective rate of thought increases Humanity's exposure to unforeseen threats emerging from the rising complexity of the universe.

Even at the current complexity of the universe, we are faced today with difficult threats to our survival that we may not have time to solve. We will not, for example, survive if we do not eliminate the threat of large scale basalt lava flows before the next one poisons our atmosphere. The same can be said of the other threats destined to befall us: the asteroid with our name on it, the massive gamma ray burst scheduled to incinerate us, the hyper-lethal global pandemic that we will accidentally unleash, the super volcanoes due to erupt, the human faction bent on catastrophic malevolence, or our sun's appointment to go nova. Our descendants, if we survive to produce them, must similarly overcome their own challenges – most of which are unimaginable by us today. The deadlines by which they must do so become all the more worrisome, if they are to be burdened as well by time limits on their study of objects disappearing behind the cosmological horizon, strict limits on available energy, and an ever worsening retardation in the growth of their civilization-wide computations, their collective capacity for thought.

Stargates to the Rescue?

What does this have to do with stargates? A stargate is a device in science fiction that uses a peculiar warping of space and time called a "wormhole" to create a shortcut between distant locations in space. A journey between two galaxies, for example, could require billions of years of travel through normal space at the speed of light. The same trip could be accomplished through a wormhole in a fraction of a second. Wormholes, then, annihilate distance. Because of this we might be able to use them to compensate for the expansion of the universe. Wormholes could nullify the ever increasing distances between points of expanding space. An ever increasing density of wormhole-exploiting stargates could, moreover, overwhelm the expansion by creating an effective contraction of space. They would shorten the distances between a growing number of computational resources, import energy from beyond cosmological horizons and parallel universes, and eliminate time limits on the study of comic objects receding out of view. The expansion would then cease to progressively retard the collective mental processes of future civilizations, thus leaving their capacity for understanding unconstrained. This stargate-enabled effective contraction will permit our descendants to perform an infinite number of processing steps before the Big Freeze. It will seem to them to be postponed to the infinitely distant future. They will have an unlimited time to develop an unlimited understanding. Stargates, then, could rescue Thales' idea and the foundation of Western culture.

What Thales Demands of Nature

In order for wormholes to more than compensate for the continued expansion of the universe, they would have to be ubiquitous and proliferative. Every microscopic region of space would require a wormhole connection to virtually every other. But we don't observe macroscopic wormholes or even mysterious disappearances or appearances of subatomic particles. So a profuse network of wormholes interconnecting all regions of space, if it exists, would have to be comprised of wormholes with exceedingly tiny mouths. The only natural scale for these mouths is that of the so-called Planck length -- the shortest possible distance defined within any conceivable theory of quantum gravity, the long sought-after marriage between quantum theory and Einstein's general relativity. The idea of tiny Planck-scale wormholes and other geometric anomalies flourishing beneath our notice originated with the pioneering physicist John Archibald Wheeler. Wheeler likened the spacetime that we perceive to an ocean viewed from an airplane at high altitude. Just as an ocean appears perfectly smooth from a cruising airliner, spacetime looks smooth at the relatively large length scales within our ability to resolve. Just as the ocean, viewed up close, becomes a roiling surface of breaking waves and foam, so it is with spacetime. Wheeler's idea – that spacetime at the Planck scale (i.e. "viewed up close") appears to be a hectic medley of every conceivable geometry – has come to be known as "spacetime foam".

If wormholes are to validate Thales' doctrine, then there must exist a profuse network of Planck-scale wormholes interconnecting the whole of spacetime. Such a network would most easily be explained as a remnant of the quasi-singular geometry of the big bang viewed within the context of spacetime foam. Instead of imagining the big bang as a tiny, perfectly smooth hypersphere (the 3-dimensional analog of sphere) of arbitrarily dense matter and energy explosively expanding to form the current universe, consider a minor variation on this theme. Imagine the universe at the instant of the big bang as a tiny, hot, dense, inchoate blob of spacetime foam. Its geometric structure and even its dimensionality are undetermined. Spatial relationships within this blob have yet to form. Every piece of it is equally connected to every other. It is the blob's explosive growth that shapes it into the smooth and familiar hypersphere of conventional cosmology. It nevertheless retains a vestige of its previous self connectedness. A dense network of filamentary, Planck-scale wormholes interconnects every location in the hypersphere to every other.

The presence of this hidden network of submicroscopic primordial wormholes would promise near instantaneous travel across the universe to any civilization advanced enough to exploit it. Physicists Michael Morris and Kip Thorne, in a paper that launched the "modern age" of wormhole physics, imagined that an "exceedingly advanced civilization" might accomplish this by somehow enlarging such wormholes for their use. If a hidden network exists, Morris and Thorne's civilization could use these wormholes to annihilate distance, to nullify and even reverse the retarding effects of the relentless expansion of the universe. This civilization could thereby demonstrate its understanding of reality by simulating it arbitrarily well. It could prove that Thales was right.

Given the reality of an ever expanding universe, then, there seems no way to escape the following conclusion: Unless stargates can harness a network of primordial wormholes, or unless some other distance-annihilating technology can exist, we must concede that Thales was wrong, some version of the supernatural exists, and the foundation of Western culture is fundamentally flawed.

The Non-Western Option: Abandon Thales

Why not admit this? Why not acknowledge that we can only understand a small and (eventually) an ever decreasing fraction of reality? Surely this is better than embracing the existence of universe-spanning primordial wormholes for which we have no evidence. The reason that we should not yet abandon Thales' doctrine is that it is always more fruitful to follow our fundamental theories to whichever conclusion they lead us. It is premature to reject any of our fundamental theories until we have rejected the conclusion derived from them. Any articulation of the epistemological principles supporting Western science would necessarily pronounce our belief in the strong form of rationalism to be fundamental. Our belief in the non-existence of primordial wormholes, by contrast, is not. If we believe in the strong form of the doctrine of rationalism, in the validity of the experiments detecting an accelerating expansion of the universe, in general relativity and in other well established physics, then our beliefs taken together lead us to conclude that it must be possible to annihilate distance. Either it is possible for stargates to accomplish this by exploiting primordial wormholes, or some other form of distance-annihilating technology must be possible. Until we are convinced that this conclusion is false, we are compelled to retain our confidence in each of the beliefs from which this conclusion follows, including our faith in the strong form of Thales' idea.

Stargates Our Only Hope?

There is reason to believe that if stargates are impossible, so is any other form of distance-annihilating technology*. If this is true, then the logical integrity of a culture inhabiting the minds of hundreds of millions of Westerners depends on whether stargates – synthetically controlled wormholes – can exist. Unfortunately, more is at stake than cultural integrity. If stargates are impossible, the consequences will be dire -- far worse than the trauma that secular Western rationalists will experience in grudgingly admitting the existence of an effec-

* Alternative means of superluminal travel such as "warp drive" share the same requirement as stargates – exotic matter.

tive supernatural, a sector of reality that they can never hope to understand. If stargates are impossible, the descendants of humanity are doomed.

Their first challenge will occur long before they notice the accelerating expansion of the universe to be a problem. They will find it impossible to communicate quickly over the vast stretches of space over which their civilization will inevitably extend. Communicating a message across a single galaxy such as our own, for example, will take 100,000 years. They will attempt to combat this time delay by concentrating their civilization within as small a region as their technology will permit. Although the maximum population density permitted by their technology will grow with time, the growth of their population (or to be more precise, that of its volume-consuming computational resources) might well be faster. Their civilization will be forced to spread outward, communication rates will slow, and the development of their civilization-wide computations will be retarded. Here, too, stargates are an answer. In the absence, however, of such a distance-nullifying technology, our descendants will eventually experience a universe expanding too rapidly for them to cope. It will thwart their best efforts to concentrate their computational resources. It will prevent them from moving brilliant (artificial) minds residing in the "provinces" to the center of their transgalactic civilization. The reason is simple. The effective rate of separation between the most distant provinces and the center will exceed the speed of light.

Well after the rate of cosmic expansion thereby becomes a problem, the increasing scarcity of local matter and energy and the ever slowing growth in the speed at which our descendants can collectively process information means that their simulation of reality will not keep pace with reality itself. In other words, the fraction of reality that they understand will continually decline. Their growing ignorance will put them at growing risk. New threats to their existence will continually emerge from the sector of reality that they do not understand. Failing to understand these threats, they will be helpless to defend themselves against them. They might survive a few of these threats, but eventually their luck will run out. Humanity, irrespective of its future physical form, will be destroyed.

But won't the expansion of the universe also slow down the rate at which its complexity grows? Yes. Then won't that allow a civilization seeking to simulate that complexity to keep pace? No. The universe merely needs to execute its physical laws, while the civilization needs to simulate them. Execution requires a single communication between any of the universe's interacting components. And, yes, the rate at which these communications occur will slow as these constituents separate. Simulation, however, requires at least two communications for each pair of interacting constituents – one to fetch the physical laws from wherever in the civilization they are stored, and at least one other to ensure that the variables in the simulation representing these constituents implement the physical laws. So the simulation is at least twice as sensitive to communication delays resulting from the expansion of the universe. Initially, the simulation will run many times faster than the reality that it simulates. As the universe expands, the growth of complexity in reality will slow, but the simulation will slow even more. Eventually, perhaps after hundreds of trillions of years, the fastest possible simulation will fail to simulate reality in real time. After that, humanity's descendants will be exposed to a genuine supernatural – a sector of reality that they will never be able to understand. And this ignorance will ultimately kill them.

What deadly threats might emerge from this supernatural sector? We cannot know. One possibility, however, would be that of malevolent beings or alien civilizations whose physical nature – based, for example, on nuclear chemistry rather than organic chemistry[†] – would allow them to think and evolve many orders of magnitude faster than the human capacity -- too fast for the descendents of humanity to ever hope to understand the descendents of such beings. Another possibility would be that of a cosmic environmental disaster -- resulting from the complex interplay of various aspects of exceedingly subtle physics – that would strike before our descendants could acquire the knowledge needed to prevent it from wiping them out. Yet another scenario for their demise could be an exceedingly intricate and rapidly mutating descendant of the computer viruses of today – a swiftly spreading disease for which our descents might have insufficient mental capacity to defeat before it terminates them. They would moreover face the threat of unfathomably malevolent factions and individuals in their midst.

They will attempt to alleviate their plight through a thorough reliance on quantum computation, effectively shifting more and more of their computational load to parallel universes. Although this will postpone their demise, it will not prevent it. The rate of communication between the quantum computers in their own universe, each of which will be performing a distinct part of a civilization-wide computation, will continue to slow. Our descendants will find themselves unable to spread the whole of this computation across parallel universes without so spreading their civilization. To do so, they would need wormhole bridges between parallel worlds. They would need stargates.

[†] In 1980 the late Robert Forward explored the possibility of such life in his novel *Dragon's Egg*.

Our descendants might similarly recognize the need to enlist the computational aid of other civilizations in distant regions of the universe in order to solve their common problem of survival. These civilizations will in general have developed independently in regions of the universe out of causal contact with our own region and with each other. To reach them, our descendants would require wormholes. They would, again, require stargates.

Stargates, then, or some other distance-nullifying technology is their only hope.

Even if our descendants are extremely lucky, and the growing complexity of the universe presents them with no new threats to their survival, they nevertheless face an ultimate challenge. They must somehow survive in a frigid universe of declining energy density, whose ever falling temperature relentlessly approaches the frigid Hawking Temperature. They must cope with the Big Freeze. Here, again, stargates could help. Our descendants could use them to import energy from less frigid parallel universes or to escape to such universes, including those corresponding to their own past.

Whether future civilizations use stargates to nullify distance, import energy, or to escape to other universes, they will use them. Or they will perish.

Why Care?

You might find it strange to be concerned with the fate of the advanced civilization that will occupy the universe a trillion years hence. But I cannot help it. Having become comfortable with the expectation of a universe inhabited by beings of ever increasing knowledge and power, it is disconcerting to discover that in the context of post-1998 cosmology such beings are doomed. I find myself disturbed by the awareness of this ultimate disaster, which unlike the Big Crunch, cannot (in the absence of stargates) be postponed indefinitely in the subjective experiences of the universe's future inhabitants. Before 1998 it was possible to regard the universe as essentially a knowledge generator – although a highly wasteful and extremely inefficient one – that was destined to continue for a subjective eternity. In its production, though inefficient, of intelligent life and in its requirement that such life rapidly and continually acquire knowledge in order to survive, the universe seemed to have an ultimate purpose. Now, with the Big Freeze as its denouement, it does not. If this realization is unlikely to cost you sleep, you will be pleased to know that wormhole physics is interesting apart from its eschatological implications. Not only do wormholes lead immediately to time travel and parallel universes, but they also enable communication with these universes and, moreover, support what has hitherto been regarded in physics as a preposterous possibility: that of travel between them.

The Bottom Line

In summary, the fact seems to remain that the fate of the distant intellectual descendants of humanity -- or that of whatever intelligent life comes to dominate the universe -- depends upon whether stargates or equivalent technology can exist. Without the effective existence of stargates our ultimate destiny is extinction – both physically *and* intellectually.

==/==

On this dramatic note we begin our survey of wormhole physics. We shall see that there are obstacles to the existence of stargates. Our aim is to understand them. We shall thereby gain insight into the validity of Thales' doctrine, the future of Western culture, and the ultimate destiny of the human project.

Questions and Answers about Wormholes and Stargates

It's normally considered good form when writing for a lay readership to dribble out scientific concepts amidst dramatic tales of discovery filled with fascinating anecdotes and vivid accounts of the emotions of the scientists involved. Personally, I hate that. Although it's nice to know what a scientist had for breakfast on the day of her big breakthrough, I'd rather just learn about the relevant science as directly and succinctly as possible. If you're of like mind, this section is for you. At the cost of a bit of redundancy, each question-answer unit is more or less independent of the others. Jump in anywhere. Read as many as you like, or, if you prefer, skip this section entirely.

What is a wormhole?
A wormhole is a short passage in spacetime that directly connects two universes or two distant regions within the same universe.

Two universes? How can there be more than one?
Understand that the word "universe" in this context does not mean "all that exists". Rather, it means "an independently existing spacetime and its contexts". To imagine two universes, think of two toy balloons floating near each other. A traversable wormhole connecting them would be like a short drinking

straw glued between them that would allow an ant to crawl from one balloon to the other.

Is there a more precise definition of a wormhole?
Yes. In normal space it is always possible to shrink any closed surface down to a point. A wormhole is a region of space containing a closed surface for which this is not the case. More correctly, a wormhole is a region of spacetime containing a "world tube" (the time evolution of a closed surface) that cannot be continuously deformed (shrunk) to a world line.

What is a black hole?
A region of intense gravitation characterized by a central point or ring of infinite energy density called a "singularity" and an enveloping surface called an "event horizon" from which nothing – not even light – can escape.

How are wormholes related to black holes?
Unlike a wormhole, a naturally occurring black hole -- one created through stellar collapse -- is not a bridge between two universes (or distant regions within the same universe). There nevertheless exist certain solutions to the Einstein equations of general relativity in which a bridge between universes – a wormhole -- appears to have a black hole at either end. This is the sense in which certain theoretically possible black holes can be said to be wormholes.

Can a wormhole with a back hole at either end be traversed?
No. The bridge between universes remains open for too short a time for any traveler to cross it.

Is there another way in which wormholes are related to black holes?
Yes. They are also related through inter-conversion. Were a black hole to exist as an untraversable wormhole, it could be converted into a traversable wormhole by dropping enough negative mass-energy into it. It is similarly possible to convert a traversable wormhole into a black hole by showering it with enough positive mass-energy. Unlike static traversable wormholes, black holes always contain singularities – regions of infinite mass-energy density, where classical physics breaks down.

Could I traverse a wormhole with a black hole at either end, if I could somehow exceed the speed of light?
Yes. Assuming, of course, that the black hole is large enough to allow you to tolerate its gravitational tidal forces as you approach and pass through. [The more massive the black hole, the weaker its tidal forces in the vicinity of its event horizon.]

What is a white hole?
A time-reversed black hole. Nothing can escape a black hole, but nothing can enter a white hole. The event horizon of a black hole prevents objects from exiting; the event horizon of a white hole prevents objects from entering. The singularity of a black hole absorbs whatever has entered its event horizon; the singularity of a white hole has emitted whatever exits its event horizon. The event horizon of a black hole is highly stable; the event horizon of a white hole is highly unstable. Black holes are believed to exist in nature; white holes are believed to exist only as parts of certain solutions to the Einstein equations.

How are wormholes related to white holes?
There are classic solutions to the Einstein equations -- called maximally extended solutions -- that describe untraversable wormholes. The mouths of these wormholes may be described as white holes that become black holes. Any external observer who views a mouth will see the white hole that it was in the past. When she approaches the mouth, she will find the black hole that it is in the present.

What is an Einstein-Rosen bridge?
A solution to the Einstein equations published in 1935 by Albert Einstein and Nathan Rosen in an attempt to create a gravity-based model for an elementary particle. This solution is today called "the maximally extended Schwarzschild solution". It describes a dynamic wormhole connecting two universes. In each universe the mouth of the wormhole appears to be a black hole that was previously a white hole. As John Wheeler and Robert Fuller showed in 1962, this wormhole cannot be traversed. Its throat constricts too rapidly.

What is an Einstein-Podolsky-Rosen bridge?
A misnomer. In 1935 Einstein, Podolsky, and Rosen proposed a famous thought experiment intended to expose what they believed to be a paradox in quantum theory. The Einstein-Podolsky-Rosen (EPR) paradox has nothing to do with wormholes and, incidentally, is no longer regarded as paradoxical.

What is the difference between classical physics and quantum physics? Which one do I need to understand wormholes?
You need both: classical physics for macroscopic wormholes, quantum physics for microscopic ones. Quantum physics is the most accurate description of physical reality known. It applies in both macroscopic and microscopic realms. Classical physics, by contrast, is merely a convenient approximation to quantum physics that is only accurate for systems of macroscopic objects (i.e. systems for which the products of measured momentum and relative

position always well exceed a particular fundamental constant of nature -- Planck's constant).

What is general relativity and what does it have to do with wormholes?
General relativity is the theory formulated by Einstein that describes gravitation as curvature in spacetime induced by the presence of matter or energy. It is used to understand physical systems in which gravity is too strong for the Newtonian theory to accurately describe. Such systems include wormholes.

What is a stargate?
In science fiction, a synthetically controlled wormhole. We shall also use the term to describe a wormhole whose throat is a planar surface in the sense that it does not appear to enclose a volume (e.g. a disk and a square are both planar, unlike the surfaces of a sphere or a cube).

What is the origin of the word "stargate"?
Arthur C. Clarke used the term "star gate" to describe the interstellar shortcut featured in his novel upon which the 1968 movie *2001: A Space Odyssey* was based. Common use of the contracted form "stargate" followed the release of the 1994 movie of the same name and its subsequent extension to television.

What is a traversable wormhole?
A wormhole through which human beings can repeatedly travel unharmed in a time short compared to the human life span.

Do wormholes have event horizons?
In general, no. They only have an event horizon – a surface surrounding a region from within which nothing can escape -- if they are black holes. Two-way traversable wormholes cannot have event horizons.

Do traversable wormholes have singularities?
Static traversable wormholes do not. Some dynamic traversable wormholes do. The singularity doesn't sit inside these dynamic wormholes. It's what the wormhole will become in the future. If the wormhole is traversable, the only way its singularity will cause a problem for a traveler is if he parks his spaceship at the throat and waits (perhaps years) for the wormhole to collapse into a singularity.

What about untraversable wormholes, do they have singularities?
Yes. These are called maximally extended black hole solutions to the Einstein equations. Unlike naturally occurring black holes, which don't connect to

anything, these form untraversable bridges to other universes or distant regions within the same universe.

Have any naturally occurring wormholes been discovered?
No. If they do exist, they are likely to be the result of primordial microscopic wormholes being inflated to macroscopic size during the inflationary phase of the universe's development.

If no wormholes have ever been discovered, why should we consider them?
The best way to test and extend our theories of nature, especially in the absence of experimental data, is to check their logical consistency in extreme hypothetical cases. This, after all, is how Einstein discovered relativity. Wormholes are examples of such cases. They are predicted by an extremely well tested physical theory, general relativity. If we believe this theory, then we believe that they can exist. If they can exist, and if circumstances conducive to their creation and maintenance have occurred, they do.

How could a macroscopic wormhole arise naturally?
The primordial universe might have spawned, through a process known as "quantum tunneling", unstable cosmic wormholes that have been expanding along with the universe. If the accelerating expansion of the universe first detected in 1998 is due to cosmic exotic matter, such matter might somehow have expanded the submicroscopic wormholes believed to be contained in the vacuum state of spacetime.

What hazards are encountered in attempting to traverse a wormhole?
A traveler attempting to traverse a wormhole might be ripped to death by gravitational tidal forces, be incinerated by high radiation emitted near a singularity, be damaged by contact with exotic matter, or die of old age in transit. He might also become trapped within the wormhole or on the distant side of it, after he inadvertently induces its collapse by attempting to pass through it. It is in principle possible, however, to engineer a wormhole that alleviates each of these hazards arbitrarily well.

Could wormholes be used for interstellar communication?
Yes. It would in fact be much easier to create a wormhole that permits the passage of message-carrying light signals than it would be to create one that ensures the safe passage of human travelers.

How difficult would it be to create a wormhole?
The ability to create a traversable wormhole is well beyond current human technology. It would require the enlargement of one of the many submicroscopic quantum wormholes believed to exist within any volume of space. The process would likely require an intense, ultra-high frequency negative energy source -- something we have no idea how to produce.

I've read that two large, ultra-dense, ultra-charged, extremely rapidly spinning rings at separate locations in the galaxy will form a wormhole connection between these locations. Is this true?
No. For sufficiently rapid rotation, each of the spinning rings would approximate a special case of the so-called Kerr-Newman solution. This describes a naked ring singularity (one not cloaked by an event horizon), passage through the center of which results in a traversal between two distinct universes. However, the actual construction of such rings would not form such a connection. Rather, as you spin up a ring, the most you could expect would be the formation of a growing bubble universe – a sort of blister or pustule formed on our own universe -- that could be entered via the ring. There is no reason to believe that two such bubbles formed in separated locations of the universe by distinct spinning rings would ever connect at all, much less form a shortcut.

What are "energy conditions" and what do they have to do with wormholes?
They are conditions once believed to be satisfied by all forms of matter. They can be expressed as inequalities that constrain the energy density of matter and its principle pressures, the pressures in each spatial direction. The matter required to hold open a traversable wormhole must violate certain energy conditions.

How many energy conditions are there?
There are five primary energy conditions that have been used over the last 50 years. Three of them are now regarded as obsolete, because there are now clear examples of their being violated by perfectly ordinary matter.

Which energy conditions matter in wormhole physics?
The two non-obsolete energy conditions are called the "Weak Energy Condition" (WEC) and the "Null Energy Condition" (NEC). [Their averaged versions are also in use.] The NEC requires that the sum of the density of matter with each of its principle pressures be non-negative. The WEC requires in addition that the matter density itself be non-negative. These conditions might some day become obsolete, as both of them are known to be violated by

quantum effects. The matter required to hold open a traversable wormhole must at least violate the WEC.

What is exotic matter and what does it have to do with wormholes?
Exotic matter is matter that violates an energy condition. A traversable wormhole requires the anti-gravitating effect of exotic matter at its throat to counter the wormhole's tendency to inwardly contract.

Is exotic matter the same as antimatter?
No. There exists a reference frame in which the energy density of exotic matter, unlike that of antimatter, is negative. Therefore exotic matter, unlike antimatter, gravitationally repels normal matter.

Is exotic matter the same as negative energy?
Strictly speaking, no. Although negative energy is a form of exotic matter, not all exotic matter contains negative energy. It can instead consist of matter with a positive energy density and a negative pressure -- the gravitating effects of the former property being exceeded by the antigravitating effects of the latter. If, however, an observer detects such positive-energy-density-with-negative-pressure exotic matter, another observer traveling sufficiently rapidly with respect to the first would detect *negative*-energy-density-with-negative-pressure exotic matter. For this reason the terms "negative energy" and "exotic matter" are generally taken to be synonymous.

What's the difference between phantom energy and exotic matter?
Phantom energy is exotic matter discussed in the context of cosmology. Specifically, it is matter that violates the null energy condition, is presumably distributed uniformly throughout the universe, is a possible explanation for dark energy, and is also known as "superquintessence".

How realistic is it to suppose that exotic matter exists?
Its existence isn't as far fetched as you might at first think. Four arguments for its likely existence are: 1) The ordinary electromagnetic field is infinitesimally close to being exotic. 2) Quantum effects are known to create negative-energy densities. 3) Something is causing the expansion of the universe to accelerate. Cosmologists have recently speculated that it might be cosmic exotic matter (which they call "phantom energy" or "superquintessence"). 4) Formerly sacrosanct energy conditions have been dying off for the last few decades. Why not a couple more?

I hear that there are two types of exotic matter. Is this true?
Sort of. Strictly speaking, there are precisely as many types of exotic matter as there are energy conditions. Because there are only two pointwise energy conditions that are still widely believed to apply to matter, their violations define the two types of exotic matter normally considered. There is exotic matter that violates the Null Energy Condition (NEC) and that that violates the Weak Energy Condition (WEC). The NEC requires that the sum of the density of matter with that of each of its principle pressures be non-negative. The WEC requires in addition that the matter density itself be non-negative. The WEC, then, despite its name, is stronger than the NEC. So all matter that violates the NEC also violates the WEC, though the reverse is not true.

What are "quantum inequalities" and what do they have to do with wormholes?
Quantum inequalities are a set of principles in quantum field theory that state that any observer's exposure to negative energy must be followed by a compensating exposure to positive energy, whose magnitude and duration exceed that of the negative energy exposure. This implies that the negative energy density at the throat of a traversable wormhole must be surrounded by a compensating region of positive energy density. Hence, quantum theory forbids wormholes whose matter is entirely negative.

What shape is a wormhole's mouth most likely to take?
A naturally occurring wormhole's mouth is likely to be spheroidal for the same reason that stars and black holes tend to be – the absence of stresses that introduce an asymmetry. In general, a wormhole's mouth must be a closed two-dimensional surface that would *seem to be* surrounding a three-dimensional volume. Although it is also possible to engineer a flat mouth (that does not seem to enclose a volume), such a configuration seems unlikely to develop naturally.

What is a thick-shell wormhole?
A wormhole whose matter is distributed throughout a nonzero (thick) volume centered about its throat, as opposed to being concentrated into an infinitesimally thin surface at its throat.

What's the difference between a wormhole's mouth and its throat?
If you think of a thick-shell wormhole as the three-dimensional analog of a two-dimensional tube that narrows at its middle, the throat is the analog of the circle of minimum circumference at the waist of this tube. That is, it is the *surface* of minimum area of the analogous three-dimensional hypertube. Its mouths are vaguely defined as the regions corresponding to the entrance and

exit of the hypertube. In a macroscopic, traversable, thick-shell wormhole, it is sometimes useful to define the mouth more precisely -- as the surface at which the acceleration of gravity is 1 g. For a thin-shell wormhole in flat spacetime, its mouth and throat are the same.

What is a thin-shell wormhole?
A wormhole whose matter is all concentrated at the infinitesimally thin surface that defines its throat. There, energy densities and spacetime curvatures are infinite. These wormholes, also known as Visser wormholes, simplify the analysis of dynamics and other aspects of wormhole physics. These contrived objects are not expected to be found in nature. [To imagine a thin shell wormhole, consider two sheets of paper. Each sheet represents a universe. Stack the sheets horizontally and use a hole-punch to create a hole through both of them. Use a thin strip of glue to attach the rim of the hole in the upper sheet to that of the hole in the lower sheet. The edge of these attached rims represents the wormhole's throat. Now an ant can crawl from the top of the upper sheet (universe A), through the hole (the thin-shell wormhole), onto the bottom of the lower sheet of paper (universe B).]

Can a traversable wormhole be used as a time machine?
Yes.

How can a traversable wormhole be turned into a time machine?
Keep one of the mouths stationary. Move the other mouth -- at speeds approaching that of light -- away from the stationary mouth for a distance of a few light years. Then return it to the vicinity of the stationary mouth. Anyone who now enters the stationary mouth will be transported years into the future. Those entering the other mouth will find themselves transported years into the past. For shorter time jumps, shorten the journey of the traveling mouth.

Is there another way to turn a traversable wormhole into a time machine?
Yes. Leave one mouth, Mouth A, in a weak (or virtually nonexistent) gravitational field. Move the other mouth, Mouth B, into a strong gravitational field, such as that near the event horizon of a black hole. Wait. Now move Mouth B from the strong gravitational field and return it to the vicinity of Mouth A. Anyone now entering Mouth A will emerge from Mouth B into the past. Anyone entering Mouth B will emerge from Mouth A into the future. For longer time jumps, wait a longer time before removing Mouth B from the strong gravitational field. Alternatively, initially move Mouth B into an even stronger gravitational field. To have an ever growing time jump, leave Mouth B in the strong gravitational field.

Isn't time travel by wormhole or any other means impossible due to the paradoxes that it implies?
Not necessarily. Dealing with time travel paradoxes by conjecturing the impossibility of time travel is only one of three ways of resolving the issue. The other ways are:
1) Impose self consistency on classical physics: A time traveler cannot change the past because he was always part of it. When he attempts to change the past, his efforts will be thwarted by an apparent conspiracy of events. 2) Impose self consistency on quantum physics: A time traveler cannot change the past because all possible pasts have already occurred in parallel universes. When he attempts to change the past, his efforts will not seem to him to be thwarted. This is because he will have entered the past of a preexisting parallel universe in which he has already made the changes that he seeks to effect.

Don't quantum effects prevent wormholes from being turned into time machines?
No. It's true that calculations seem to show that a wormhole-destroying feedback loop of virtual particles appears whenever a wormhole is configured as a time machine. But these calculations don't assume the existence of the parallel universes that would have prevented the feedback. Even without this assumption, it's possible to make the feedback arbitrarily small by replacing the single wormhole with a Roman ring of wormholes.

What is a Roman ring?
An arrangement of several wormholes that functions collectively as a time machine, even though no subset of the wormholes functions as one.

Are wormholes the only means of creating a time machine?
No. Any method of creating time loops – called "closed timelike curves" – will produce a time machine. Other methods include those that involve the generation of exceedingly high angular momentum densities.

Is a wormhole always a shortcut?
In practice, yes. In principle, no. Consider a wormhole that connects two regions within the same universe – an intra-universe wormhole. It is possible for the path through the wormhole to be longer than the shortest path through normal space between the wormhole's mouths (Figure 1.1). Intra-universe wormholes have in practice, perhaps due to the influence of science fiction, come to be synonymous with shortcuts. Inter-universe wormholes, by contrast, cannot be described as shortcuts. There is no path through normal space between distinct universes in comparison to which the path through an inter-universe wormhole appears shorter.

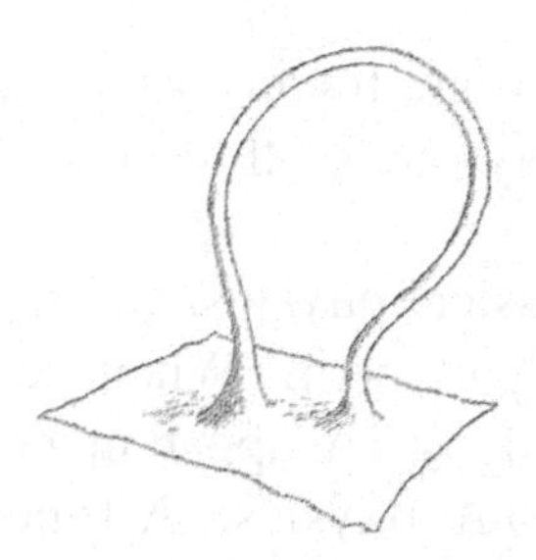

Figure 1.1. A wormhole that is not a shortcut. Such wormholes are even less stable than those usually considered.

Is there such a thing as a quantum wormhole?
Yes. A quantum wormhole is one whose complete description requires quantum theory. An example of a quantum wormhole is the ephemeral, submicroscopic wormhole of the sort that characterizes Wheeler's spacetime foam. Physicist John Wheeler was the first to imagine the vacuum state of a quantized gravitational field. He believed that the corresponding spacetime would at tiny length scales be a roiling froth (analogous to sea foam) that momentarily permits every conceivable geometry -- including wormholes. Unfortunately, the effort to create a theory of quantum gravity has not yet advanced sufficiently to permit anything approaching a complete description of spacetime foam or of the quantum wormholes it is presumed to contain.

If quantum wormholes exist, why don't we see elementary particles disappearing from one place and appearing elsewhere or jumping backward and forward in time?
The mouths of quantum wormholes are many orders of magnitude smaller than the smallest elementary particle.

How does a wormhole differ from a space warp?
A wormhole is a special case of space warp.

Are wormholes accurately depicted in science fiction films and television programs?
Not usually. Wormholes are normally depicted as swirling drain holes in space. Sometimes they are shown to have luminous event horizons.

What would a wormhole look like?
A wormhole would likely appear to be a bubble or window through which unfamiliar stars are visible. If the wormhole is massive, dense, rotating, and in

the proximity of luminous matter -- such as that found in stellar atmospheres -- it could be surrounded by a visible "accretion disk" much like those surrounding black holes.

Can a wormhole's mouth be a flat disk instead of a spheroidal surface?
Yes. Some people refer to such wormholes as "stargates". Others call them "portals".

Would a flat "stargate-style" wormhole require less exotic matter than a spherically symmetrical one?
Yes.

What is a ringhole?
A ringhole is a wormhole whose mouth and throat have the topology (rough shape) of a torus.

Can a wormhole possess an electric charge?
Yes. The maximally extended Reissner-Nordstrøm solution to Einstein's gravitational field equations, which describes a charged black hole, contains a one-way wormhole. A charged, two-way wormhole – no longer surrounded by an event horizon -- could be created by dropping sufficient amounts of negative-energy matter into this charged black hole. Solutions to the Einstein equations for such charged traversable wormholes have been found.

Is there such a thing as a rotating wormhole?
Yes. The Kerr black hole solution to Einstein's equations, which describes a rotating black hole, contains a one-way wormhole. By dropping a sufficiently large amount of negative-energy matter into a rotating black hole, one could produce a spinning, two-way wormhole that is no longer surrounded by an event horizon. Such a solution was found in 1998.

How stable are wormholes?
It depends on a feature of the exotic matter that supports it. Specifically, it depends on what's called the "equation of state" of the exotic matter – how its pressure depends on its density. For exotic matter modeled more or less conservatively (as massless particles of negative energy), traversable wormholes are unstable. Although it is always possible to choose an equation of state that guarantees stability, this stability is not unlimited. Any wormhole will collapse to a black hole after a sufficiently large influx of positive energy. Conversely, any wormhole will expand ceaselessly after a suitably large influx of negative energy.

Could wormholes be stabilized artificially?
Yes. When a wormhole begins to contract, an artificial stabilizer would halt the contraction by injecting negative energy. The stabilizer would similarly halt expansions by injecting positive energy.

Can a wormhole have more than one throat?
Yes. The neck – the inter-universe/intra-universe passage -- of a spherically symmetrical wormhole can narrow and widen in several locations. The spheres of locally minimal area at which the neck is locally most narrow are throats. Such a wormhole is highly unstable, but it could be stabilized artificially to create a static wormhole. Dynamic wormholes, whose geometric throats are either expanding or contracting, acquire multiple throats in a different way. The wormhole's functional throat, which coincides with its geometric throat in the static case, splits in the dynamic case into two throats -- one on either side of the geometric throat.

What is the difference between a wormhole's geometric throat and its functional throat?
Imagine an incoming spherical wave front of light that converges (contracts) on a spherically symmetrical wormhole, enters it, and diverges (expands) as an outgoing spherical wave front in the other universe. The functional throat of the wormhole is the surface at which the area of this incoming-outgoing wave front is minimal. A wormhole's geometric throat, by contrast, is its surface of minimum area -- irrespective of the behavior of light passing through it. The geometric and functional throats of a static wormhole are the same. This is not true for dynamic wormholes, however, because the motion of the traversing light is affected by the expansion or contraction of the wormhole. In this case there are two functional throats -- one for each of the two directions in which the wave front could have traversed the wormhole.

When a wormhole forms a connection to another universe is this other universe a "parallel" universe?
Yes, but only if we assume that reality admits a single system of fundamental physical laws. In that case, the set of all parallel universes is the set of all universes that are physically possible according to this single system. Because, by assumption, no universes external to this set exist, a wormhole necessarily connects to a parallel universe.

What is a *Tolman* wormhole?
A Tolman wormhole, while mathematically similar to a wormhole, is not a true wormhole. It is a cosmological solution to the Einstein equations -- that is, it is a universe -- that contracts to a point of maximum finite density and re-

expands. This contraction followed by a re-expansion (as opposed to a big crunch) is called a "bounce". Hence, a Tolman wormhole is also known as a universe with a bounce.

What is a Krasnikov tube and how does it compare to a wormhole?

A Krasnikov tube is a theoretical alteration of spacetime that would permit one-way, superluminal travel. Imagine a tube connecting our solar system and the Rigel system. Outside of this tube, spacetime is ordinary Minkowski space. Inside the tube spacetime has been altered so that spaceships traveling *from* Rigel *to* Sol at sufficiently high *sub*luminal speeds appear to external observers to be traveling backward in time. If a ship travels from Sol to Rigel *without* using the tube, but uses the tube on its return trip, the round-trip travel time -- according to clocks on Earth -- could be arbitrarily short. Hence, the first outbound ship to a distant star would need to construct a Krasnikov tube on its way. This would allow itself and future ships to enjoy short round-trip times. A pair of Krasnikov tubes – one outbound and the other inbound – would permit a traveler to the Rigel system to return to the Sol system *before* he left. Unlike wormholes, Krasnikov tubes do not require a change in the macroscopic topology of space and only allow effectively faster-than-light travel in one direction. Like wormholes, Krasnikov tubes require exotic matter, are effectively superluminal shortcuts through space, permit time travel, and are utterly beyond our civilization's current capacity to construct.

A wormhole solution was known in 1916. Why did a thorough investigation of wormhole physics not begin until the late 1980s?

Beyond the fact that wormhole and black hole solutions were universally regarded as physically uninteresting until about 1935, the development of wormhole physics was retarded by sociological factors. Physicists are not in general financially independent. To survive, they must be employed. To be employed, they must be credible. To be credible, they must eschew topics of great interest to crackpots. [Crackpots are those engaged in the promulgation of irrationally held ideas typically involving the occult, fantasy, or science fiction.] One indication of the stigma associated with the subject is that the first wormhole paper of the "modern era" when published in 1988 was not billed as a research paper (which it was) but as "a tool for teaching general relativity". It was only because its lead author was an established physicist of unquestioned credibility that less established physicists subsequently felt it professionally safe to publish traversable wormhole papers as well. They were further reassured when in the same year other established physicists happened by sheer chance to be publishing important papers involving *Euclidean* wormholes, which – due to their irrelevance as interstellar shortcuts – never invoked the same stigma as their Lorentzian counterparts. Another reason for the delay in wormhole research was that most practitioners of general relativity did not appreciate the existence of viable methods for creating exotic matter. Many had undoubtedly plugged a wormhole or other interesting geometry into the Einstein equations, discovered that the geometry required exotic matter, and consequently dismissed the geometry as unphysical.

How much exotic matter is needed to hold open a traversable wormhole?
A Morris-Thorne wormhole with tolerable tidal forces -- due to having a mouth that is 600 times the earth-sun distance -- requires about 10^8 solar masses of exotic matter. A thin-shelled (Visser) wormhole with a 1-meter wide mouth requires about 1 Jupiter mass of exotic matter. While it has recently been shown that a wormhole traversable in principle can be held open with arbitrarily small quantities of exotic matter, it still appears that enormous quantities of exotic matter are required to hold open wormholes that are traversable by humans in a timely fashion.

Does the exotic matter in traversable wormholes have anything do to with dark matter or dark energy?
It has nothing to do with dark matter. Dark matter is *non*-exotic matter hypothesized to exist because the assumed primordial inflation of the universe requires its density to be many times greater than that due to observed luminous matter. Dark energy, by contrast, is an attempt to explain the apparent acceleration of the expansion of the universe that was first detected in 1998. A possible explanation for dark energy is the supposition that the universe is awash in dark *exotic* matter (albeit at a very low density). This is precisely the sort of matter required to sustain traversable wormholes.

If a wormhole is created or destroyed, won't that require a change in the topology of space? I thought that was impossible.
Wormhole creation and destruction need not involve topology change. When wormhole creation is discussed, it's generally understood that it means enlargement of preexisting microscopic wormholes. Wormhole destruction connotes their conversion into black holes or their contraction and absorption into the spacetime foam of virtual geometries.

Can a wormhole be used as a weapon?
Yes. But then, what can't be used as a weapon?

What's the connection between wormholes and string theory?
Not much, although the brane world concept that has emerged from string theory might provide a framework through which we might begin to understand how wormholes connect to parallel universes. In the brane world

scenario the universe is a 3+1-dimensional membrane, or "3-brane" for short, embedded in a higher-dimensional spacetime called the "bulk" – rather like a sheet of paper floating in space. Parallel universes can be imagined as a stack of slightly separated parallel sheets (3-branes) floating in space (the bulk). It's clear, then, that a wormhole that links different regions in the bulk would necessarily link different parallel universes. Another connection is the slight possibility that the projection of a specific string theory of the bulk onto a 3-brane representing our universe might have an interesting effect. Specifically, it might contribute to the stress energy of the 3-brane in a way that simulates the presence of exotic matter. Or to state the matter in a Wheeleresque fashion, there is still a chance that wormholes in string theory might enjoy the benefit of *exotic matter without exotic matter.*

How could an advanced civilization ensure that the mouths of their intra-universe wormholes are in desired locations?

They would proceed as follows. Step 1: At the desired location of the first mouth, enlarge a virtual wormhole extracted from the spacetime foam. Step 2: Determine the location of the second mouth by traversing the newly enlarged wormhole and studying the sky from the vantage point of the second mouth. Step 3: If this location is undesirable, return through the wormhole, collapse it, and begin again at Step 1. Otherwise, continue to Step 4. Step 4: Charge the second mouth by showering it with charged particles. Step 5: Use electrostatic attraction to precisely position the second mouth by suitably dragging it.

Can a wormhole be used as a source of energy?

Generally not. It could, however, be a *font* of energy. If it is a bridge to a hotter universe or to a hotter region of our own universe, a wormhole could funnel energy to us. In the special case of certain rotating wormholes, it would be possible to extract energy from them through the Penrose process. This is a well known method through which matter suitably injected into a rotating black hole by an object increases the energy of the object, while the black hole loses an equal amount of mass.

What is a geon and what does it have to do with wormholes?

A geon or "gravitational-electromagnetic entity" was conceived by John Wheeler as a divergence free, quasi-stable solution to the Einstein-Maxwell equations that might serve as a model for a charged particle. The idea was to model a negatively charged particle as the mouth of a wormhole into which electric lines of force entered. The other mouth, from which the same lines of force emerged, was to model a corresponding positive charge. Wheeler's 1955 paper entitled "Geons" contained the first freestyle drawing of a wormhole ever to appear in the physics literature.

If wormholes that are black holes are not traversable, in what sense do they form an inter-universe or intra-universe connection?
They do in the sense that people from different universes (or from distant parts of the same universe) can both enter the black hole and meet each other. The meeting will likely be interrupted, however, by the violent deaths of both parties in the black hole's singularity.

I've heard that some wormhole solutions are called "self-consistent". How can a solution *not* be?
The self-consistency of a solution does not pertain to its logical consistency. It means instead that a solution is complete or self-contained. Such a solution not only specifies the curvature of spacetime due to the wormhole, but also explicitly specifies all of its matter. For example, if the wormhole's matter includes an electromagnetic field, a self-consistent solution includes a solution to the combined Einstein-Maxwell equations that govern *both* gravity and electromagnetism. Solutions to these interwoven equations describe 1) the geometry of spacetime that is produced by the stress energy of the electromagnetic field, *and* 2) the electromagnetic field that produces this stress energy and is affected by this geometry. Self-consistent solutions are difficult to find.

What would it be like to pass through an event horizon?
You wouldn't notice. For a huge, ultra-massive black hole, you wouldn't notice a thing. For a somewhat smaller black hole there would be significant tidal forces in the vicinity of its event horizon. Assuming that you could tolerate these, you would find that they wouldn't have changed in any sudden way as you passed through the horizon. Human beings could not, however, survive the approach to the event horizon of a "typical" black hole, whose mass would only be a few times that of the sun. The tidal forces there would be too great.

Does a traveler passing through a traversable wormhole have to come into contact with exotic matter?
No. Two methods have been proposed to protect travelers. The first was simply to punch a hole in the exotic matter through which travelers could pass. This violates the spherical symmetry of the wormhole solution. But it was assumed that if the hole was small enough, the perturbed wormhole solution would be very close to the original symmetric one. The other proposal, due to physicist Matt Visser, was to abandon spherical symmetry and consider polyhedral thin-shell solutions. In Visser's wormholes exotic matter is confined to the edges and vertices of a polyhedral throat. This allows travelers to pass through the faces of the polyhedron without being exposed to exotic matter.

If I stick my arm into an event horizon, can I pull it back out again?
No.

What does an event horizon look like?
It is completely black.

If a wormhole connects distant regions A and B, can its mass measured in region A differ from its mass measured in region B?
Yes. Residents of these regions can determine the mass of their end of the wormhole by putting an object in orbit around the wormhole mouth in their region. For a given orbital radius, the orbital period determines the wormhole mass that they measure. This value depends on how the wormhole curves their local region of spacetime. There is nothing requiring this curvature to be the same at both ends of the wormhole. Hence the wormhole masses measured at it ends need not agree.

Can a wormhole's total mass be negative?
Yes, if classical exotic matter exists. However, the wormhole would be gravitationally repulsive. This would require travelers to expend additional fuel to approach the wormhole, because doing so would be an "uphill" trip. If exotic matter is due exclusively to quantum effects, a purely negative-mass wormhole would be forbidden by the quantum inequalities.

What other nonspherical wormhole topologies are possible?
Klein bottles, disks, and tori with any number of handles have been discussed in the literature. There are no restrictions on wormhole topology imposed by classical physics. However, quantum physics, in particular that governing the weak interactions, requires spacetime to be orientable. At every point it should be possible to distinguish left from right.

How would the presence of wormholes in the cosmos be detected?
If its total mass is negative, it could be detected through the unusual way in which it distorts light. Unlike objects with positive mass that act as "convex" gravitational lenses, a negative-mass wormhole would, by contrast, function as a sort of "concave" gravitational lens.

Does wormhole physics suggest a preference among the competing interpretations of quantum theory?
Perhaps. If wormholes exist, then we must either explain: a) why it is impossible to turn them into time machines, or b) how it is possible to resolve the paradoxes that result, when they are turned into time machines. The latter seems most naturally accomplished in the Many Worlds interpretation.

Do wormholes behave qualitatively the same in theories of gravity other than general relativity?
Yes, at least in theories that become general relativity in the low-energy limit. It was once hoped that such a theory might permit traversable wormholes to exist in the absence of supporting exotic matter. This turned out not to be the case in nearly all of the commonly considered alternative theories of gravity.

If I toss a spinning object into a wormhole, will that cause the wormhole to spin?
Yes. However, if it is a traversable wormhole, any additional angular momentum imparted to it will be lost, when the spinning object emerges from the other side.

How could I destroy a wormhole?
If it's an artificially stabilized wormhole, turn off its stabilizer. The wormhole would then collapse and become a black hole. If it's a naturally stable wormhole, dumping into it an amount of positive matter that well exceeds that of its intrinsic exotic matter will similarly destroy it. To destroy the wormhole without leaving a black hole remnant, you would have to neutralize the matter in its throat region. You would have to repeatedly inject normal or exotic matter in just the right amounts to ensure that an observer at the throat would measure an ever decreasing stress-energy there. This would effectively destroy the wormhole by shrinking it to a microscopic size.

Can a black hole be used as a wormhole that is traversable in only one direction?
Possibly, if it is charged or rotating. As I mentioned before, there are certain well known solutions to Einstein's field equations describing charged and rotating wormholes that might be one-way traversable. A traveler could in principle enter the black hole sector of these solutions and emerge from its white hole sector in another universe. From there he could never return to his universe of origination. There are a few problems with this scenario, however. The real universe is full of light and other radiation. This energy accumulates on certain surfaces in the traveler's path. Crossing these surfaces could kill him. Moreover, these energy accumulations would likely destroy the wormhole. In particular, the wormhole's white hole sector, from which the traveler would emerge, would be destroyed by the accumulation of energy on its outer horizon. Lastly, there is no reason to believe that this sort of wormhole would work as a shortcut – as a path to a distant region within a traveler's own universe. Even if it somehow did, it would only be a path to the distant past of that universe.

Is it possible for travelers to simultaneously move through a traversable wormhole in opposite directions?
Yes. The only restriction is that the throat needs to be large enough to simultaneously accommodate at least two travelers.

Is there any way in which a wormhole might require a source of power to hold it open?
Yes. A synthetic wormhole's stabilizer unit, which would inject normal or exotic matter as needed, would require power. Holding a wormhole's exotic matter in place might also require power. One could imagine exotic matter that is self-repulsive both gravitationally and electrically. A wormhole's designers could have decided to use electromagnetic fields to confine this matter at the throat. Generating such fields, even using superconducting coils, requires power. Turning such a wormhole "off" would amount to deactivating these fields. This would allow the exotic matter to spread out from the throat. Upon doing so, the matter's density and pressure would decline sufficiently to enable physical barriers to confine it. The wormhole's throat diameter would constrict, its traversal time would increase, and it would cease temporarily to be a viable means of human transport.

Could I use a wormhole to escape from the inside of a black hole?
Yes. Classically, there appears be to nothing to prevent the existence of a wormhole that connects the inside of a black hole with the region exterior to its event horizon. However, to an observer within the horizon, the outward direction points to the past, i.e. backward in time. So escaping a black hole via a wormhole is only possible, if time travel by wormhole is possible.

Are elementary particles really just tiny classical wormholes?
No. This is an idea that dates back to the 1930s. No one has gotten it to work in the intervening time. One problem is that the wormholes would have to be held open by a ubiquitous negative-energy field. We don't observe such a field. In the absence of this field, the wormholes would become tiny black holes. Because of the large charge-to-mass and spin-to-mass ratios typical of elementary particles, the corresponding black holes would have to expose their singularities. Naked singularities are bad. The laws of physics break down there. Predictability is lost. Anything could fly out of them or be absorbed by them. We don't observe this.

Are elementary particles really just quantum wormholes?
Not very likely. The main problem appears to be the discrepancy between the mass of the quantum wormhole and that of elementary particles. A quantum wormhole can have a mass of zero, or it can have a mass that is at least 10^{19}

times that of the proton. This makes it impossible to reproduce the observed mass spectrum of elementary particles.

Can the Large Hadron Collider, the most powerful particle accelerator in history, produce detectable quantum wormholes?

Possibly. It depends on two things: 1) whether, as string theorists believe, there are hidden extra dimensions beyond the usual four, whose existence would sufficiently lower the mass of quantum wormholes to fall within the LHC's energy range, and 2) whether the production of such wormholes is a sufficiently likely gravitational outcome of a high-energy collision in comparison to other possibilities such as quantum black holes or bursts of gravity waves. If quantum wormholes are possible but too unlikely, the LHC could run for centuries without producing any.

Do quantum wormholes explain quantum teleportation?

No.

Can a wormhole periodically disappear and reappear?

Yes. It can do so in two ways: through external manipulation or internal dynamics. To externally manipulate a wormhole for this purpose, we would carefully inject it with matter in such a way as to ensure that its throat becomes microscopic. We would then reverse the process, by changing the sign on the matter injected, until the wormhole returns to its original size. For a wormhole to be periodic by virtue of its internal dynamics, its exotic matter need only possess a suitable equation of state. This equation determines how the matter's pressure depends on its density. For certain pressure-density relationships, the wormhole will oscillate between being microscopic and macroscopic and thus appear to periodically disappear and reappear. We don't know whether the type of exotic matter required to accomplish this exists.

Will the mouths of a periodic wormhole always appear in the same place?

Yes. True topology change is impossible within general relativity, as it is conventionally formulated. So a periodic wormhole never actually ceases to exist. It merely shrinks until it is microscopic. Its mouths, irrespective of their sizes, retain their positions relative to co-moving observers.

What would happen if two wormhole mouths were to collide?

If one mouth were much larger than the other, the smaller mouth would simply pass through the wormhole with the larger mouth, as any other object would. If the colliding mouths and their associated wormholes were identical, there would be two possible outcomes: 1) If the effective mass of both mouths is negative, they will approach, possibly coalesce briefly (if their relative momentum is high enough), emit gravitational waves, and separate. 2) If the effective mass of both mouths is positive, they will approach, coalesce, emit gravitational waves, and remain joined. In neither case would the topology of space have changed. When the mouths of the two wormholes coalesce, they do not become a single wormhole. Rather, they become a three-mouthed system roughly resembling a stethoscope: A mouth in one universe branches internally to connect to two separate mouths in separate universes (or distant regions within the same universe). Unfortunately, no one has yet taken the trouble to perform calculations that would confirm or refute this speculation.

What would happen if two mouths of the same wormhole were to collide?
After the collision identical mouths would 1) coalesce, if their masses are positive, or 2) separate, if their masses are negative. Upon coalescing, the wormhole's shape would be something like the three-dimensional generalization of the surface of an eye screw (Figure 1.2). A traveler could enter the single mouth of the wormhole, travel the length of its interior region, and exit the mouth only to discover that she is back where she started.

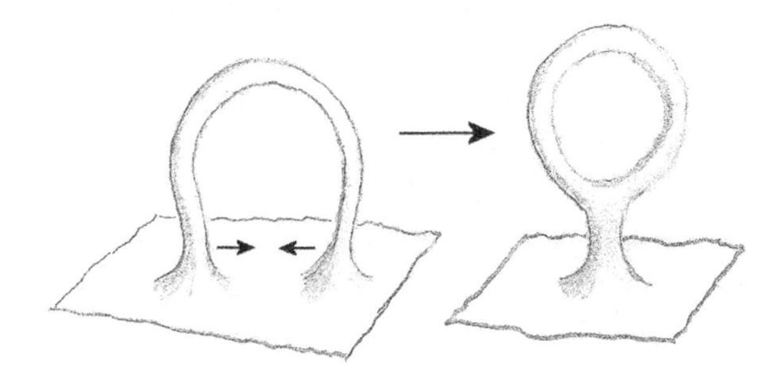

Figure 1.2. A wormhole Self Collision. Mouths will likely coalesce.

What would it be like to travel through a wormhole?
As you enter a deep spherical wormhole, you will see stars concentrated in a sphere directly ahead. After awhile you peer out of your spaceship's port window. You will notice another ship on a parallel course off in the distance. Upon inspecting it with your telescope, you will notice that it looks exactly like the starboard side of your spaceship. That's because it is. Looking through your starboard window similarly reveals a distant port view of your ship. The sky will show relatively few stars in directions at right angles to your inward trajectory. The more forward your angle of view, the more stars you will see, until you see the aforementioned spherical concentration straight ahead. As you approach this concentration -- the wormhole's throat -- you will

notice that the parallel image of your ship is much closer. You wave and, after a momentary delay, can (through your telescope) see yourself waving. As you pass through the throat, you will find yourself gazing upon a normal sky of unfamiliar stars. The parallel image of your ship will recede and soon vanish. In contrast to this experience, traveling through a maximally benign Visser (thin-shell) wormhole would be just like stepping through a doorway. You could even straddle this doorway and be at once in two universes or vastly separated regions of the same universe.

Do the mouths of a rotating wormhole have to rotate at the same rate?
No. The wormhole's exotic matter would merely need to have an angular momentum density that is distributed asymmetrically about the wormhole's throat.

Do the mouths of a charged wormhole have to be equally charged?
No. But the wormhole's exotic matter would need an asymmetrical charge density distribution.

Can a wormhole look like a black hole at one mouth and a traversable wormhole at the other, i.e. can only one of its mouths have an event horizon?
Yes. This, however, would result in a one-way traversable wormhole. Travelers entering from the horizon (black hole) side could reach the other mouth. But they would have to leave in time to avoid being crushed in the singularity that forms within this wormhole, as it pinches off. Travelers who innocently enter from the horizon-free mouth would invariably meet their doom at this singularity.

Can the size of a wormhole's throat oscillate?
Yes. This would just be a periodic wormhole whose minimum throat size remains macroscopic. All comments regarding periodic wormholes apply, especially those emphasizing the importance of a suitable exotic matter equation of state to the realization of this phenomenon.

Is it always obvious that a traversable wormhole is not a black hole?
No. It's possible for a traversable wormhole to be arbitrarily close to being a black hole. It's an actual black hole only if it has an event horizon. You might try to test for the presence of an event horizon by attempting to detect Hawking radiation. That won't work. For a wormhole that's large enough to traverse, the Hawking radiation will be imperceptibly weak and drowned out by background radiation. You might try lowering a probe into the wormhole and trying to pull it back out. Unfortunately, gravitational time dilation

ensures that this experiment could take centuries. The easiest method to test for a horizon in a short subjective time is to pilot your spaceship into the wormhole and try to reverse out of it. If you can't, you've entered a black hole.

What is the black hole information paradox and what do wormholes have to do with it?

In 1974 Stephen Hawking discovered that quantum theory requires black holes to emit radiation, which causes them to eventually evaporate away leaving nothing behind. This means that any information that fell into the black hole will vanish along with it. That's a problem, because quantum theory also insists that information cannot be destroyed. That quantum theory applied to black holes seems to require information to both be destroyed and not be destroyed is called the black hole information paradox. Most physicists suspect that the paradox will be resolved by understanding how Hawking radiation actually carries away information. For example, when an encyclopedia is dropped into a black hole, physicists hope that the Hawking radiation emitted by the black hole will be perturbed in a manner that encodes the information of the encyclopedia. This preserves its information after the black hole evaporates away. However, if traversable wormholes can exist, this idea won't work. A traversable wormhole that connects regions interior and exterior to the black hole will allow some of its contents to escape. The encyclopedia could fall into the black hole, fall into the mouth of a small wormhole beneath the black hole's event horizon, and thereby escape the black hole. If Hawking radiation now encodes the encyclopedia's information, we now have perfect information cloning exterior to the black hole – i.e. the encyclopedia together with its information encoded in radiation are simultaneously accessible to a single external observer. This is unacceptable, however, because perfect information cloning is also forbidden by quantum theory.

Does the motion in space of one wormhole mouth affect that of the other?

No. Because there is no way to establish an absolute frame of reference within a universe (or between universes), the mouths of a wormhole normally *are* in motion relative to each other. This translational motion of the mouths is completely independent.

So I can station the mouths on separate planets and not worry about the relative motion of the planets interfering with the wormhole's operation?

Yes.

Is it possible to create a network of wormholes in which any wormhole mouth could be used to reach multiple destinations directly?

Yes. Assuming that it's possible to create a single wormhole, the creation of a network of many is not particularly far fetched. You would proceed as follows: 1) Create a wormhole that connects locations *A* and *B*. 2) Create another wormhole connecting locations *A* and *C*. There are now two wormhole mouths at the location *A* – one that leads to *B* and another that leads to *C*. 3) Move the *A* mouth of the *AC* wormhole into the *A* mouth of the *AB* wormhole until the *A* mouth of *AC* is at the throat of the *AB* wormhole. 4) You now have a wormhole network in which there is a non-stop connection between any of *A*, *B*, and *C*. Repeat the procedure to add other destinations. 5) Ease congestion at the throat nexus by expanding it (with additional exotic matter) and by adding an automated system that properly routes spacecraft based on their transmitted destination codes.

What would happen if one of the mouths of a traversable wormhole fell into the sun?

The wormhole's stabilizer (if present) would be overwhelmed, causing the wormhole to collapse and become a black hole. This black hole would make its way to the center of the sun, where it would feed on the sun's mass. As the black hole grows, the time dilation effects near its horizon would effectively retard the rate of nuclear fusion there. I would speculate that the resulting loss of outward pressure would cause the sun to contract. If this contraction occurs too rapidly -- which I think unlikely -- it could cause an inward-going high-pressure wave that would temporarily boost the rate of fusion. The sun could then explode, or it might merely expand until it returned to a size somewhat smaller than its original radius. Assuming that the sun did not explode, its radius would then decrease for the same reason as before. A sudden shrinkage would continue the sporadic cycle just described. Or it might continue to contract smoothly. In either case, the sun, or its remnant, will continue to shrink until the black hole completely devours it. Depending on the size of the black hole, the process could take seconds or millions of years.

What can you tell me about the future of wormhole physics?

Very little. However, John Wheeler once explained to me that "Our goal in physics is to make our mistakes at the fastest possible rate." The degree to which investigators take this to heart, is that to which I expect the rapid progress of the field to continue.

What can you tell me about the history of wormhole physics?

Everything that you'll find in Chapter 2, to which we now turn.

2. Wormhole History

The history of wormholes began the year after Albert Einstein published his theory of general relativity. In 1916 Ludwig Flamm, in attempting to work out the geometry given by general relativity for the interior of a sphere of incompressible fluid, produced an embedding diagram that extended the 1915 solution of Karl Schwarzschild. His embedding showed an incomplete version of what is now known as a Schwarzschild wormhole.

In 1924 Hermann Weyl suggested that electric charges are manifestations of topological features of space that are essentially wormholes.

Eleven years later Albert Einstein and his long-time collaborator Nathan Rosen further elaborated the features of the Schwarzschild wormhole. Their "bridge", as they called it, was an attempt to model an elementary particle in a way that was free of singularities.

Twenty years after this, John Wheeler invented his "geon", another attempt to devise a singularity-free particle model. Wheeler's geons – a word he coined to stand for "gravitational-electromagnetic entities" – were hoped to be quasi-stable solutions to the Einstein-Maxell equations, the equations that describe gravitational and electromagnetic fields as well as their interaction. Wheeler believed, in accordance with Weyl's old suggestion, that it was possible to avoid the point charges of classical electrodynamics by modeling a charged particle as a wormhole threaded by an electric field. A negatively charged particle would be manifested as a wormhole mouth into which electric field lines entered. A positively charged particle would be the companion mouth from which the same lines emerged. Thus charges could effectively exist in the absence of an actual charge density. John Wheeler's 1955 paper featured the first appearance in the physics literature of a free-style drawing of a wormhole.

In 1957 Wheeler, in a paper that he co-authored with his graduate student Charles Misner, further elaborated his geon concept as part of a larger speculation that the whole of physics can be explained as a manifestation of geometry. In explaining his "charge without charge" concept that I described above, he introduced the word "wormhole" into the physics literature.

In another 1957 paper Wheeler speculated that at tiny length scales the quantum vacuum of the gravitational field must be characterized by turbulent fluctuations in the geometry of spacetime. He would later liken this turbulence to that of the ever changing structure of sea foam.

In 1960 John Cramer and Dieter Brill elucidated the properties of the Reissner-Nordstrøm wormhole. The Reissner-Nordstrøm solution, discovered in 1917, described the geometry of spacetime induced by a charged, massive spherically symmetrical object. They discovered that, unlike the Schwarzschild wormhole studied by Einstein and Rosen, the Reissner-Nordstrøm wormhole has a throat radius that "pulsates periodically in time". This made the wormhole potentially traversable.

In 1962 John Wheeler and Robert Fuller eliminated the Schwarzschild wormhole as a possible shortcut, by showing that it contracts too rapidly to be traversed.

Another potentially traversable wormhole was discovered in 1963, when New Zealander Roy Kerr, then working at the University of Texas, solved the Einstein equations for the case of a rotating black hole.

In 1968 Roger Penrose, then at Birkbeck College in London, discovered that the inner horizon of the extended Reissner-Nordstrøm wormhole solution is unstable. It would henceforth be classified as untraversable.

It was 1973 when a wormhole next appeared in the literature. Homer G. Ellis, in another attempt to develop a singularity-free particle model, proposed what he called a "drainhole". This was a coupling of the gravitational field to a massless, negative-energy scalar field that resulted in a traversable wormhole – the first to appear in the literature.

A few years later Stephen Hawking revived, in the context of Euclidean quantum gravity, Wheeler's view of the quantum fluctuations of spacetime geometry. Widespread use of the term "spacetime foam" for this idea can be traced back to Hawking's 1978 paper of the same name.

In 1985 Carl Sagan, who was finishing his novel *Contact*, famously contacted his friend Kip Thorne of Caltech for advice regarding a plausible means of interstellar travel for his story. Sagan had originally considered a black hole as the means of accomplishing this. Thorne, well aware that black holes are not traversable, began working on an alternative device. He suggested a wormhole, which Sagan ultimately incorporated into his novel. Thorne continued to develop his ideas and with his graduate student, Michael Morris, submitted in 1987 the seminal paper, "Wormholes in spacetime and their use for interstellar travel: A tool for teaching general relativity". Its publication in 1988 launched the modern age of Lorentzian wormhole physics.

Meanwhile, Stephen Hawking had published "Wormholes in Spacetime", in which he attempted to calculate the contribution to Euclidean path integrals of the joining and splitting off of baby universes. This and similar work by Andrew Strominger and Steven Giddings lead Sydney Coleman of Harvard to use Euclidean wormholes to attempt to solve the cosmological constant problem, that is to explain its unreasonable smallness. His famous attempt, "Why Is There Nothing Rather Than Something: A theory of the cosmological constant", was published in 1988 and spawned a cottage industry of refutation, attempted modification, and extension.

By early 1988 Thorne understood that traversable Lorentzian wormholes could be used as time machines. With his students, Morris and Ulvi Yurtsever, he published this result later that year, thus breaking the taboo against papers with "time machine" in the title. This encouraged others, including I.D. Novikov and V.P. Frolov of the Lebedev Physical Institute in Moscow, to publish their own results concerning time travel.

In 1989 Matt Visser broke new ground by departing from the spherical symmetry that characterized the wormhole solutions of Thorne and his collaborators. In so doing, he was able to eliminate a drawback of spherical symmetry – exposure of wormhole travelers to dense exotic matter, which would have probably resulted in unpleasant consequences for them. Visser's thin-shell wormholes were also free of gravitational tidal forces. Unlike Morris-Thorne wormholes, however, there seemed to be no way of constructing them or of even pulling them out of the quantum vacuum, as Morris and Thorne had envisaged.

Within a few years virtually every aspect of wormhole physics had been touched upon, including wormhole stability, quantum wormholes, supersymmetric wormholes, wormholes in alterative theories of gravity, wormholes and chronology protection, wormholes and topological censorship, primordial wormholes and inflation, and wormholes as a means of understanding black holes. The annual production of wormhole-related papers continued to trend upwards, reaching a peak in 1994.

At this point Matt Visser began to write what would become the definitive monograph summarizing the field. Completed in 1995 and published by the American Institute of Physics, his *Lorentzian Wormholes – From Einstein to Hawking* marked the maturation of wormhole physics.

Later in 1995 a team of collaborators -- including John G. Cramer, Robert Forward, Visser, and Morris -- suggested a means of detecting cosmic wormholes through their induced microlensing effects.

1996 saw the application by Lawrence Ford and Thomas Roman of quantum constraints to wormholes. These constraints, which are also known as quantum inequalities, restricted the geometry of wormholes. They could be submicroscopic, or they could have all of their exotic matter confined to a submicroscopically thin shell at their throats. It appeared, then, that naturally occurring macroscopic wormholes were virtually impossible, and that the technological sophistication required of a wormhole-constructing civilization would have to be even greater than the "absurdly advanced" level originally supposed.

The first inkling that this gloomy state of affairs need not be permanent seems to have appeared in a 1997 paper by Dan Vollick. He showed that normal (non-exotic) matter interacting via a normal scalar field can have negative interaction energy. He used this negative energy to hold open a wormhole. Because this interaction energy was classical, rather than of quantum origin, the Ford-Roman constraints did not apply.

The following year Vollick, while not the first to consider wormholes in alternative theories of gravity, appears to have been the first to explicitly obtain Lorentzian wormhole solutions to low energy string theory. Meanwhile David Hochberg and Visser turned their attention to wormhole dynamics. They were able to prove that NEC violations at or near the throat were required in any traversable wormhole. They also improved the definition of a wormhole. That same year Edward Teo discovered a rotating wormhole solution. 1998 also saw measurements of supernova luminosities that indicated that expansion of the universe is accelerating. One proposed explanation was the prevalence of cosmic exotic matter (superquintessence) – precisely the sort of matter required to support wormholes.

During this time Sean Hayward -- then at the Yukawa Institute for Theoretical Physics in Kyoto, Japan -- had extended his study of black hole dynamics to wormholes. This lead to a definition of a wormhole essentially equivalent to that independently proposed by Hochberg and Visser. In 1999 Hayward was the first to explain that black holes and wormholes are merely different dynamical states of the same object.

In the final year of the twentieth century Sergei Krasnikov showed how the Ford-Roman constraints could be met in a wormhole with quantum-based exotic matter by relaxing the condition of asymptotic flatness. Meanwhile, Carlos Barceló and Matt Visser were pointing out that every known energy condition had classical violations. This meant that precious theorems of general relativity whose proofs relied on these conditions – including the singularity, positive mass, and topological censorship theorems – do not in general apply. It also meant that wormhole-supporting exotic matter immune to the constraints of Ford and Roman could in principle exist.

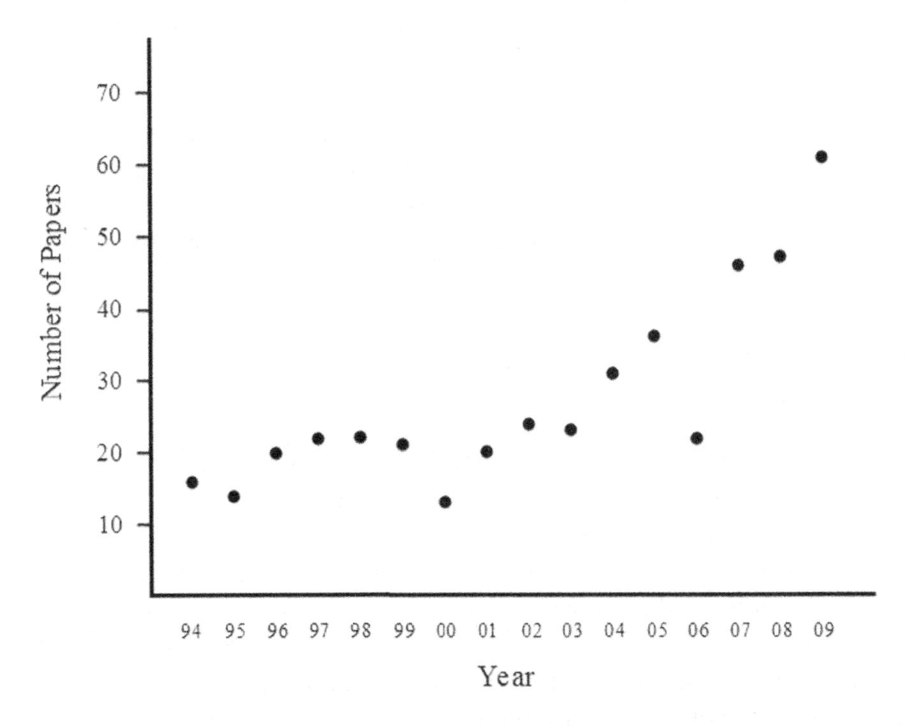

Figure 2.1. Rise in wormhole papers near turn of the 21st century. Number of papers in the physics archive (arxiv.org) with the word "wormhole" in the title for each year shown.

At the turn of the century there was a brief lull in the production of wormhole-related publications. By 2002, however, production returned to pre-millennial levels. Thick-shell wormhole stability studies tended to bear out the belief that such wormholes are unstable. The following year Visser, Sayan Kar, and Naresh Dadhich found that wormholes could be held open with arbitrarily small quantities of exotic matter. Their result was improved in 2004 by its being re-expressed in terms of a coordinate invariant measure of the quantity of exotic matter. Hayward and Koyama found an analytic solution describing the conversion of a black hole into a wormhole by an intense pulse of exotic radiation. A study of wormholes in R^n gravity confirmed and extended the results of the early 1990s. In 2005 previous results were also confirmed, when wormholes supported by "phantom energy" – a special case of exotic matter -- were shown to display no conceptual surprises.

The first decade of the twenty-first century ended with scientific interest in wormholes reaching an all-time high. Investigations included wormhole-based models of dark matter, new violations of the energy conditions by scalar fields, the discovery that wormholes in Einstein-Gauss-Bonnet gravity -- a higher-dimensional theory of gravity that reduces to General Relativity in 4 dimensions – need not require exotic matter, and the birth of wormhole thermodynamics.

In 2010 the intuition of wormhole pioneer John Wheeler seemed to be vindicated, when Alexander Balakin and Alexei Zayats of Russia's Kazan State University working with José Lemos of Portugal's Technical University of Lisbon showed that a Wheeler geon solution is possible for a suitable coupling of gravity to electromagnetism.

Some of the highlights of wormhole history are summarized below.

1916 Ludwig Flamm publishes first paper containing a two-dimensional embedding diagram of a Schwarzschild wormhole.

1924 Hermann Weyl publishes his ideas on the topological nature of electric charges, implying that they are manifestations of tiny wormholes.

1935 Albert Einstein and Nathan Rosen describe their bridge – an untraversable Schwarzschild wormhole.

1955 John Wheeler illustrates his geon concept with the first drawing of a wormhole to appear in the physics literature.

1957 John Wheeler introduces the word "wormhole" to physics.

John Wheeler introduces the idea of quantum fluctuations in spacetime geometry and thus the possibility of virtual wormholes.

1960 John C. Graves and Dieter Brill determine that the Reissner-Nordstrøm wormhole, unlike the Einstein-Rosen bridge, contracts to a minimum diameter and re-expands in an endless cycle.

1962 John Wheeler and Robert Fuller show that the wormhole in the maximally extended Schwarzschild solution (Einstein-Rosen bridge) constricts too rapidly to be traversed.

1963 Roy Kerr finds a rotating black hole solution. Like the Schwarzschild, and Reissner-Nordstrøm solutions, it also contains an apparent bridge to other universes.

1965 Ezra Newman extends Kerr solution to the case of a charged rotating black hole, finding similar wormhole bridges as in the Kerr and Reissner-Nordstrøm solutions.

1968 Roger Penrose shows that the inner horizon of a Reissner-Nordstrøm wormhole is unstable, thus making the wormhole untraversable.

1973 Homer G. Ellis describes his "drainhole", the first traversable wormhole solution.

1985 Carl Sagan asks Kip Thorne to devise a scientifically accurate means of interstellar travel for Sagan's novel *Contact*, thus catalyzing Thorne's investigation of wormhole physics.

1987 Stephen Hawking uses Euclidean wormholes to describe a non-quantum-coherence-preserving quantum field theory for macroscopically flat spacetime.

Steven Gidding and Andrew Strominger use Euclidean wormholes in a system coupling an axion to gravity in order to argue that quantum coherence might be lost through information being swallowed by baby universes.

1988 Michael Morris and Kip Thorne launch the modern age of wormhole physics with detailed descriptions of traversable wormhole solutions.

Morris, Thorne, and Ulvi Yurtsever show that traversable wormholes can be converted into time machines.

Sidney Coleman uses Euclidean wormholes to attempt to explain why the cosmological constant is zero, as it was then believed to be.

1989 Matt Visser devises thin shell wormholes that protect travelers from exposure to exotic matter and tidal forces.

1990 Matt Visser describes a quantum wormhole as a solution to the Wheeler-DeWitt equation in one-degree of freedom.

David Hochberg attempts to determine whether traversable wormholes free of exotic matter exist in R^2 gravity.

Peter D'Eath and D. I. Hughes consider supersymmetric quantum mini-superspace wormholes.

1991 Kazuo Ghoroku and Teruhiko Soma fail to find WEC-respecting wormholes in $R + R^2$ gravity.

1992 Thomas Roman demonstrates the possibility of naturally occurring macroscopic wormholes by showing that quantum primordial wormholes can inflate.

Alty, D'Eath, and Dowker find massless supersymmetric minisuperspace quantum wormholes.

Stephen Hawking's "Chronology Protection Conjecture" suggests the impossibility of traversable wormholes.

1995 Eric Poisson and Matt Visser determine that thin-shell wormholes can be stable depending on the equation of state of exotic matter.

Matt Visser writes the landmark monograph, *Lorentzian Wormholes*, which summarizes the state of the art.

John Cramer et al. specify wormhole microlensing signatures and recommend their use in a search for cosmic wormholes.

1996 Lawrence Ford and Thomas Roman apply quantum inequalities to wormholes and conclude that macroscopic wormholes are only possible if their exotic matter is confined to an ultra-thin shell at their throat.

Pedro González-Díaz finds a "ringhole" solution – a wormhole with toroidal symmetry.

1997 Dan Vollick shows that a wormhole can be held open by the negative interaction energy of a classical scalar field.

David Hochberg, Arkadiy Popov, and Sergey Sushkov find a self consistent wormhole in semiclassical gravity.

1998 Edward Teo finds a rotating wormhole solution.

Dan Vollick finds a wormhole solution in low-energy string theory.

David Hochberg and Matt Visser prove that any traversable wormhole must violate the NEC at or near its throats.

The acceleration of the expansion of the universe is detected, thus raising the possibility of cosmic exotic matter.

1999 Peter Kuhfittig shows that the exotic matter shell of a wormhole can be arbitrarily thin.

Sean Hayward conjectures that black holes and wormholes are different states of the same dynamical object: black holes can be converted into wormholes and vice versa.

2000 Sergei Krasnikov relaxes the asymptotic flatness condition and is able to maintain wormholes using vacuum stress-energy of neutrino, electromagnetic, or scalar fields.

Carlos Barceló and Matt Visser demonstrate classical violations of all energy conditions.

2001 Hayward, Sung-Won Kim and Hyunjoo Lee find analytic solutions that demonstrate black-hole-wormhole interconversion in a model of 2-d gravity coupled to exotic radiation.

2002 Bronnikov and Grinyok, Shinkai and Hayward, and Armendáriz-Picón separately study the stability of traversable wormholes, mostly concluding that they are unstable and might collapse or inflate.

2003 Visser, Sayan Kar, and Naresh Dadhich show that wormholes can be supported by an arbitrarily small quantity of exotic matter.

Marc Kamionkowski and Nevin Weinberg realize that a universe dominated by cosmic exotic matter is doomed to rend itself asunder in a "Big Rip".

2004 Kamal Nandi, Yuan-Zhong Zhang and K.B. Kumar build on Visser et al. to propose a coordinate invariant measure of the quantity of wormhole matter that violates the Averaged Null Energy Condition.

Hiroko Koyama and Sean Hayward find an analytic solution describing the conversion of a black hole into a wormhole via an exotic radiation pulse.

Pedro González-Díaz proposes the "Big Trip" scenario in which the universe of the distant future is swallowed by a giant wormhole and travels through it.

2005 Sergey Shuskov and Francisco Lobo independently show that cosmic exotic matter (phantom energy) can support static wormholes.

Francisco Lobo studies stability of phantom energy wormholes finding them stable under certain conditions.

2006 Pedro González-Díaz finds that the Big Trip is only possible in the context of a multiverse of parallel universes.

2007 Elias Gravanis of Kings College London and Steven Willison of the Center for Scientific Studies in Chile show that wormholes in Einstein-Gauss-Bonnet gravity need not require exotic matter.

A. Kirillov and E. Savelova from Russia's Uljanovsk State University propose that dark matter might be due to a gas of wormholes.

2009 Lobo and Miguel Oliveira find extensions of general relativity that support wormholes in the absence of exotic matter.

Sean Hayward and at Shanghai Normal University and separately Prado Martín-Moruno and Pedro González-Díaz at the Spanish National Research Council elucidate the laws of wormhole thermodynamics.

Douglas Urban and Ken Olum of Tufts University notice that arbitrarily great violations of the averaged null energy condition are possible for particular quantum states of a scalar field.

2010 Alexnder Balakin, Alexei Zayats, and José Lemos find that geons are possible for a suitable coupling of gravity and electromagnetism.

3. Preliminaries

The study of wormholes relies on the two great pillars of twentieth-century physics – relativity and quantum theory. Because these concern situations beyond normal human experience, they were discovered late in our history. In particular, the branch of relativity known as special relativity describes objects traveling at inordinately high speeds – near that of light. General relativity considers in addition regions of exceedingly high gravity such as near the surface of a dense star. Quantum theory describes objects too small to be seen with our most powerful optical microscopes. The ideas of relativity – special and general – and those of quantum theory have been elaborated in many a weighty tome penned over the last century. Here we will wade into these subjects only as far as needed to support subsequent chapters.

Special Relativity

Special relativity applies where gravity is too weak to noticeably curve spacetime. An example of such a place is the surface of the earth. Although gravity is of course present here, it is too weak for us to notice the tiny deviation from Euclidean geometry that its associated curvature has induced. The sum of the angles of any human-scale triangle differs only imperceptibly from its flat-space value of 180 degrees. However, for larger triangles such deviations begin to be noticeable. The triangles formed by lines interconnecting you and any two satellites of the Global Positioning System are large enough for such deviations to begin to matter. For this reason GPS software relies on *general* relativity. As a rule, for gravity of any particular strength there exists a particular length scale below which spacetime is approximately flat, and special relativity applies. This is perfectly analogous to the surface of the earth being noticeably curved at large scales, but being approximately flat on the scale of a few miles.

Special relativity follows completely from two postulates and an understanding of the concept of an "inertial reference frame".

Inertial Reference Frames

An inertial reference frame is the frame of reference of any observer who experiences no acceleration, i.e. that of any observer whose motion is uniform. "A frame of reference" is just another way of saying "a coordinate system". A coordinate system is just a systematic way of assigning a set of numbers to every point. For example, any point on the earth can be assigned a meaningful pair of numbers – a latitude and longitude. In three dimensions we assign a triple of numbers, e.g. latitude, longitude, and altitude. In spacetime we assign four numbers – three for location in space, one for location in time.

Suppose that you convert a room-sized wooden box into a laboratory. You paint its interior white. You then paint a blue coordinate system grid on the white floor of the box – the x-y plane. You ensure that each line in your grid has only one of two orientations – North-South or East-West. You also paint horizontal coordinate lines on the walls – the z coordinates. You choose one corner of the laboratory to be the origin of the coordinate system. You can now assign a triple of numbers (x, y, z) that specify the position of any particle in the lab. The first two numbers locate its position on the floor relative to the origin (the special corner) of the lab. The third number assigns a height above this position. Lastly, you mount a clock on a lab wall. You may now conduct experiments in which you keep track of the positions of the particle involved and the times t at which they occupied these positions. [Recording video of your experiments from 3 different angles (at least one of which includes a view of the clock) would be a way to do this.] Your experiments allow you to rediscover Newton's laws, which you express mathematically in terms of the blue coordinate system.

Now you use a crane to lift your box laboratory and rotate it by, say, 17 degrees about a vertical axis through the lab's origin, and set the lab down again. Because of this rotation the grid of blue lines on the floor of your lab no longer runs North-South and East-West. So you overlay the previous blue coordinate grid lines with a freshly painted red coordinate grid, whose lines do run North-South and East-West. To be thorough, you also replace the blue grid lines on the walls with red ones. You perform the same experiments. This time, however, you keep track of the position of the particles using the red coordinates system. You are now able to express Newton's laws in terms of the red coordinates system. Your experiments have the same qualitative results as before. Yet you find that your numerical expression of the regularities described by Newton's laws differ between the blue coordinates and the red ones. This is to be expected, because the coordinates of any point on the lab bench – on which you are colliding ball bearings and performing other such experiments – differ between the red and blue coordinate systems.

Newton's laws do not change, simply because you rotated the laboratory. There must be some sort of mathematical entity that reconciles your expression of Newton's laws in the blue coordinates and that in the red coordinates. There is. It's called a *tensor.* A tensor is like a character in a play, such as Othello in *Othello, the Moor of Venice.* In any particular staging of the play, Othello's particular characteristics will be unique. The actor playing him will possess unique physical attributes and deliver a unique performance. The character, however, remains the same -- Othello. Similarly, the actual numbers representing a tensor, say the force tensor, in any particular coordinate system are unique to that system. Yet the tensor remains the same – the force. [The tensors that appear in Newton's laws are tensors of rank 1, more commonly known as *vectors*.]

Each orientation of the laboratory defines a reference frame. Newton's laws are clearly the same in each of these. In other words, these laws are *covariant* under rotations, a particular type of coordinate transformation. Just as a staging of *Othello* in New York will have the same form as when another company stages it in London, so it is with Newton's Laws. When expressed in terms of tensors Newton's laws have exactly the same form in each reference frame, the precise meaning of covariance.

In special relativity we consider another type of rotation. Instead of a rotation that changes the coordinates for the floor of our laboratory, the *x-y* plane, we consider rotations involving time that change the coordinates of the *x-t* plane (or *y-t* or *z-t*). Each "orientation" of our laboratory with respect to this new type of rotation also defines a reference frame – an *inertial reference frame*. We find that the tensor expressions of Newton's laws do *not* retain their form from one inertial reference frame to another. They are *not* covariant under this type of rotation. Einstein believed in essence that the laws of physics (expressed in terms of tensors) should retain their form not merely under ordinary rotations, but also under rotations involving time. The set of coordinate transformations consisting of both types of rotations – the usual ones and those involving time – is called the *Lorentz transformations*. Einstein proceeded to replace Newton's laws, with new laws that retain their form under these transformations. Just as the physical characteristics of Othello changes for each staging of the play, so do the values of the components of the tensors in these new laws change for each inertial reference frame. Special relativity consists of nothing more than an elaboration of these changes, especially those involving lengths, durations, and velocities. It turns out that an observer's velocity determines the "angle" associated with any of the aforementioned rotations involving time. The spatial reference frame of an observer facing North is related to that of an observer facing East by a purely spatial rotation. But the *inertial* reference frame of a stationary observer is related to that of an observer moving at 50% of the speed of light by a spatio-temporal rotation. A particular angle (defined with respect to an arbitrary choice for the spatial frame corresponding to 0 degrees) specifies a particular spatial reference frame. Similarly, a particular velocity (defined with respect to an arbitrary choice for the inertial reference frame corresponding to 0% of the speed of light) specifies a particular inertial reference frame. As stated above, the concept of the inertial reference frame allows the whole of special relativity to follow from only two postulates, to whose explicit statements we now turn.

The Two Postulates

1. *The Principle of Relativity: The laws of physics are the same in all inertial reference frames (i.e. their tensorial expression retains the same form in each such frame).*

2. *The Constancy of the Speed of Light: That the speed of light is 3.00×10^8 meters per second in a vacuum is a law of physics.*

Consequences of the Postulates

These consequences are most easily described from the point of view a particular observer, who we will call "the stationary observer". Another observer, "the moving observer", travels uniformly in a straight line at a speed v relative to the stationary observer. Some of the consequences of the postulates of special relativity are as follows.

• No absolute definition of simultaneity: Events that appear to occur simultaneously to the stationary observer will not so appear to the moving observer.

• Time dilation: The moving observer's watch appears to the stationary observer to tick more slowly than his watch.

• Length contraction: The moving observer's meter stick oriented in the direction of her motion appears to the stationary observer to be shorter than his meter stick.

• Equivalence of mass and energy: The mass of a stationary object can be converted to a quantity of energy given by the product of the object's mass and the speed of light squared ($E = mc^2$).

• Mass divergence: The mass of the moving object will appear to the stationary observer to increase in a manner that diverges toward infinity as the speed of the object approaches that of light.

• No superluminal speeds: No finite impulse exists that can accelerate an object to a speed beyond that of light.

• Counter-intuitive velocity addition law: A bullet fired with speed u by the moving observer, who moves with speed v relative to the stationary observer, will not appear to the stationary observer to have a total speed of $v + u$; it will have a lesser speed, thus ensuring that the total speed of the bullet does not exceed that of light.

Although special relativity will be useful in understanding certain peripheral aspects of wormholes, these objects are actually manifestations of general relativity, a subject without which they cannot be understood.

General Relativity

The Equivalence Principle

Newtonian physics has a strange feature. It defines two types of mass – "inertial mass" and "gravitational mass". Inertial mass is measured by applying Newton's Second Law of Motion: (inertial mass) = (Force) / (acceleration of the mass due to the force), or in symbols

$$m_i = \frac{F}{a} \qquad (1)$$

In other words, you obtain the inertial mass of an object by applying a known force to it – *any* force. You then measure its resulting acceleration. Lastly, you take the ratio of these values.

The gravitational mass of the same object is a measure of the strength of the gravitational force between it and another object. If, for example, the gravitational mass of the object is m_g and that of the earth is M_g, the gravitational force F_g between the object and the earth, which is also know as the object's weight, is according to Newton's Universal Law of Gravitation

$$F_g = -\frac{GM_g m_g}{r^2} \tag{2}$$

where r is the distance between the center of the object and that of the earth, G is a constant, and the minus sign indicates that the force points downward, toward the center of the earth. A little algebra tells us that this implies that the gravitational mass of the object is

$$m_g = \frac{F_g}{-GM_g/r^2} \tag{3}$$

Now here is the strange part. Even though the definition of the inertial mass of an object (equation 1) differs from the definition of its gravitational mass (equation 3), we always find that the inertial mass and the gravitational mass are equal,

$$m_i = m_g \,. \tag{4}$$

Why should this be? Why should two seemingly distinct quantities, obtained in two completely different ways have the same value? Why should a measure (m_g) of the strength of an object's attraction to other bodies via the gravitational force be the same as a measure (m_i) of the object's resistance to acceleration by *any* force? Or, to put it another way, why is gravity unlike any other force in that it accelerates objects at a rate independent of their intrinsic properties? Why should a grain of sand, a bowling ball, and a grand piano all accelerate toward the ground at the same rate?

Newton took this equivalence of gravitational and inertial mass as a postulate.

PP1. Newton's Equivalence Principle: Inertial mass is equal to gravitational mass.

Einstein took Newton's principle further. He in effect considered two situations.

Situation 1: A woman stands on a scale in a windowless, hermetically sealed box that is sitting on the surface of the earth (Figure 3.1a). The woman's *gravitational* mass is given by her weight, as measured by the scale on which she stands, divided by the geophysical quantity shown as the denominator of equation (3) with r being the radius of the earth. This geophysical quantity is the familiar acceleration of gravity at the surface of the earth.

Situation 2: A woman stands on a scale in a windowless, hermetically sealed box that is being accelerated in empty space so that the scale reads exactly as it did in Situation 1 (Figure 3.1b). The woman's *inertial* mass is given by the accelerating force, as measured by the scale on which she stands, divided by the acceleration of the box.

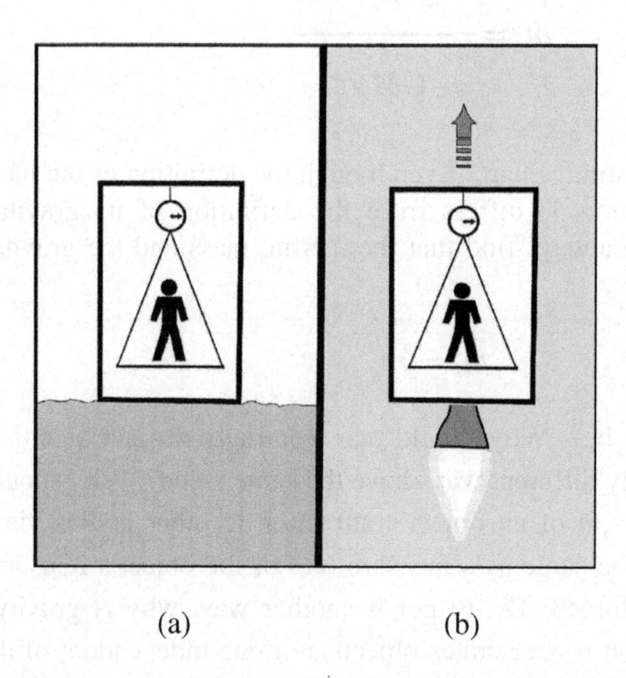

Figure 3.1. Equivalence of inertial and gravitational mass. Weight of an experimenter on the surface of the earth (a) is the same as her weight in a space capsule accelerating at 1 g (b).

We know that the woman's gravitational mass measured in Situation 1 will be the same as her inertial mass measured in Situation 2. Einstein explained this by postulating that Situation 1 and Situation 2 are fully equivalent. In other words, we have

PP2a. *Einstein's Equivalence Principle – Specific Form: Any experiment performed within a sealed laboratory in a uniform gravitational field with acceleration of gravity g will have the same result when it is performed within any other sealed laboratory undergoing an acceleration g through empty space.*

In other words, the laws of nature do not depend on how the weight is generated. Whether it is due to a planet's gravity or the acceleration produced by a rocket engine's thrust, the laws of nature are indifferent. This idea can be stated more generally.

PP2b. *Einstein's Equivalence Principle – General Form: Any experiment performed within a sealed laboratory, within which a particular apparent acceleration of gravity is measured, will have the same result when it is performed within any other sealed laboratory, within which the same apparent acceleration of gravity is measured, irrespective of this laboratory's motion or location within the universe.*

The reason that the laboratories mentioned in these statements of the equivalence principle must be sealed is to prevent experiments that probe the exterior of the laboratory (e.g. by peering through a window). The results of such experiments could easily differentiate between an accelerating laboratory and a stationary one and thus violate the spirit of the equivalence principle: The local physics within two laboratories is the same, as long as the locally measured gravitational fields are the same, *and there are no other observable differences.*

What the Equivalence Principle Does Not Say – Mach's Principle

One might be tempted to extend the equivalence principle to cover the case in which the laboratories are not sealed. Let us do this by considering transparent laboratories. We might then suppose that the equivalence principle could be strengthened to handle this case, as follows:

Hypothetical Equivalence Principle – Most General Form: The forms of the laws of nature do not depend on the motion or location within the universe of the ***unsealed*** *laboratory in which these laws are demonstrated.* [*WRONG!*]

Here is a consequence of this hypothetical most general form. Consider a transparent laboratory that is being rotated about a distant central axis to create an effective 1-g gravity field, rather like a section of the rim component of the wagon-wheel-shaped space stations typical of older science fiction. If an occupant of this space station were to look through its transparent floor, he would see distant stars moving past as the laboratory rotates. An observer in a transparent laboratory on earth would, of course, not see this when he looks downward through the transparent floor of his laboratory. He would just see the ground. Yet the laws of physics (according to our hypothetical principle) would retain their form, because the same equations would be used to calculate the 1-g field. In one case the matter source in the equations would be all of the matter in the universe moving precisely as seen from within the moving section of the space station. In the other case the matter source would simply be the earth. The point is that the equations – the *form* of the laws – would be the same despite different values for their matter related variables. Or would they?

The answer, surprisingly, is "no". Einstein found this surprising as well. He was inspired by an idea tantamount to our hypothetical most general form of the equivalence principle. This idea is called Mach's Principle after Ernst Mach, the nineteenth century philosopher who was its most ardent promoter. Mach's Principle is notoriously vague. This explains why fringe controversies regarding it have survived into the twenty-first century. To the extent that his principle can be made concrete, we have according to Mach the following assertion. If the distant stars revolve *en masse* about your position, it will cause you to feel a centrifugal force. That force, moreover, will be indistinguishable from what you would feel if, instead, the stars were to remain stationary, while you rotate your body at the same angular rate. This seems to make sense. How, after all, could you possibly distinguish between the stars revolving around you and your body rotating at the same rate? Would you not see the same thing in both cases?

Yes, you would see the same thing, but you would not feel the same thing. Were you to close your eyes, you could feel the rotation of your body, but not the revolution of the stars about you. The situations, then, are not identical. Mach's principle, it turns out, is false.

Return to the comparison of the terrestrial observer and the observer in the rotating space station. Because the terrestrial observer (correctly) obtains his local gravitational field *solely* by considering the matter distribution he observes below his feet (the earth), we supposed (incorrectly) that the space-based observer can do the same. He cannot. The acceleration-induced force that he experiences is not due to distant stars. The terrestrial observer need not consider acceleration-induced forces, because he is not accelerating (except negligibly due to the planet's rotation). In short, our hypothetical most general form of the equivalence principle fails, because, unlike the specific and general forms above, it attempts to describe a global symmetry (unsealed labs) that does not exist, rather than the local one (sealed labs) that does. So it is with Mach's Principle. Despite its motivating effect on Einstein, Mach's principle is not part of general relativity.

PP2c. *Mach's Principle – the idea that an observer cannot distinguish between his own acceleration and the (visually indistinguishable) acceleration of distant masses – is not part of general relativity.*

There is, however, a sort of ghost of Mach's principle that does in fact survive within general relativity. According to the theory the revolution of the distant stars *en masse* – were that somehow to occur -- would have a measurable local effect. It would be miniscule, however. It would be far less intense than what you would feel were you to spin your body at the same angular rate. The effect would be limited to the slow precession of local gyroscopes in the direction of the stellar revolution. This is called the *Lense-Thirring effect.* It has been confirmed experimentally by satellite-based laser ranging measurements in the late 1990s. In order to check these confirmations, NASA launched in 2004 *Gravity Probe B* – a satellite containing the world's most precisely fabricated gyroscopes. In this experiment the rotating mass that induces the Lense-Thirring "frame dragging" is not that of the distant stars, but that of the nearby earth. Though the distributions of spinning mass differ, the procedure for calculating their effect in general relativity is the same.

The Modern View of the Equivalence Principle

Since about 1960 the term "Equivalence Principle" has come to mean one of three things:

1) The Weak Equivalence Principle – *the trajectory of bodies falling in a gravitational field does not depend on what the bodies are made of.*

This is tantamount to Newton's equivalence principle: inertial mass = gravitational mass. This version is what experimentalists usually use.

2) The Strong Equivalence Principle – *Any experiment performed within a free-falling sealed laboratory will have the same result,*

when it is performed within an unaccelerated sealed laboratory in empty space.

This is just the general form (PP2b) specialized to the case of a measured acceleration of gravity of zero.

3) The Einstein Equivalence Principle – *Any non-gravitational experiment performed within a free-falling sealed laboratory will have the same result, when it is performed within an unaccelerated sealed laboratory in empty space.*

This is merely the Strong Equivalence Principle (version 2 above) specialized to the case of non-gravitational experiments. These modern statements of the principle all involve free fall. Einstein's original statement of it -- gravity = acceleration – did not. What is the connection? It is this. If gravity = acceleration, then the absence of gravity = the absence of acceleration. Or, to be more specific, the absence of gravity within a laboratory in a gravitational field is equivalent to the absence of acceleration within a laboratory in empty space. Put another way, physics within a free-falling laboratory is equivalent to physics within an unaccelerated one in empty space.

One way to summarize these ideas is to say that within a sufficiently small free-falling laboratory, *spacetime* – the union of space and time into a single geometrical object -- is flat, just as it is in the absence of gravitating bodies. This is true, even if it is obscured by one's choice of coordinate system. Within this laboratory it is always possible to choose a simplifying rectilinear coordinate system through which the local flatness of spacetime is obvious. That such a choice is always possible is a consequence of what is called the *diffeomorphism invariance* of general relativity. Diffeomorphism invariance (more commonly known as *general covariance* before about 1975) is simply this: The laws of physics do not depend on the coordinate system in which they are expressed.

Gravitational Field Equations

Once Einstein had his equivalence principle, he could deduce that spacetime is curved. He considered a light beam in an accelerating elevator, shown in Figure 3.2. As the light beam moves horizontally toward the opposite wall, the wall accelerates away from its previous speed and position. The resulting beam appears to curve downward. By the equivalence principle, the gravitational field of a star or planet must also cause light to bend. Einstein knew that light follows Fermat's Law – it seeks the shortest path between two points.

The only way in which the shortest path between two points is curved is if space itself is curved.

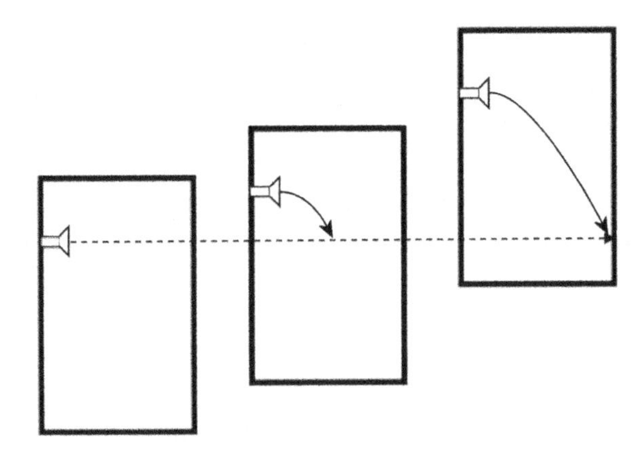

Figure 3.2. Acceleration bends light. A pulse of light is emitted within an accelerating glass elevator. An external observer (whose constant velocity matched that of the elevator at the moment of emission) sees the pulse travel in a straight line. A passenger within the elevator sees the light bend. [Of course, no human passenger could survive the extreme acceleration depicted.]

Now Einstein had all of the concepts that he needed to work out the equations of general relativity. He knew that the equations must have the same form, irrespective of the motion of the frame of reference. He knew that space must be curved by gravitational fields, which are in turn produced by the presence of matter. He also new that time must enter on the same footing as space, so that he could recover his special theory of relativity in the limit of vanishing gravitational fields and unaccelerated motion.

The field equations that he ultimately derived may be roughly translated as follows. At any point in spacetime

$$\text{“essential” curvature of spacetime} = \text{stress-energy of matter.} \qquad (5)$$

Einstein's friend, Marcel Grossman, had introduced him to Riemannian geometry. So Einstein was aware of a well known method of defining the curvature of spacetime. He was also keenly aware that curvature defined in this way – i.e. using the Riemann tensor – was not directly related to the stress energy of matter. Just as it is possible to have a nonzero electric field in the absence of charge, it was clear that it was possible to have a nonzero curvature (as determined by the Riemann tensor) in the absence of matter. The big problem facing Einstein was that of figuring out which part of the curvature of spacetime – what I've called the "essential" part – is produced by the stress energy of matter. By late 1915 he had solved this problem. Essential curvature is quantified by an object that we now call the Einstein tensor.

For the stress-energy of matter Einstein used a mathematical object (the stress-energy tensor) that specifies the mass-energy density of matter, its pressure (stress) in each of the spatial directions, and any currents of mass-energy that the matter might contain. We use the plural in describing the field equations, because there are actually several closely related equations – one equation for each possible pairing of two of the four spacetime dimensions (e.g. *tx*, *xx*, *yx*, *tt*, etc. for the spacetime dimensions *t*, *x*, *y*, and *z*).

The spacetime underlying Einstein's equations is an example of a mathematical entity called a *manifold*. To understand what a manifold is, we must first become familiar with the concept of R^n. R^n is simply the *n*-dimensional generalization of the real number line R. R, also known as R^1, is the infinite set of numbers – integers and nonintegers – between negative and positive infinity. With R we can define a real number plane, R^2, as RxR. This is the familiar plane in which points are identified by their *x* and *y* coordinates. R^n is the continuation of this procedure from 2 to *n* dimensions. A point in R^n, then, is identified by *n* coordinates. A manifold, roughly, is any set that can be divided into subsets that can each be brought into one-to-one correspondence with a subset of R^n (for some particular value of *n*). As an example of a set that is *not* a manifold, consider the set of all types of fruit. Unlike the case in R^n, we cannot always find in this set a third element (called a "point" in R^n) that lies between any two other elements. What lies, for example, between banana and orange? Here is an example of a set that is a manifold: the set of all points on an idealized "rubber" sheet. This sheet is idealized in the sense that it is not made of rubber molecules but of some indivisible substance. For any two points on the sheet, there is always a third point (actually an infinite number of points) between the two points. Just like R^n. In short, you may think of a manifold as any set that locally resembles R^n.

For the curvature of spacetime Einstein used in his equations another mathematical object (also a tensor) invented by Riemann in the mid nineteenth century. This object, the Riemann tensor, depends on another object – the metric tensor.

The metric tensor specifies the distance between two infinitesimally close points in spacetime. Imagine a rubber sheet. Draw a rectilinear *x*-*y* coordinate grid on the sheet (Figure 3.3a). Consider two nearby points on this sheet. Let the difference between the *x* coordinates of the points be *dx* and that between the y coordinates be dy. Then the distance ds between these points is given by the Pythagorean theorem

$$ds^2 = dx^2 + dy^2. \tag{6}$$

Imagine now that the rubber sheet is stretched and distorted in some arbitrary manner (Figure 3.3b). The coordinate grid is no longer rectilinear. We can no longer use Pythagoras' theorem to obtain the distance between nearby points. We must instead use a generalization of Pythagoras' theorem,

$$ds^2 = g_{11}dx^2 + g_{12}dxdy + g_{21}dydx + g_{22}dy^2. \tag{7}$$

This expression is called a *line element*. The four numbers – $g_{11}, g_{12}, g_{21}, g_{22}$ – are components of the metric tensor. [Actually, there are only three components, because g_{21} must equal g_{12}.] The metric tensor is an example of what is called a tensor of rank 2. Such a tensor is a two-dimensional matrix, the value

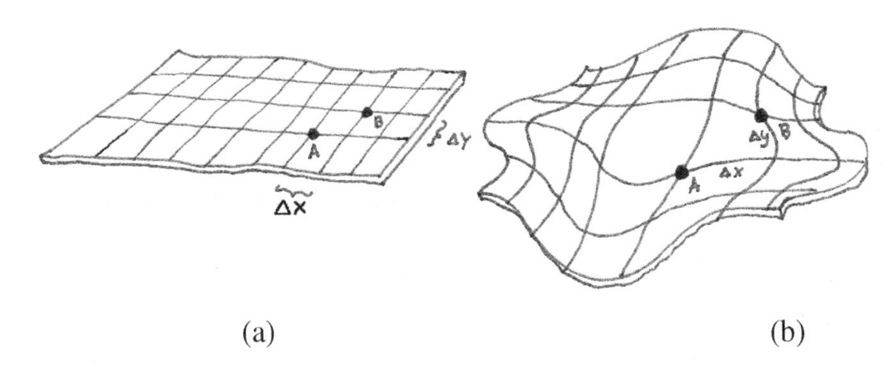

Figure 3.3. The role of the metric tensor. (a) The distance Δs between Point A and Point B can be obtained using the Pythagorean Theorem: $\Delta s^2 = \Delta x^2 + \Delta y^2$. (b) The Pythagorean Theorem no longer applies. Instead, we must use the formula, $\Delta s^2 = g_{xx}\Delta x^2 + 2g_{xy}\Delta x\Delta y + g_{yy}\Delta y^2$.

of whose components change according to a particular rule when the matrix is expressed in different coordinate systems. These rules ensure that field equations written exclusively in terms of tensors retain the same form in different reference frames. This is precisely the property that Einstein required to mathematically implement his equivalence principle.

Instead of describing a two-dimensional rubber sheet the metric tensor of general relativity describes four-dimensional spacetime. There are a few quick facts that it helps to know about the line element for spacetime. It is always possible to find a coordinate system so that within a sufficiently small local region the line element looks the way it does in flat spacetime, namely

$$ds^2 = g_{tt}dt^2 + g_{xx}dx^2 + g_{yy}dy^2 + g_{zz}dz^2 \tag{8}$$

with

$$g_{tt} = -1,\ g_{xx} = g_{yy} = g_{zz} = 1. \tag{9}$$

The metric tensor components that appear in equations (8) and (9) are called diagonal components. There is one for each coordinate of spacetime. Notice that all of the spatial components of the metric tensor are positive, but the temporal component is negative. Time is the odd man out, which agrees with our experience. These metric tensor components are in general functions of space and time. In certain regions of spacetime an amazing thing occurs. The coordinate that is the odd man out changes. For example, instead of equation (9), we could have

$$g_{xx} = -1,\ g_{tt} = g_{yy} = g_{zz} = 1. \tag{10}$$

The x and t coordinates have exchanged their meanings. Our sense of moving from the past to the future would in this case be associated with moving in the x direction. While movement back and forth in the t direction would be no different than moving in any other spatial direction. When g_{tt} goes from being negative in one region (normal spacetime) to being positive in another, it is possible to find the points in spacetime at which $g_{tt} = 0$. These points define a closed surface called an *event horizon** which encloses a region from which nothing can escape. The reason that nothing can escape is precisely the same as the reason that you cannot move backward in time. When the time coordinate and a spatial coordinate exchange roles in the region bounded by the event horizon, the one-way nature of movement in time is transferred to the spatial coordinate. The event horizon can only be crossed in one direction (inward). We have, then,

PP3. *The coordinate that functions as a local time coordinate is that whose diagonal metric tensor component is negative.*

---

* Actually, it defines a closely related surface called an apparent horizon, which may be thought of as a "predicted" event horizon. In the case of a rotating black hole this condition defines a surface called the static limit.

***PP4.** A change in the local role of a coordinate from temporal to spatial is signaled by its diagonal metric tensor component becoming zero.*

The Riemann tensor, the measure of local curvature, depends on how the metric tensor changes (and how the rate of change of the metric tensor changes) in each of the spacetime dimensions.

There are two ways to use the field equations, whose meaning is summarized in equation (5). The traditional method is to globally specify a stress-energy of matter, and then find a metric tensor that generates a curvature that solves the equation. This is difficult. Obtaining exact solutions in this manner has only been possible in a few cases, some of which we will discuss in Chapter 5. The other method -- favored by curious physics undergraduates and circa-1988 wormhole researchers -- is to choose a metric, insert it into the left side of equation (5), and thus obtain the corresponding distribution of stress-energy. This is easy.

Why do we care about the curvature of spacetime induced by matter? We care because the curvature of spacetime determines the motion of particles of matter. When a particle moves solely under the influence of gravity, it travels along the special paths in curved spacetime. Just as a particle in flat space moves in straight lines, a particle in curved space moves along *geodesics*. Geodesics, examples of which are the straight-line trajectories in flat space, are the shortest paths between two points. Other examples of geodesics are the lines of longitude on the surface of the earth. These are the shortest paths on the surface of the earth between the north and south poles. Like the geodesics of an arbitrary spacetime, they are curved.

There is another aspect of general relativity that is important to know. The presence of a gravitational field can cause time to slow down. A person living in close orbit around a massive black hole or wormhole ages less rapidly than a person living in deep space. If you are reading this on the surface of the earth, you are aging slightly less rapidly than are astronauts in orbit overhead. The gravitational fields of black holes, wormholes, and the earth are each a special case of a family of solutions to the Einstein field equations for which we can say the following.

***PP5.** The stronger the gravitational field, the lesser the amount of time that elapses in its presence relative to that that elapses beyond the field.*

While this is true for the gravitational fields produced by objects whose matter is entirely confined within a spherical region of some size – planets, stars, and black holes – it is not true of all gravitational fields, including those of certain wormhole solutions that have been considered. This is because the slow down of time is due to a particular feature of the gravitational field. The field must be the result of the curvature of both space *and* time. Gravitational fields resulting solely from the curvature of space do not slow the elapse of time.

Matter in General Relativity

As mentioned above, matter enters general relativity through the right-hand side of the Einstein equations, which is caricatured by equation (5). Physicists have long believed that for the Einstein equations to describe a spacetime that corresponds to reality, the stress-energy inserted into the equation's right-hand side had to be "physically reasonable". Their definitions of physical reasonableness are called *energy conditions.* For example, they once believed that no observer could ever be confronted by a chunk of negative energy density sitting at rest in his laboratory. Though seemingly reasonable, this idea (the weak energy condition) turns out to be false.

Energy conditions are not part of general relativity. They are external ideas about the nature of matter imported for the purpose of taming the stress-energy tensor and thereby eliminating unphysical spacetimes. These ideas have had to be revised several times over the last few decades. This occurred whenever new theoretical or experimental results demonstrated the invalidity of the reigning energy condition. If we assume that the matter of interest can be characterized by an equal pressure p in each direction and an energy density ρ, then the various energy conditions are defined as restrictions on the possible values of p and ρ. The conditions are of two types – pointwise conditions and averaged conditions. Pointwise conditions must hold at each point of spacetime. Averaged conditions need only hold on average over an observer's path in spacetime, her so-called "world line". In each of the expressions listed below the speed of light is in "natural" units in which its value is precisely 1. The density ρ that appears below is actually ρc^2, which makes it clear that the ρ that appears below and the pressure p have the same dimensions.

The commonly considered averaged conditions are defined as follows.

Averaged SEC (ASEC):	SEC holds on average over the path of any observer.
Averaged WEC (AWEC):	WEC holds on average over the path of any observer.
Averaged NEC (ANEC):	NEC holds on average over the path of any light ray.

Pointwise conditions are summarized in the table below.

Pointwise Energy Condition	**Meaning**	**Definition for Perfect Fluid**	**Violations**
Trace (TEC)	Density must equal or exceed sum of principal pressures	$\rho - 3p \geq 0$	Neutron stars, Slowly expanding/contracting universe with rapidly accelerating contraction
Strong (SEC)	Matter must gravitate toward matter	$\rho + p \geq 0$ $\rho + 3p \geq 0$	Accelerating expansion of universe, Inflation, Very "explosive" Big Bang, Scalar field (minimally coupled)
Dominant (DEC)	Energy must not flow faster than light	$\rho \geq 0$ $-\rho \leq p \leq \rho$	Casimir effect, Squeezed vacuum, Near black hole horizon, Scalar field (nonminimally coupled), Slowly expanding/contracting spatially open universe, Very "implosive" Big Crunch
Weak (WEC)	Local mass-energy density must not be negative.	$\rho \geq 0$ $\rho + p \geq 0$	Casimir effect, Squeezed vacuum, Near black hole horizon, Scalar field (nonminimally coupled), Slowly expanding/contracting spatially open universe with rapidly accelerating contraction
Null (NEC)	Stress-energy experienced by light ray must not be negative	$\rho + p \geq 0$	Casimir effect, Squeezed vacuum, Near black hole horizon, Scalar field (nonminimally coupled)

The pointwise definitions are graphed in Figure 3.4 and their relationships mapped in Figure 3.5.

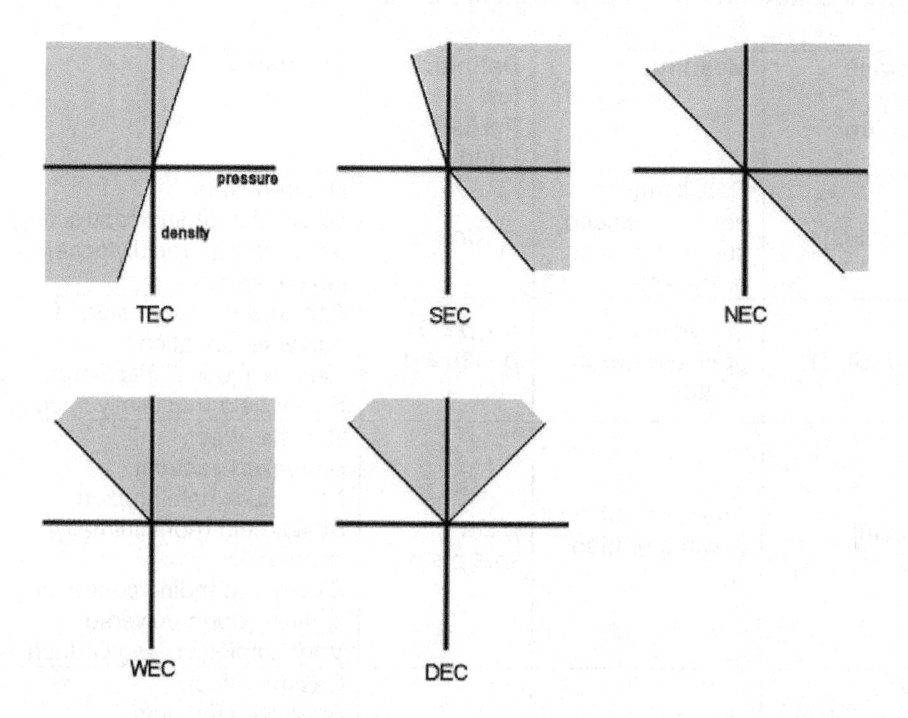

Figure 3.4. Energy Conditions. Density (vertical axes) is plotted against pressure (horizontal axes). Shaded regions show density-pressure combinations deemed to be reasonable by each energy condition. [When the TEC was in favor (before the mid 1960s), it is likely that density and pressure were both tacitly assumed to be positive.]

Notice that a violation of the NEC -- the weakest (least stringent) of all the pointwise energy conditions -- does not necessarily imply a negative matter density. This is because pressure can be negative, as we shall see in subsequent chapters. It is the matter's equation of state – how its pressure depends on its density – that determines whether negative pressures occur for positive densities.

The trace energy condition was called into doubt when it was shown in 1961 to be violated macroscopically by an ordinary quantum field. It soon fell into disuse, when it was understood to be inconsistent with the equation of state of neutron stars. The massive scalar field that drives the cosmological inflation proposed in 1979 violates the strong energy condition. The SEC was dealt a mortal blow when in 1998 astronomers discovered the ongoing acceleration of the universe's expansion, which is driven presumably by such a scalar field. By the late 1980s physicists had come to acknowledge that quantum effects

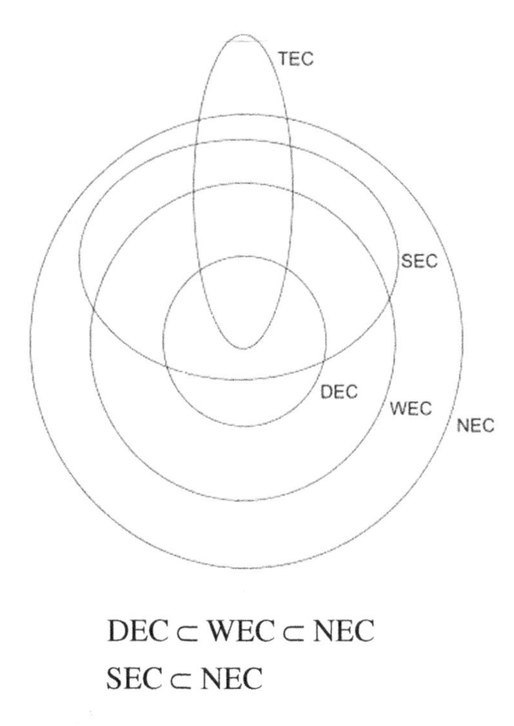

Figure 3.5. How the energy conditions relate. Each point on this plane represents a particular pair of matter descriptors (pressure, density). Points contained within a particular closed curve are those that satisfy the corresponding energy condition. Note the DEC is contained within the WEC, which is contained within the NEC. The SEC is contained within the NEC. The TEC is not contained within (i.e. is not a special case of) any single energy condition. Alternatively, one can say that adherence to the DEC implies adherence to the WEC, which implies adherence to the NEC, i.e. DEC ➔ WEC ➔ NEC. Similarly, SEC ➔ NEC.

violate the dominant, weak, and null energy conditions. They had by this time embraced a fallback position. The view emerged that the averaged versions of the least stringent of the energy conditions – the AWEC and the ANEC – would apply to all matter. They had hoped that quantum violations of the pointwise DEC, NEC, and WEC were local anomalies that did not endure on average. This turned out not to be the case. We now know that the same quantum effects can also violate the AWEC and the ANEC. Moreover, certain widely held classical field theories violate them as well. In short,

PP6. *Matter in general relativity has been regulated traditionally by externally imposed ideas about physical reasonableness called "energy conditions". All of these energy conditions are now known to be violated.*

General Relativity in a Nutshell

The conceptual content of general relativity can be summarized as follows.

PP7. *The laws of physics do not depend on the state of motion or the location in which they are experienced, i.e. there is no preferred state of motion or location.*

PP8. *Matter curves spacetime, in a manner specified by Einstein's field equations, in order to produce what is known as a gravitational field.*

PP9. *Matter solely under the influence of a gravitational field moves along geodesics of spacetime.*

As John Wheeler put it, "Matter tells spacetime how to curve; spacetime tells matter how to move."

Quantum Theory

It is difficult to describe what quantum theory is without immediately delving into controversy. While there are several rival schools of thought, three of these "interpretations" of the theory are of the greatest consequence. They are the Copenhagen interpretation, the Many Worlds Interpretation, and the Consistent Histories interpretation. The reason that we need to consider them at this point, is that I want to tell you what quantum theory *is*. In other words, I want to make a statement about physics itself. Such a statement is by definition metaphysical. It is precisely in the realm of metaphysics that the interpretations of quantum theory differ.

PP10. *The major interpretations of quantum theory differ on metaphysical grounds, i.e. on their assumptions concerning the nature of reality and its relation to the theory.*

The Copenhagen Interpretation

An Outgrowth of Logical Positivism

This is the interpretation to which the majority of physicists subscribe. It holds the greatest sway amongst the older generations of physicists. It is a manifestation of *logical positivism*, an influential philosophy that arose in Vienna during the late 1920s. This anti-metaphysical doctrine influenced Niels Bohr and his collaborators at the Bohr Institute in Copenhagen, who were at that time laying the foundations of quantum theory. Logical positivism's fundamental tenets are 1) there is no reality beyond our sense experiences, and 2) any statement that is not experimentally verifiable is meaningless. Its application to quantum theory is known as *instrumentalism*: Quantum theory does not describe an underlying reality, it is only an instrument that can be used to predict particular sensory experiences, namely the results of experimental measurements.

This interpretation of quantum theory, as appealing as it was to the "tough-minded" physicists of the day, engenders problems that have not been overcome despite nearly a century of attempts. The problems arise because the output of quantum theory is *not* a prediction of experimental results. The way in which the actual output of quantum theory is converted into a prediction of experimental results is called an *interpretation*. The Copenhagen interpretation is the particular way in which most physicists have been taught to perform this conversion.

The State Vector and Its Interpretation

The output of the quantum mechanical description of a physical system is a mathematical entity called a *state vector*. The state vector is a collection of numbers called components. There is one component for each possible state in which the system can exist. This poses an immediate problem. When I perform a measurement that determines the state of the system, the result is just *one* of the possible states. Yet the state vector – the quantum description of the system – does not single out any particular state. It merely assigns numbers to *all* of them. If the state vector were to assign a zero to all of the possible states except for one of them, there would be no problem; it would be singling out a state just as an experimental measurement does. But the state vector does not do this.

The Copenhagen interpretation conjectures that the meaning of the state vector is statistical. It supposes that the number assigned to a particular state in the state vector (the value of a state vector component) determines the *probability* that the system is in that state. More precisely, it determines the probability that a measurement will find the system to be in the state of interest. This interpretation assumes that 1) there exists an experimenter external to the system who is able to perform measurements on it, and 2) it is possible to create an ensemble of identical systems that give meaning to the statistics. The ensemble could either be a collection of simultaneously existing, identically prepared systems, or it could be a single system that is able to be reset to its original state after each measurement.

Incompatible with Quantum Cosmology

Quantum cosmologists, who treat the entire universe as a quantum system, have a big problem with the Copenhagen interpretation. Where, they would ask, is the experimenter who is external to the system – that is, *external to the universe* – who can perform measurements on it? How, moreover, is it possible to prepare an ensemble of universes or repeatedly restart the universe in its original state?

Incompatible with Quantum Computation

A quantum computation requires the simultaneous, *parallel* execution of its component calculations. Rudimentary quantum computations have been performed in laboratories. The observed experimental results can only be explained if parallel calculations had in fact happened. If parallel universes do not exist, exactly *where* did these parallel calculations occur? This question becomes particularly acute for quantum computations that require more parallel operations than there are particles in the universe.

The Measurement Problem

The interpretation has another problem. Before the state of the system is measured, it can be in any of its possible states. The Copenhagen interpretation assumes that after the measurement, the system is in precisely one state – the outcome of the measurement. This measurement-induced elimination of all but one of the system's possible states is called the "collapse" of the state vector. It turns out to be impossible to unambiguously define exactly when this collapse occurs. This is called the *measurement problem.*

That Darn Cat

To understand the measurement problem, recall the plight of Schrödinger's cat. It is sealed in a transparent box together with a sample of a radioactive isotope, a Geiger counter, and a canister of poisonous gas. When an alpha particle is emitted by the radioactive sample, it sets off the Geiger counter, which in turn opens the electric valve of the canister, thus flooding the box with instantaneously deadly gas. A quantum mechanical calculation of the state vector of the system indicates that 15 minutes after the box is prepared there is (according to the Copenhagen interpretation) a 50% probability that the alpha particle has been emitted.

The transparent box is prepared as described above and covered with an opaque tarp. Fifteen minutes later, the state vector describing the box indicates two equally likely outcomes – the cat is alive, or the cat is dead. At this point our experimenter, Bob, removes the tarp and observes the state of the cat. Suddenly, two possibilities collapse to a single possibility. The system can now only be in the state observed by Bob. Bob either sees that the cat is alive, or he sees that it is dead.

It appears, then, that the state vector has collapsed unambiguously to a single possibility. But has it? Consider Alice, who is standing outside of the closed door of the laboratory in which Bob has made his measurement (by removing the tarp and looking). As far as Alice is concerned there are still two equally likely possibilities. When she enters the lab she will either see Bob staring at a dead cat or Bob staring at a live cat.

For Bob the system to be measured consists of the contents of the box. But for Alice the system to be measured consists of the contents of the *laboratory*, which includes the box together with Bob.

Different Rules for Systems Containing People?

It is difficult to argue cogently that when Bob collapses the state vector of the box, that he also collapses the state vector of the laboratory. To so argue, would be to assert that a human being cannot be part of a quantum system as it is normally defined. In other words, Alice's observation of a system containing a human being would have to follow different rules than Bob's observation of a system containing a mere cat. When the cat performs a measurement by either choking on gas or not, this does not collapse the state vector of the system observed by Bob. But when Bob performs a measurement by observing a dead cat or not, this *does*, according to this argument, collapse the state vector of the system observed by Alice. This asymmetry is often attributed to the special role of consciousness in physics. Bob is conscious, the cat is not. Hence, the results differ. Assigning, however, an ill defined, nebulous, and quasi-supernatural concept such as "consciousness" a central role in the formulation of a physical theory seems, at least to me, to be a great backward leap toward mysticism.

No Guarantee that the State Vector Will Ever Collapse

This measurement problem cannot be escaped by supposing that Bob's observation does not collapse the state vector of the laboratory observed by Alice. Because Alice cannot be known to be the last observer, this supposition leads to an open-ended regress. We are tempted to say that when Alice walks into the laboratory, she makes an observation that eliminates one of two possibilities and thus collapses the laboratory's state vector. End of story. But is it? What about Joe? He will later ask Alice what she saw in the lab. He can expect two answers. The state vector of the system that he is studying – the Alice-Bob-cat system – has not yet collapsed. Surely, though, we can suppose that Joe or *someone* will be the final observer. Surely we can at some point collapse the state vector as required by the Copenhagen interpretation. No, we cannot, because no one is guaranteed to be the final observer. There remains, for example, the possibility of an archeologist ten thousand years in the future, who will reconstruct the entry in Joe's diary that describes his conversation with Alice. There could be another archeologist *twenty* thousand years hence, who will piece together the remnants of the notes left by the first one. There could in principle be an observer in the distant future observing a particle of dust that will float to the left or to the right depending on whether an early twenty-first-century cat was gassed or not. As long as the consequences of a measurement can in any way and to the slightest degree affect the future, the measurement will not have eliminated any possibilities by inducing a collapse of a state vector. This lack of an unambiguous collapse violates the implicit assumption that such a collapse is possible and thus renders the Copenhagen interpretation inconsistent.

PP11. *The Copenhagen interpretation is inconsistent with quantum cosmology, cannot explain quantum computation, and has not satisfactorily resolved the measurement problem.*

PP12. *The deficiencies of the Copenhagen interpretation are inconsequential to the daily operations of the overwhelming majority of physicists.*

The Many Worlds Interpretation

A Return to Metaphysical Realism

The Copenhagen interpretation leads to inconsistencies. It assumes that quantum theory cannot be describing a reality beyond our experiences, because – by a fundamental tenet of instrumentalism – there is no such reality. But what if instrumentalism is wrong? What if there *is* a reality beyond our experience? In that case we would be free to take the output of quantum theory – the state vector – as a description of this larger reality. To make the connection with our experience, we would only need to explain how it emerges from this underlying reality. In other words, we would no longer be forced to directly convert the state vector into a description of our experiences. There would no longer be a need for a mechanism, such as the collapse of the state vector, whose sole purpose is to perform this conversion. We would then be rid of the inconsistencies engendered by such a mechanism.

In 1957 Hugh Everett, then a graduate student under John Wheeler at Princeton, incorporated these ideas in a novel formulation of quantum theory. His "Relative State" formulation has since come to be known as the Many Worlds Interpretation.

The Many Worlds Interpretation assumes that reality consists of an amalgamation of parallel universes that has come to be called the *multiverse*. In Hugh Everett's original formulation, the number of parallel universes in the multiverse increases each time an observation occurs. In the early 1980s Oxford physicist David Deutsch proposed a slight variation on this theme. In his variation the number of parallel universes remains constant. What changes with each observation is the apportionment of universes amongst the various possible results of the observation. The brief description that follows will also assume a constant number of parallel universes.

The concept of parallel universes is a familiar one, due to the influence of numerous popular fictional stories, whose initial appearances predate Everett by at least twenty years. Everything that can happen does happen. Each universe is a particular sequence of possibilities. Each possible history of events has occurred in at least one parallel universe. In some universes you are wealthy, in others you are impoverished, in most you do not exist.

Back to the Lab

Let us return to the laboratory, where Bob is about to lift the tarp off of the transparent box containing Schrödinger's cat. For the sake of simplicity, let us suppose that this scene is replicated in all parallel universes. When the Bobs lift the tarp, the multiverse changes. The state vector describing the multiverse will indicate that half of its universes contain a Bob who is staring at a dead cat, the other half a Bob staring at a live cat. The Alices, who are standing outside of the laboratory door in each universe, do not yet know whether they exist in a dead-cat universe or a live-cat universe. After they open their respective doors, they do. The Joes, who later talk to their respective Alices, are similarly informed. In other words, the state vector never collapses. The multiverse is forever divided into two classes – dead-cat and live-cat.

This eliminates the measurement problem, because it is no longer necessary to attempt to determine when the state vector collapses. It never does. The problem of inapplicability to quantum cosmology also vanishes, because there is no longer a need for an external observer. Nor is there a need to construct an ensemble of universes. The multiverse itself serves this purpose. Quantum computation is no longer mysterious. Its parallel calculations occur within parallel universes. The ensemble of parallel universes results in the probabilistic character of the Copenhagen interpretation being retained for any particular observer, while the multiverse as a whole is fully determined.

In fact Many Worlds is the only interpretation of quantum theory that is uncontroversially known to be entirely free of inconsistencies. It, moreover, yields the advantage, as we will see in Chapter 10, of resolving all known time travel paradoxes.

Too Expensive a Solution?

For the great majority of physicists – including many of those of the highest order -- the conceptual cost of the Many Worlds view is too high. They prefer to live with the inconsistencies of the Copenhagen interpretation rather than believe in the reality of an infinite number of parallel universes. Others prefer even to embrace interpretations that require the possibility of superluminal communication.

Were I to guess the reason for the prevalence of such positions, I would say that it results from a misapplication of *Ockham's Razor*. Ockham's Razor, as you will recall, is the dictum that advises that the simplest of competing ideas of equal explanatory power is most likely to be the correct one. But how should we interpret the word "simplest"? Should we take it to mean *conceptual* simplicity or *physical* simplicity? A measure of the conceptual simplicity of an idea is the number of words required to express it. The fewer the words, the simpler it is. An indication of the physical simplicity of an idea is the ease with which it can be physically implemented. Many Worlds is conceptually simpler than Copenhagen (because Copenhagen requires long explanations about, for example, the role of consciousness in physics). Copenhagen is physically simpler than Many Worlds (because MW requires the physical construction of an infinite number of parallel universes). Those that believe that Ockham's Razor pertains to physical simplicity, will opt for the

Copenhagen interpretation. Those who think that the Razor concerns conceptual simplicity will favor the Many Worlds view.

For my part it seems that in considering an idea we should only be concerned with the ratio of its explanatory power to its complexity. The cost of its physical implementation is irrelevant. Conceptual simplicity, in this view, should be all that matters.

Analogous to the Problem of Time

For another perspective on the matter consider the problem of time. We all experience a sense of a special time that we call "now". Yet nowhere in the equations of physics is this instant singled out for special treatment. We have come to regard our special awareness of "now" as a manifestation of our natures rather than a global characteristic of the universe. "Now" is just another instant, no more or less special than any other.

We experience similarly a sense of specialness regarding the outcomes of our measurements of a quantum system – a system in a state of superposition. The actual readouts that we see on our measuring devices are more special, it seems to us, than all of the other possible values that these numbers could have assumed. Nowhere, however, are these special outcomes – the actual readouts on our measuring devices – singled out for special treatment in the equations of quantum physics. Just as there is no special time, there is in general no special measurement outcome. The Many Worlds Interpretation merely raises this observation to the status of a metaphysical principle. To summarize,

PP13. *The Many Worlds Interpretation is the only interpretation of quantum theory that is free of inconsistencies. It also resolves all time travel paradoxes that arise in general relativity. It requires, however, a belief in the reality of the multiverse – an infinite number of parallel universes.*

PP14. *In the Many Worlds Interpretation our observations – the results of our experiments – tell us where we are in the multiverse; they restrict our particular universe to membership in a certain class of universes.*

The Consistent Histories Interpretation

A Hybrid

One could argue that the Consistent Histories, also known as Decoherent Histories, is a hybrid of the Copenhagen and Many Worlds Interpretations. It inherits from its Copenhagen father a belief in the primacy of human experience over theoretical abstraction. From its Many World's mother it obtains its emphasis on the inclusion of all possible histories in the calculation of experimental outcomes. The latter inheritance is weak, however. Consistent histories is best considered a modernized version of the Copenhagen interpretation. It was born in 1984 of the work of Robert Griffiths, now at Carnegie Mellon University. It has been developed subsequently by Nobel-prize-winning particle theorist Murray Gell-Mann of the Santa Fe Institute, quantum cosmologist James Hartle of the University of California at Santa Barbara, and French theorist Roland Omnès of the University of Paris, among others.

The Basic Idea

To understand the idea behind consistent histories, we must understand that quantum theory says nothing about probabilities. The theory is couched in terms of what are known as *amplitudes.* An amplitude is a complex number that has traditionally been *assumed* to have a simple relationship to probability. Like all complex numbers, an amplitude is a vector in the complex plane. The square of the length of this vector is the probability that is *assumed* to correspond to this amplitude. This assumption is called the "Born rule", named after physicist Max Born who proposed it in 1926*. The reason that we care about the link between amplitudes and probabilities is that we can (at least in most cases) experience probabilities, but we cannot directly experience amplitudes. The outputs of quantum theory are amplitudes. Linking amplitudes to probabilities is the way that we connect the theory to our experience.

Probabilities follow certain rules that have been known since the seventeenth century. One of them is that the sum of the probabilities for each possibility is one. Another is the rule by which probabilities add: P(A or B) = P(A) + P(B) – P(A and B). [In other words, the probability that event A or event B occurred is the sum of the probability that A occurred with the probability that B occurred minus the probability that events A and B both occurred.] We can use quantum theory to work out the amplitude for every conceivable possibil-

* It has been recently shown that this assumption can be extracted from the other axioms of quantum theory. See D. Deutsch, *Proceedings of the Royal Society of London* **A455**, 3129 (1999).

ity or *history* of the system under study. We can use the Born rule to obtain probabilities that correspond to each of these amplitudes. We will discover, however, that some of these probabilities violate the aforementioned rules of probability. The histories that correspond to these outlaw probabilities are deemed *inconsistent*. These inconsistent histories are excluded from any calculations.

For an example of an inconsistent history, consider the two slit experiment. A barrier with two parallel slits in it is placed between an electron gun and a phosphor screen. In calculating the probability that an electron emitted by the gun will reach a particular point on the phosphor screen, one would naively consider every possible way in which this could occur, i.e. every history. These (one-event) histories are: a) the electron went through slit #1, b) the electron went through slit #2, c) the electron went through slit #1 or slit #2. One can show that the inclusion of history (c) violates the rules of probability. It is therefore banned as an inconsistent history. Through the explicit exclusion of superpositions such as history (c), the Consistent Histories approach seeks to solve the conundrum posed by Schrödinger's cat.

The Role of Decoherence

The other pillar of the consistent histories approach is the concept of *decoherence*, defined as follows,

> **decoherence** is the collapse of the wave function through subtle interactions with the environment in advance of any conscious attempt at measurement.

In order to understand this, let us return to the laboratory, where Bob is about to remove the tarp covering the apparatus containing Schrödinger's cat. Although Bob doesn't yet know whether the cat is alive or dead, there is actually subtle evidence available to him from which he could, given the proper instruments, determine the state of the cat. For example, if the cat is dead, the laboratory environment would differ subtly from the contrary case. The room temperature would be slightly lower, the concentration in the air of carbon dioxide would be lower and that of certain molecules associated with necrosis would be higher (assuming that the box containing the cat is not hermetically sealed). The sound of the cat's breathing and heartbeat, though never audible to Bob, would nevertheless be absent. Adherents to consistent histories argue that the presence of this subtle evidence acts like a cat state detector, just as Bob's eyes do. Hence, the action of this subtle-evidence-cat-state detector caused the cat to enter one of its possible states. The cat would no longer be in a *coherent* state – a superposition of the quantum states "alive" or "dead". It would instead, through the action of this detector, *decohere* into one or the other. Before Bob lifts the tarp to reveal the cat's state, the argument goes, the cat is already either definitely alive or definitely dead as a consequence of its interaction with the subtle-evidence-cat-state detector. One solves the paradox of Schrödinger's cat by stating that in the real world it could never occur. Subtle interactions with the cat would always cause the superposition of cat states to decohere before any observation by a human could occur. In the real world cats are either alive or dead. Never both.

In other words, supporters of coherent histories believe in banning certain histories if they are sufficiently improbable. They would in particular discard the following history and its converse: subtle evidence points to the cat being dead, but Bob lifts the tarp and finds the cat to be alive. By discarding this possibility, Bob's observation will always match the pre-detection performed by the subtle-evidence-cat-state detector. He will never face the possibility of two outcomes to his observation due to quantum indeterminacy. Coherent states will thus be banished from macroscopic experience.

The Cat Cannot Really Exist

Where this argument succeeds is in its demonstration that the line of demarcation between the microscopic-quantum world and the macroscopic-classical one is for all *practical* purposes drawn in the microscopic world at the site of the occurrence of subtle interactions. For all practical purposes Schrödinger's cat – a sustained macroscopic superposition – cannot exist. The importance, however, of this observation to the conceptual content of the theory is unclear. It is possible to engineer an *im*practical experimental setup that eliminates extraneous subtle interactions arbitrarily well, thus allowing direct macroscopic observation of coherent states. In other words, we can always devise a situation that is conceptually identical to the Schrödinger cat setup through sufficiently careful engineering. Although such engineering might currently exceed human capabilities (which I do not believe to be the case), this is irrelevant. Schrödinger's cat is a *thought* experiment. All that matters, then, are physical principles, not whether we can engineer them.

Decoherence No Match for the Cat

Even in the absence of such engineering as in the original setup, there will still be two cat states irrespective of whether Bob pre-detects them through subtle interactions. The arguments above regarding Bob, Alice, and the cat proceed exactly as before, if we replace their optical detectors (their eyes) with subtle-evidence-cat-state detectors. Ultimately, it is still the case, whether through decoherence or other means, that a quantum superposition gets mapped to a single output value of a macroscopic detector. One of the problems, then, raised by Schrödinger's cat -- that of knowing how to collapse the wave function -- remains. The Copenhagen interpretation cannot solve it. Many Worlds solves it by refusing to collapse the wave function. And Consistent Histories attempts to solve it by retaining all possibilities – ascribed to separate universes in Many Worlds – as counterfactual histories of a single universe. A counterfactual history is, of course, a history that did not occur. That these histories are counterfactual and not real does not emerge from the mathematical formalism of Consistent Histories. It is an ontological assumption grafted onto it. "Bob viewing a live cat" and "Bob viewing a dead cat" occur in separate, equally real universes of the Many World's multiverse. In Consistent Histories, one of them is real, and the other is part of a ghostly counterfactual history. The Consistent Histories approach to the question of how the wave function should be collapsed is valid only to the degree that prominent elements of its mathematical formalism – the consistent histories themselves – can be denied reality. This situation is unlike that of classical stochastic models. There the probabilistic component can be reduced arbitrarily by injecting the model with additional information. Here the quantum indeterminacy of the uncertainty principle prevents a similar injection. One cannot argue, then, that there exist classical precedents for a *fundamentally* probabilistic theory within a single universe.

Inconsistent Metaphysics

In its attempt to retain the single universe and probability interpretation of Copenhagen while exploiting the utility of what are in effect the multiple universes of Many Worlds, Consistent Histories appears to be mathematically consistent but also seems to be metaphysically unclear. To summarize,

PP15. *The Consistent Histories interpretation excludes histories that are inconsistent with the laws of probability.*

PP16. *The Consistent Histories interpretation emphasizes that macroscopic detections of quantum superpositions (Schrödinger's cat scenarios) do not, as a result of decoherence, occur in practice.*

PP17. *The Consistent Histories interpretation ascribes no reality to all but one of the consistent histories, regarding these others as mere counterfactuals.*

PP18. *The Consistent Histories interpretation, despite its emphasis on decoherence, does not solve the problem of providing an unambiguous means of collapsing the wave function.*

Quantum Mechanics

Quantum theory is a theoretical framework for a particular type of physical theory. This framework includes axiomatic conjectures about the fundamental elements of the theory and their correspondence to reality. The state vector is, for example, a fundamental element of quantum theory whose correspondence to reality was described above. Quantum *mechanics* is a collection of techniques for calculating the state vector in a wide variety of circumstances.

One circumstance of great interest is that of a particle subjected to the influence of a force that varies with spatial position. To work out the likely positions of such a particle, one begins by considering the state vector as a vector in an infinite-dimensional space. Each dimension of this infinite-dimensional space corresponds to a particular position in physical, three-dimensional space. The state vector becomes in effect a function of position in physical space. This function is called the *wave function.* It is calculated by solving the *Schrödinger* equation. Quantum mechanics is largely a collection of techniques for solving this equation for various physical systems.

When to Use Quantum Mechanics

Let us suppose that you are interested in describing a system that consists of a single particle being acted on by a spatially varying force. Suppose further that you are able to measure the momentum of the particle to an accuracy of Δp and its position to an accuracy of Δx. Then quantum mechanics is the only accurate way of describing the system, if the following is *not* true (">>" means "is much greater than"),

$$\Delta p \Delta x >> h. \qquad (8)$$

In other words, ordinary *classical* mechanics accurately describes the physics of the system if the product of the uncertainties in the particle's momentum and position is much greater than Planck's constant h. Because Planck's constant is exceedingly small (6.63×10^{-27} erg seconds), equation (8) will be satisfied by measurement accuracies that we would find adequate for the study of virtually any macroscopic system. Suppose instead that the particle of interest is constrained to move within a tiny spatial region, such as within an atom. If we are interested in locating its approximate position within the atom, then the accuracy Δx of our position measurements must be much smaller than the size of the atom. A similar comment applies to the particle's momentum. The uncertainty Δp of the particle's momentum in this case must be less than the maximum momentum that the particles can possess without escaping the atom. This is an example of a case in which the inequality (8) does not hold, and quantum mechanics would have to be used.

Where Quantum Weirdness Comes From

The deviation of systems described by quantum mechanics from the classical behavior with which we are all familiar has come to be termed "quantum weirdness". Yes, there is weirdness, but quantum mechanics is not unintelligibly weird. Two or three generations of physicists grew up within a positivist intellectual environment that discouraged the metaphysical contemplation required to make sense of this weirdness. As a result, a quasi-mystical attitude toward the foundations of quantum theory came to inhabit the minds of most working physicists. These attitudes were over a period of decades transmitted to the general public and repeatedly reinforced. As a consequence, most people are surprised to learn that many physicists, particularly those working within certain specializations, find quantum theory fully intelligible from the bottom up. "Spooky action at distance" is neither spooky, nor action at a distance. Wave-particle duality is a confusing description of a crystal-clear phenomenon. The paradox of Schrödinger's cat has long ceased to be paradoxical. Unfortunately, a paradox-free exposition of the fundamentals of quantum theory well exceeds our purview. I shall instead resort to the usual shortcuts in my thumbnail sketch of the theory. My point, however, is that any confusing ideas encountered should be regarded as one regards such ideas in classical mechanics – as perhaps difficult to comprehend but not fundamentally incomprehensible.

All of the weirdness in quantum mechanical systems stems from one rule,

$$\Delta p \Delta x \geq \frac{h}{4\pi} \qquad (9)$$

the famous Heisenberg Uncertainty Principle. If h, Planck's constant, were zero, all quantum weirdness would vanish. Using certain mathematical tools, this rule can be derived from the Schrödinger equation. Let us see instead whether we can come to an intuitive understanding of its meaning. In what follows I shall refer only to the Many Worlds Interpretation.

Consider a particle I have measured to have been at position $x=a$. Suppose that I perform a second measurement and determine with arbitrarily low uncertainty that the position of the particle is now $x=b$. The section of the multiverse that I now know that I occupy is the one that contains only those parallel universes in which the particle has gone from position a to position b. I no longer have to consider those parts of the multiverse containing parallel universes in which the particle did not go from a to b, or those in which I never took the measurements, or those in which I was never born, or those in which life did not evolve on Earth. What are the different ways in which the particle could have gone from a to b? Its only other property is its momentum. So the parallel universes of interest are the ones in which the particle's position went from a to b with a momentum distinct to each universe.

PP19. *The only parallel universes of consequence to an observer in a particular parallel universe are those whose history matches that of the observer's universe prior to her current experiment.*

The State Vector from a Superposition of Interfering Universes

A key fact about the multiverse that I have yet to mention concerns that manner in which the state vector is obtained from the parallel universes. Recall that in the Copenhagen interpretation a component of the state vector determines the probability of our experiencing a particular possibility. In the Many Worlds Interpretation such a component determines instead the fraction of universes (of the universes of consequence as in PP19 above) in which a particular possibility actually occurs. Here is how. As mentioned above, each component of the state vector corresponds to a particular possibility. A component of the state vector is calculated by adding together contributions from all of the universes of consequence. These contributions interfere with each other. If the interference is constructive, it increases the fraction of universes in which the possibility occurs. Destructive interference has the opposite effect.

For the sake of simplicity, the universe's that we shall consider each contain a single particle executing one of an infinite number of possible motions. Each universe may be assigned a particular number called a *phase angle* or *phase* for short. The phase of a universe is obtained by calculating what's known as the "*action*" for the particle's motion in it. Roughly speaking, the action is the product of the particle's momentum and the length of the path traveled by the particle. So to determine the fraction of universes (the probability) that a particle has reached position b from position a, we add together the phases of all the universes representing the various ways (momentum values) in which this travel could have occurred. Because these phases depend on the distance traveled (in general on the *path* traveled), the fraction of universes in which the particle has reached position *b* differs from the fraction that reached some other position *b'*. There is a familiar example that demonstrates this. Place an opaque card with single slit in it between a monochromatic light source (e.g. a laser) and a projection screen. The screen will show a bright central band bordered by dark bands. These dark bands are bordered by light bands that are dimmer than the central band. This alternating pattern of dark and light bands of diminishing intensity continues outward from the central band until it is too dim to be discerned. This familiar diffraction pattern on the projection screen results from the possibilities represented by various universes constructively and destructively interfering with each other. The light bands occur at positions on the screen where photons, particles of light, have impacted in many universes between which there is constructive interference. Dark bands occupy positions where photon impacts occur in universes between which there is destructive interference. The position-dependent sums of complex numbers that determine the result of this interference at each position – which each in turn determine the fraction of universes in which photons have reached particular positions on such a projection screen -- is an example of a state vector.

PP20. *In the Many Worlds Interpretation each state vector component determines the fraction of universes (of consequence) in which a particular possibility occurs.*

If the fraction of universes in which the photon strikes the central position on the screen approaches 100%, and if our particular universe is a typical member of the multiverse, then we are overwhelmingly likely to see the photon strike this position. The distribution of possible outcomes in the multiverse determines the statistics of the outcomes we experience in our universe.

The reason that we are forced to invoke parallel universes to explain this diffraction pattern is that it is clear that photons are not interfering with other photons in the same universe. In other words, photons cannot be interfering in the ordinary way in which ripples in a pond interfere. We know this, because we can replace the projection screen with photographic film. Film records the positions of photon impacts on it. Upon putting the film in place, we will discover an amazing fact. An identical diffraction pattern appears on the film, when the photons are sent through the slit toward the film *one at a time*.

Lest we think that the pattern results from the photon somehow interfering with itself, consider the two-slit experiment. The card now contains two parallel slits. When both slits are open, light travels from its source on one side of the card, through slits, and onto the projection screen. In doing so, it produces a distinctive two-slit interference pattern on the screen. When one of the slits is blocked, we get the one-slit diffraction pattern that we considered before. The natural explanation for the two-slit pattern is that the light that passes through one of the slits interferes with the light that passes through the other. This explanation cannot be right, however. As before, let us replace the projection screen with a plate of photographic film. Once again, we will be shocked to observe that the two-slit pattern appears on the film, even when the photons are emitted from their source one at a time. We know, because of the interference pattern, that the photons are interfering with *something*. We know, because each photon traveled in isolation, that they are not interfering with other photons. We know that they are not interfering with themselves, because they can only pass through one slit or the other but not both. We are driven, then, to a startling conclusion.

Destructive and Constructive Interference between Universes

Such photons traveling solo can only be interfering with their counterparts in other universes. This interference results from adding together the phase-determined complex numbers associated with each universe. The sum of any two such numbers whose phases differ by 180 degrees is zero. They cancel each other out. This also occurs for the sum of many complex numbers in which every phase (a direction in the complex plane) is equally represented. The sum of all directions cancels to no direction. In situations well described by classical physics, such as that of an extremely wide single slit, this sort of cancellation is the norm. In these situations, for every state vector component *but one*, the sum of these complex numbers is zero. All of the universes, except for one, cancel each other out. The one state vector component whose value is not zero corresponds to the state that would emerge from a classical treatment of the system.

Classical physics, then, occurs when there is virtually perfect cancellation. Quantum weirdness occurs when the cancellation is imperfect. The degree to which the weirdness occurs is the degree to which cancellation is poor.

The Uncertainty Principle from Imperfect Destructive Interference

To see this, let us return to our particle located at some position x with arbitrarily small uncertainty. If we perform the sum described above, we see that the only nonzero state vector component is that associated with the state of the particle having position $x = a$, its original position. This is because all directions and magnitudes of momentum are equally represented in the parallel universes. They cancel, leaving an effective momentum of zero. The other state vector components -- which correspond to the movement of the particle to some new position -- are zero due to a perfect cancellation in their defining sums. This is the classical case: A particle not known to have a nonzero momentum stays put.

To summarize: In the case we are considering, the state vector component for $x = a$ is non-zero; the state vector component for any x other than a is zero (Figure 36a). This results from perfect cancellation occurring in the aforementioned sums. We can introduce quantum weirdness by corrupting these sums and thus ruining the perfect cancellation. Recall that each parallel universe in these sums is identified with a particular value of the particle's momentum. I could make the sum imperfect in the following way. Instead of summing over parallel universes with every possible value of the momentum, I will only sum over those universes whose momentum is less than some maximum value Δp. The smaller Δp, the larger the imperfection.

How would an imperfection in the sums manifest themselves? Components of the state vector whose value was zero in the perfect cancellation case, would become nonzero in the case of imperfect cancellation. The components nearest to the $x=a$ component are (due to their smaller phases) affected the most (Figure 36b, where a is the center of the central fringe). Let Δx be the largest x for which the state vector now acquires an appreciably nonzero value. We know that perfection occurs ($\Delta x = 0$) when Δp_x is infinite. We know that the smaller Δp_x is, the less perfect the cancellation, and the larger Δx must be. So it is reasonable to guess that Δx must be inversely proportional to Δp. We also know that Δx is a measure of quantum weirdness, and that such effects vanish when $h = 0$. We can reasonably guess, then, that Δx is proportional to h. Combining these guesses, we have

$$\Delta x > h/\Delta p_x. \qquad (10)$$

We need the "greater than" sign, because Δp_x is the *maximum* value for the uncertainty in the horizontal momentum. Which means that $h/\Delta p_x$ is the minimum value for Δx. When multiplied by Δp_x, equation (10) is (within a factor of 4π) the Heisenberg Principle that we were hoping to understand.

How might the limit Δp_x on the value of the momentum of the parallel universes arise? One way occurs when monochromatic light shines through a single slit. The momentum component of the light in the plane of the slit cannot exceed a certain maximum value. Were it to do so, the lateral motion of photons traveling from their emitter would cause them to miss the slit. They would instead travel too far to the left or too far to the right and hit the surface of the opaque card in which the slit has been cut. The limit on Δp_x, the lateral component of the momentum, is easy to work out. It turns out to be proportional to the ratio of the width of the slit to the photon wavelength (see Fig. 3.6). Another example of such a limit results when a particle is contained within a potential barrier of finite depth – a barrier that is only able to contain particles whose momentum does not exceed a certain value. This describes and electron within an atom, exactly the system for which quantum mechanics was invented to explain.

Cause of the Weirdness: Reduced Destructive Interference due to Missing Universes

The purpose of this section has been to show that quantum effects (weirdness) arise when our observations place us in a section of the multiverse from which certain universes have been excluded. For example, we might have, as above, light of a given wavelength passing through a slit of unknown width. If the slit's width is much larger than this wavelength, there is a nearly total cancellation of universes corresponding to deviations from the classical path of the light ray. As we shrink the slit's width, some of these deviant paths cease to be available (because the opaque card now blocks them). The universes that correspond to them no longer contribute to the wave function. And the aforementioned cancellation weakens. Think of a cockroach crawling into a house through an open doorway. As we magically shrink the size of the doorway, there are progressively fewer paths available for the cockroach to enter. When the diameter of the doorway becomes that of the roach, it is limited to little more than a single path. Replacing cockroaches with photons and associating each path with a universe, we can restate our conclusion as follows. The relatively poor cancellation between universes is due to our entering (as a consequence of measuring the width of the slit to be small compared the wavelength of our light source) a region of the multiverse in which certain universes are excluded.

This exclusion prevents the nearly total cancellation of universes that characterizes classical physics. This explains, for example, why the lowest energy state of a pendulum described by quantum mechanics is not zero, as it is in classical physics. In the quantum treatment there is imperfect cancellation. The universes, that in classical physics would have cancelled the universes causing the lowest energy state to be non-zero, have been excluded by our observation (of a tiny upper limit on the pendulum's momentum) from the section of the multiverse that we occupy. Such exclusions are tantamount to a narrowing of the known ranges of the positions or momenta of the particles of a system. When the products of these ranges, better known as uncertainties, are no longer large compared to Planck's constant, quantum effects ensue.

The thrust of this section is summarized in the following preliminary principle, a restatement of the Heisenberg principle:

PP21. *A system of particles need only be described by quantum mechanics, if our observations place us in a section of the multiverse from which sufficiently many universes have been excluded, so as to prevent the products* $\Delta x \Delta p$ *-- of the uncertainties in position with those of momentum -- from being large compared to Planck's constant h.*

I should mention in passing that there also exists a Heisenberg uncertainty relation between energy and time. It is often claimed to be precisely analogous to the position-momentum uncertainty principle mentioned above. This is not quite true. While it is possible to measure the position of a particle, one cannot measure its time. Its time is either always "now", or it is an interval that begins at the particle's creation and ends at its destruction. One describes this lack of symmetry between time and position by saying that a particle's position is an "observable", while its time is not. Hence, the energy-time uncertainty principle,

$$\Delta E \, \Delta t \geq \frac{h}{4\pi}, \tag{11}$$

has the following slightly different meaning. When an isolated system enters a particular state, the minimum uncertainty in the energy of the system is inversely proportional to the time that it remains in that state. Suppose, for example, that an atom enters an excited state and remains unperturbed in that state for a time Δt, before it returns to its initial state. The uncertainty ΔE in the energy of the excited state – the spread in measurements of the state's energy in many similarly prepared atoms -- is related to the state's duration Δt by the inequality (11). As before, certain observations induce weirdness. In this case it is those whose duration and energy resolution exclude certain universes from one's region of the multiverse. Inter-universe destructive interference diminishes. Weirdness ensues.

The ideas touched upon in this section – quantum superposition, sum over histories, uncertainty principle – are by no means peculiar to the Many Worlds Interpretation. They are very old ideas that are normally presented in the context of the traditional Copenhagen orthodoxy.

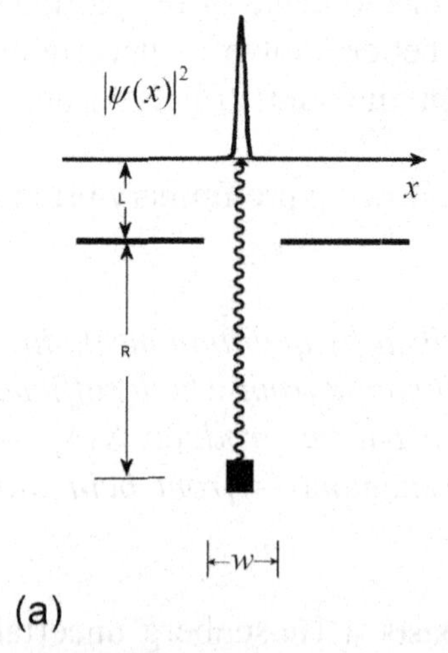

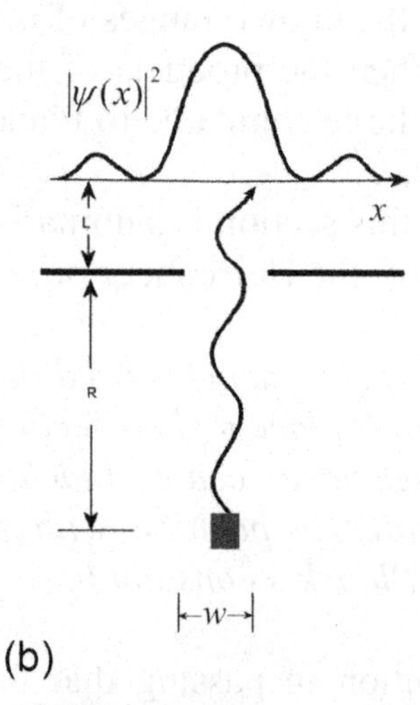

(a)

Classical Case:
- **Wavelength much smaller than slit**
- **Action varies rapidly with position**
- **Huge uncertainty in lateral momentum**
- **Nearly perfect cancellation away from center**
- **Small impact region**

(b)

Quantum Case:
- **Wavelength the size of slit or larger**
- **Action varies slowly with position**
- **Small uncertainty in lateral momentum**
- **Weak cancellation away from center**
- **Large impact region**

$$\Delta p_x = \frac{hw}{2R\lambda}$$

$$\textit{width of central peak} = \frac{\lambda}{w}L$$

$$\psi(x) = \sum_{\substack{\text{all paths to x with} \\ |p_x| < \Delta p_x}} e^{\frac{i}{\hbar}Action(x)}$$

Figure 3.6. Ordinary Diffraction, an example of Quantum Weirdness. Photon with momentum p ($=h/\lambda$) impinge upon a screen after passing through an opaque card with a slit. The intensity of light at a point x on the screen is given by the squared modulus ($|\psi(x)|^2$) of the wave function ψ. ψ is obtained by summing complex numbers each of whose phase corresponds to the so-called "action" of a particular way in which the photon could have reached the screen. Only photons from the source with a horizontal momentum p_x satisfying $|p_x| < \Delta p_x = \frac{1}{2}(h/R)(w/\lambda)$ hit the slit.

(a) In the classical case the wavelength λ is much shorter than the slit width w. The maximum horizontal momentum Δp_x of the possible paths, which is proportional to ratio of w to λ, is high. This leads to virtually perfect cancellation away from the center and only one impact location (the center). (b) In the quantum case, a long wavelength relative to the slit width results in poor cancellation and many possible impact locations. This fuzziness in the impact region, which is just ordinary diffraction, is an example of quantum weirdness.

Quantum Field Theory

Consider a circular rubber sheet. Suppose that I nail the edges of the sheet to a circular wooden frame so that I now have something that resembles a tambourine. Imagine the sheet being held in place while I simultaneously deform the surface in a few places by pressing the sheet with my fingers. I can describe this deformation as a function of position on the sheet. This function would give at each point on the sheet the deviation in millimeters from the sheet's equilibrium position. This deformation function is an example of a field. A particular deformation is called a configuration of the field.

This deformation function is subject to a force that depends on the degree of deformation. The more I press my fingers, the greater the deformation, the stronger the force attempting to restore the rubber sheet to its equilibrium position. This is an example of a field that is subject to the influence of a force. Just as the state of a particle may be specified by its position, the state of a field may be specified by its configuration.

Quantum mechanics is concerned with describing a particle acting under the influence of a force. Quantum field theory, similarly, is concerned with describing a field acting under the influence of a force. In quantum mechanics the likely positions of a particle are given by the wave function, which is a function of the particle's position. In quantum field theory the likely configurations of a field are given by a *wave functional*, which is a function of the field's configuration (i.e., it is a function of a function -- the definition of a functional).

The point of quantum field theory is that it addresses a shortcoming in quantum mechanics. In quantum mechanics the number of particles in the system under study must be fixed. If the system of interest allows particles to be created or destroyed, its description requires quantum field theory. In quantum field theory the strength of the field in some region of space (the degree of deformation in the above example) corresponds to the number of particles there.

Physicists have assigned to each of the various fundamental particles in nature a particular type of field. The type of field assigned depends on spin, mass and other intrinsic properties of the particle. For example, photons are states of the quantized electromagnetic field – a massless, spin-1 field. Electrons are states of a massive spin-½ field called the Dirac field.

Although it is possible to write down, in direct analogy to quantum mechanics, a Schrödinger-like "variational" (as opposed to differential) equation for the wave functional, this equation is too hard to solve in nearly all cases. Instead, the field (the deformation function) is expressed as a series (i.e. a sum) of *operators*. Each operator creates or annihilates a particle (which might be thought of as a unit of deformation) that has a particular momentum.

Just as the ground state energy of a quantum mechanical system is not zero, so it is with quantum fields. In the rubber sheet analogy it would be as if tiny deformations were to be constantly bubbling on the sheet, even when it is in its most relaxed state. These bubbling deformations are the analogs of *virtual particles* of a quantum field. The lowest energy state of a quantum field is called the *vacuum state* of the field or *vacuum* for short.

The energy of the vacuum is infinite. The calculation of virtually any measurable quantity through the use of quantum field theory yields an infinite result. This is because such calculations must include contributions from virtual particles of all possible momenta – no matter how high. For some types of fields the infinites can in effect be subtracted out through a process called *renormalization*. For other fields, such as the gravitational field, this cannot be accomplished.

PP22. *Quantum field theory, unlike quantum mechanics, can describe systems in which particles can be created or destroyed. Infinities are endemic to nearly all calculations. They can be benign or malignant.*

Quantum Theory in a Nutshell

The conceptual content of quantum theory can be summarized as follows.

PP23. *Reality consists of a deterministic multiverse of interfering universes in which every possibility is represented.*

PP24. *For each possibility there exists a particular number of universes (a particular fraction of the multiverse) in which it is realized.*

PP25. *Our observations result from the interference between universes that occurs according to a particular rule.*

PP26. *Universes can be excluded from participating in inter-universe interference by experimental setups that eliminate their corresponding possibilities.*

PP27. *The smaller the number of possibilities represented by the universes participating in inter-universe interference, the greater will observations deviate from classical behavior.*

PP28. *Due to the nature of the interference rule, observations performed in experimental setups that exclude universes representing particular possibilities – those for which the products $\Delta p \Delta x$ or $\Delta E \Delta t$ are large compared to Planck's constant -- reveal "quantum weirdness".*

Quantum Gravity

quantum gravity -- a theory of gravitation that is fully consistent with quantum theory and reduces to a theory similar to general relativity in its classical limit.

Major Approaches

Why Gravity Must be Quantized

There are two reasons. The first is to avoid an *inconsistency*. Consider an electron. Suppose that I attempt to measure its position by illuminating it with a photon of light. The uncertainty in my position measurement will be the wavelength λ of the photon. When the photon hits the electron, it will impart some of its momentum. The uncertainty in the electron's momentum is the maximum momentum that the photon with wavelength λ can impart, namely, its entire momentum, which by quantum theory is h/λ. The product of the uncertainty in position with that of momentum is $\lambda \times h/\lambda = h$, which is consistent with Heisenberg's Uncertainty Principle.

Suppose that instead of using a photon of light, I attempt to measure the electron's position by using a classical (i.e. unquantized) gravitational wave pulse. The uncertainty in my position measurement will be the length of my gravitational wave pulse. However, because the wave pulse is not quantized, its momentum need not be inversely proportional to its pulse length. The momentum of the gravitational wave pulse of a given length can be made arbitrarily small (by reducing its intensity). So the maximum momentum imparted to the electron, the uncertainty in its momentum, can also be arbitrarily small. The product, then, of this uncertainty with that in the electron's position can be arbitrarily small as well. This violates the uncertainty principle. Put another way, quantum theory and classical gravity are inconsistent. Taken together, they just don't make sense.

The second reason is to avoid a *breakdown*. General relativity breaks down at spacetime singularities. There spacetime curvature reaches infinity, in effect eliminating spacetime as a stage upon which physics can unfold. In short: no space, no time, no physics. One generally tends to regard the total breakdown of physics as bad. It would not be such a problem, however, if we could be assured by the classical theory that singularities would never arise. That way, breakdowns would never occur. Unfortunately, we have no such assurance. We have, on the contrary, theorems that -- if their assumptions are valid -- *guarantee* the existence of singularities.

Two Traditional Approaches – Canonical and Operator

As in quantum field theory there are two major, traditional approaches to quantum gravity. The first is the *canonical* approach. It is non-perturbative. This means that it does not attempt to describe quantum gravity as a classical theory augmented by a sum of progressively smaller quantum-induced corrections. This non-perturbative approach involves, as above, the writing down of a Schrödinger-like equation for gravitational field configurations. In the case of gravity, such configurations are actually three-dimensional geometries. The output of the canonical approach, then, is the prevalence in the multiverse (or probability in Copenhagen language) of various three-dimensional geometries. As in non-gravitational field theories, the problem is that this output can never be obtained because the equation is far too difficult to solve. As in elementary quantum mechanics, this differential version of the approach has an equivalent integral version. However, the integral to be calculated – a special type known as a "path integral" – is at least as difficult to evaluate as it is to solve the aforementioned equation.

The second approach is the *operator* approach. It is explicitly perturbative. As in non-gravitational field theories, this involves reducing the gravitational field to a sum of operators that create and destroy hypothetical quanta of the gravitational field called *gravitons*. These operators must act on a special space of states of the gravitational field. This space of states assumes that any state of the gravitational field may be expressed as a superposition of graviton states. This allows approximate calculations to be performed. If we are lucky, the infinities that inevitably appear in such calculations can be removed. And it will be possible, moreover, to construct the desired perturbation expansion –

a series of quantum corrections of progressively less importance to the value of the calculation.

The Problem with the Canonical Approach

The Schrödinger-like equation at the core of the canonical approach, which is called the *Wheeler-DeWitt equation* is infinitely more difficult to solve than the Schrödinger equation of quantum mechanics. The former involves a finite number of degrees of freedom. If, for example, the quantum system of interest consists of a single particle, this system's Schrödinger equation involves three degrees of freedom – one for each spatial dimension in which the particle can move. For a system of two particles, the equation would involve 2 x 3 = 6 degrees of freedom, and so on. The greater the number of degrees of freedom the more difficult it is in general to solve the equation. The reason that the Wheeler-DeWitt equation is particularly difficult to solve is that it involves an *infinite* number of degrees of freedom. Each degree of freedom corresponds to the value of a particular component of the metric at a particular point in space. The traditional approach to solving the equations is to drastically reduce the number of degrees of freedom – usually to one. This is a ridiculously drastic assumption. It is a bit like modeling the ocean with a single parameter, the mean sea level. Such an ocean model would not permit the description of ocean waves or currents. It might not even be useful in describing tides, as they result from bulges -- a spatial non-uniformity in sea level. A drastic assumption is not a problem, however, if it is part of a perturbative expansion. If successive terms in the expansion provide corrections to the previous terms, the drastic nature of the initial term can be mitigated arbitrarily well. Unfortunately, no one has discovered a means through which one can calculate a correction to a partial differential equation in n degrees of freedom in order to obtain a solution for such an equation in $n+1$ degrees of freedom. In short, the Wheeler DeWitt equation is too hard to solve except in one degree of freedom, and there is no way to use this solution as a starting point for solutions with more degrees of freedom. The same applies to the path-integral variant of the canonical approach.

The Problem with the Operator Approach

The treatment of electromagnetism as a quantum field theory was the greatest achievement of mid-twentieth-century physics. No theory before or since has matched the degree of precision with which it agrees with experimental results. It was natural, then, for physicists to use this theory -- known as *quantum electrodynamics*, or *QED* for short -- as a template for quantizing gravity. Their attempts, however, immediately failed. The reason for this was not hard to find. Gravity, unlike electromagnetism (or even the strong nuclear force) interacts with itself in a particularly powerful way. A gravitational wave, because it carries energy, generates a gravitational field. The more energetic the wave, the stronger the field it generates. This doomed attempts to follow QED and to reduce gravity to quantized bits – *gravitons* analogous to the photons of electromagnetism. The way that one calculates observable quantities in QED makes this clear. For example, a set of observables of great interest in QED are the results of experiments that collide electrons against each other. To calculate the results of such scattering we must consider every possible way in which the electrons could interact. When we consider two electrons interacting through the exchange of a photon, we must consider every possible photon energy – from zero to infinity -- at which the exchange could have occurred. The effect of considering all of these possibilities in our calculation is to make the result infinite. Fortunately, however, this infinity is benign. It can be isolated from the rest of the calculation and effectively subtracted away.

If, by contrast, we attempted to calculate the results of an interaction between two electrons mediated through the exchange of a graviton, things are not so easy. When we attempt to consider the contributions to the calculation due to the exchange of gravitons at all energies, the calculation explodes. A high-energy graviton must interact with *itself* through the exchange of gravitons. The gravitons involved in the mediation of this self interaction must themselves self interact, which involves more gravitons that must self interact by means of more gravitons that must self interact in a endlessly escalating chain of self interaction. It is not merely the self-interaction that makes this explosion cataclysmic. It is that the degree of the self interaction increases with the energy of the gravitons involved, and we must consider gravitons at all energies. This sort of *energy-dependent* self interaction resulted in infinite contributes to calculations, that unlike the QED case, could not be subtracted out. In short, the stupendously successful techniques of QED could not be applied to gravity.

Breakthrough for the Operator Approach? Superstrings

This was the situation in the late 1950s. Quantum gravity remained at impasse until the mid 1980s, when two breakthroughs occurred. The first involved a theory originally invented in about 1970, to explain interactions between nuclear particles. This *string theory*, as it came to be called, was realized later in that decade to contain a particle fitting the description of the graviton. About this time a new hypothetical symmetry, known as *supersymmetry* was being applied to gravity in the hope of dealing with the malignant infinities described above. Unfortunately, the resultant theory, *supergravity*, was shown to possess certain internal inconsistencies known as *anomalies*. These were symmetries in the classical version of the theory that did not survive after the theory was quantized. In the early 1980s Michael Green, then at Queen Mary College, and John Schwarz of Caltech worked together to apply supersymmetry to string theory. The resultant theory, *superstrings*, was expected to possess the same anomalies as those contained in supergravity. In 1984 Green and Schwarz showed that these anomalies vanish, when certain other symmetries are also assumed to hold in the theory. Thus *superstrings* suddenly became the leading contender for a theory of quantum gravity. Not only did it purport to unite gravity with quantum theory, it also seemed to be able to incorporate the *standard model* – the unified theory of all particles and non-gravitational interactions that was established ten years earlier. Superstrings promised to be, as the cliché goes, a theory of everything.

Breakthrough for the Canonical Approach? Loop Quantum Gravity

The second breakthrough, totally unrelated to the first, occurred in 1986. Abhay Ashtekar, then at Syracuse University, reformulated gravity in terms of a new set of variables that enabled general relativity to take the form of a gauge theory. A gauge theory is the standard type of a quantum field theory in which the invariance of theory under the (position-dependent) action of some symmetry group is enforced by a field called a gauge field. The electromagnetic field is an example of a gauge field. The quantum field theories of all non-gravitational forces are expressed as gauge theories. By casting gravity in this form, Ashtekar enabled gauge-theory techniques to be applied to gravity. One such technique was that of the Wilson loop. The Wilson loop is an observable quantity that, roughly speaking, is a measure of the amount of "magnetic" flux through a small loop in space. This "magnetic" flux is related to the gauge field in a manner analogous to the way in which the familiar magnetic flux is related to the gauge field associated with electromagnetism (the so called vector potential). Founders of the loop quantum gravity approach -- who include Lee Smolin and Ted Jacobsen -- used Wilson loops as a means of specifying the state of the gravitational field. Unlike the traditional specification in terms of the three-dimensional metric, their loop-based specification allowed them to find a large class of solutions to the Wheeler-DeWitt equation. They were, however, still unable to solve it in general.

Superstring's Status as a Theory of Quantum Gravity

Since 1984 two additional superstring models were found, bringing the total to five. Certain members of this set were shown to be equivalent, the nature of these equivalences being specified by what are known as *dualities*. A unity of all superstring theories together with 11-dimensional supergravity within a theory featuring higher-dimensional extended objects was proposed. Certain of these extended objects were conjectured to serve as termination points for strings. This allowed the entropy of a class of extremal black holes – those with the least mass consistent with their charges -- to be explained.

Unfortunately, the theory has yet to make a testable prediction. It, moreover, continues to be a theory formulated on a fixed background geometry. This is the geometry on which strings and related higher-dimensional objects (p-branes) move. Dependence on such a background would seem to be a peculiar feature for a theory of quantum gravity to possess. Background independence might, however, arise in the hoped-for duality-based unification of the whole of string theory known as *M-Theory*. [No one quite knows what the M stands for, but "mother", "matrix", and "magic" are often suggested.] This theory has yet to be fully formulated. A non-perturbative version of the theory, then, has yet to form. Nevertheless, it is possible to extract from the theory an equation of motion, valid at macroscopic length scales, that is essentially the Einstein equations with quantum corrections. Such a perturbative expansion is precisely what one would require from a theory of quantum gravity. A non-perturbative theory of quantum gravity has yet to emerge from theory, however.

String theory's greatest leap toward a relevant theory of quantum gravity occurred in 1997. That year Argentinean physicist Juan Maldacena, then at Harvard, discovered something remarkable. He was able to show, within the context of string theory, that a quantum field theory defined on the boundary of a particular type of spacetime is equivalent to a quantum theory of gravity within this spacetime. The quantum field theory defined on the boundary, which is called a *conformal quantum field* theory or CFT for short, does not involve gravity. Consequently, its quantum properties are well understood, being free of the aforementioned problems that arise in mixing gravity with quantum theory. Yet the CFT is miraculously equivalent to quantum field theory *containing gravity* defined on a particular solution of the Einstein equations known as an *anti-de Sitter space* or AdS for short. Anti-de Sitter space may be thought of as an empty space with negative curvature and a negative cosmological constant, which means that this spacetime is in effect contracting. Unfortunately, Maldacena's accomplishment, known as *AdS/CFT correspondence,* has lately become an embarrassment of riches. It now seems that virtually any quantum system can be used similarly to define an equivalent quantum theory of gravity. The problem now is to determine which of these many theories properly describes our world. This problem is exacerbated by the fact that string theories tolerate a huge number of "vacua". These vacua are possible states of minimal energy (vacuum states) that determine the properties of the theory. Currently, there is no physical principle that selects among the vacua other than the *Anthropic Principle.* This principle asserts that the vacua selected must result in physical laws consistent with the evolution of the human race. So far, no single choice among these many vacua has yet to be found that is known to match our world. In particular, no single string theory yet agrees with *all three* of the following observations: There are only three macroscopic spatial dimensions, there is no supersymmetry observed, and the cosmological constant is exceedingly small and positive.

Loop Quantum Gravity's Status as a Theory of Quantum Gravity

Since 1986 Wilson loops were replaced in the formulation of loop quantum gravity by more general entities known as spin networks. The theory has succeeded in explaining the black hole entropy -- at least to within a multiplicative factor. That factor, known as the *Immizri parameter*, is the only undetermined parameter in the theory. It arose when the variables of the theory were changed in order to exploit known mathematics and thus allow the theory to advance. Unlike the theory of superstrings, an approximation of loop quantum gravity for macroscopic length scales has yet to be derived. So we have no way of checking the validity of this non-perturbative theory by seeing whether it reduces to some sort of quantum-corrected form of general relativity. The theory might be correct. But until classical gravity emerges from theory, few will believe this. Fortunately, loop quantum gravity makes at least one prediction that we can test. It predicts that the effective speed of light increases very slightly with the frequency of light. By the time you read this, satellite observations might already have confirmed or refuted this prediction.

Do We Have a Theory of Quantum Gravity?

Consider the following comparison of the two leading contenders for a theory of quantum gravity.

	superstring theory	**loop quantum gravity**
Bekenstein-Hawking entropy	yes$^+$	yes*
background independence	no	yes
perturbative approximation	yes	no

non-perturbative theory	no	yes
testable prediction	no	yes
shown to reduce to general relativity	yes	no

* manually selected Immirzi parameter. + extremal black holes only.

Notice that contents of neither column are exclusively "yes". Notice further that I have given loop quantum gravity a "yes" as a non-perturbative theory, even though we have no idea whether this theory reduces to general relativity at long length scales. I have, moreover, also generously awarded superstrings a "yes" for having a perturbative approximation, even though this approximation is not unique. So what is the answer to our question? Do we have a theory of quantum gravity? Technically, the answer is, "Sure. String theory has tons of them." The problem, of course, is that we do not seek *a* theory of quantum gravity but *the* theory, the one that corresponds to our world. More importantly, neither string theory nor loop quantum gravity (LQG) is an actual physical theory in the sense that Maxwell's electromagnetism and Einstein's general relativity are. Unlike the theories of Maxwell and Einstein, neither strings nor LQG explain observed phenomena as manifestations of an underlying pattern. Currently, strings and LQG are merely *approaches* for obtaining a proper theory. They are more accurately described as research directions or research programs than as physical theories. We should also keep in mind that one or both of these approaches might well lead (or might already have led) to dead ends. In short, the answer to our question is, "no". We do not yet have a theory of quantum gravity.

Quantum Gravity in a Nutshell

PP29. *Gravity must be quantized in order to ensure consistency of physics and to prevent its breakdown at singularities.*

PP30. *The two traditional approaches to quantum gravity – the operator and the canonical – spawned respectively the efforts in superstrings and loop quantum gravity.*

PP31. *Despite successes neither approach has yet to achieve a complete theory of quantum gravity.*

4. Wormhole Classification

What is a wormhole -- *exactly*?

A wormhole is normally defined as a tunnel that connects distinct universes or distant regions within the same universe. Figure 4.1 shows the standard diagrammatic representation of this statement. However, it turns out that it is unexpectedly difficult to define precisely what we mean by the terms "distinct universes", "connect", and even "distant regions". Fortunately, it is rather easier to define the concept of "tunnel". It is just the wormhole's *geometric throat*, which we can define as a surface of minimal area. That is, a surface is a throat, if *any* small deformation of that surface increases its area. The resultant deformed surfaces are only of interest with respect to this comparison, if they are contained within the hypersurface that contains the original surface.

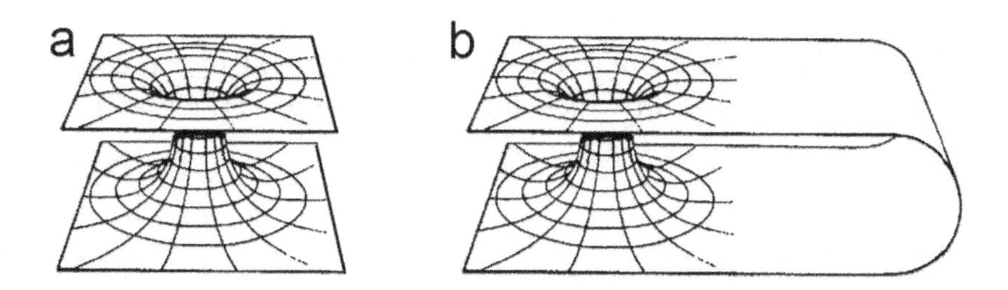

Figure 4.1. Standard wormhole diagrams. 3-dimensional space is depicted as a 2-dimensional surface. **(a)** inter-universe connection. **(b)** intra-universe connection.

A hypersurface is a three-dimensional subset of spacetime. An example of a hypersurface is the three-dimensional geometry of the universe at precisely 3:01 pm tomorrow according to your watch. This might seem like an unambiguous concept, but it is not. Ambiguity arises when we attempt to establish exactly how observers on Mars or in the Alpha Centauri system will know that your watch reads 3:01 pm. One way of addressing the problem is to establish a network of clocks. These clocks would be synchronized with your watch and slowly (to minimize time dilation) transported to observation posts throughout the region of interest. The resulting coordinate system for this region of spacetime would look like Figure 4.2a. Suppose now that we replace the clock on Mars with one that runs more slowly than the original. On Alpha Centauri we replace the original clock with one that runs too fast. Suppose further that these two clocks become less accurate with the passage of time. The slow clock runs slower, the fast clock faster. In this case the coordinate system would resemble that shown in Figure 4.2b. Even though all of its clocks are not synchronized, this is a perfectly useful coordinate system, however. The reason is simple. For a coordinate system to be useful it merely needs to assign, in an unambiguous way, time and space coordinates to every point within a particular region of spacetime. This means that we are free to use unsynchronized clocks in our network. In other words, we are free to slice spacetime however we like. Each slice is a hypersurface of simultaneity – a three-dimensional geometry of the universe at a particular time as defined by our network of unsynchronized clocks (Figure 4.2c). These arbitrary slicings – stacked collections of sequential hypersurfaces of simultaneity -- are called *foliations*. We are only restricted in choosing foliations that ensure that the imagined clock at each point of space run independently from all other clocks. The clock at any point *B* within any hypersurface of simultaneity should not have had time to receive a signal from the clock at any other point *A* within the same hypersurface. Hypersurfaces with this property are called *spacelike*.

With these definitions we can define a wormhole in the following way.

> A **wormhole** is a spacetime for which a foliation exists that contains a spacelike hypersurface in which there is a closed surface of minimal area.

Where

> A **surface of minimal area** is one whose area is positive and for which there exist no local deformations confined to the hypersurface that have less area.

And, as stated above,

> The **geometric throat** of a wormhole is its surface of minimal area.

Which leads us to,

P1. *A wormhole is a spacetime containing a geometric throat.*

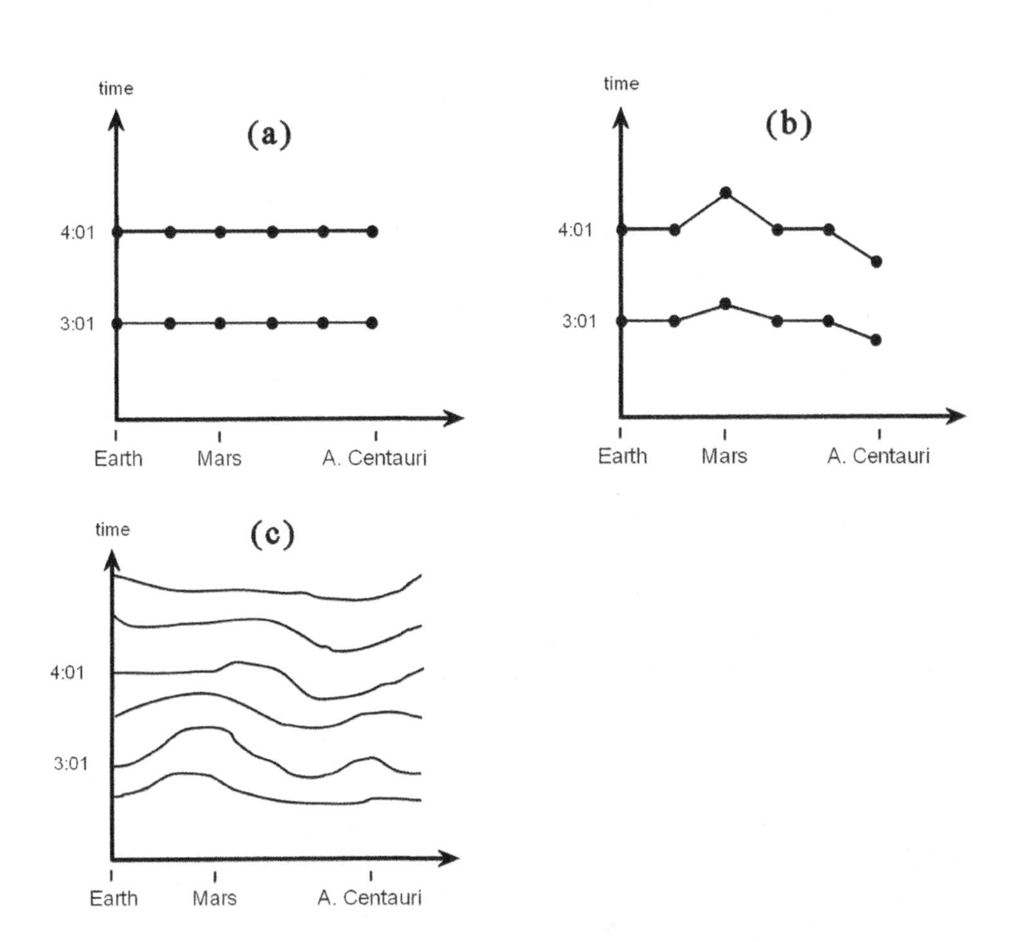

Figure 4.2. Foliation of spacetime. (a) Clocks within a network of outposts have been synchronized by slowly moving to each outpost a clock synchronized to a master clock on Earth. Hypersurfaces of simultaneity are shown for 3:01 and 4:01 tomorrow. **(b)** The hypersurfaces of simultaneity that result when the clock on Mars has been replaced by one that runs more slowly and the clock on Alpha Centauri has been replaced by one that runs more quickly. Points shown are times at which clocks on the outposts now read 3:01 and 4:01 according to the coordinate system established in part (a). **(c)** The hypersurfaces of simultaneity that result when clocks at each point in spacetime are chosen arbitrarily. Such a slicing of spacetime (i.e. an assignment of clocks to each spatial point) is called a "foliation".

Note that a wormhole is an entire spacetime. We will, nonetheless, continue in certain contexts to refer to the throat-and-mouth-containing subset of this spacetime as "the wormhole".

It is also important to notice that the above definition is purely geometrical. The appearance of a wormhole does not necessarily alter the topology of spacetime. Although the geometry of spacetime determines its shape, its topology determines what might be called its shape category. Examples of such categories are the 4-dimensional versions of the sphere, torus, pretzel, Klein bottle, plane etc. No member of any such category can be continuously deformed into a member of another category. The rules for continuous deformation forbid any surface (or its higher-dimensional analog) from being pierced, severed, or joined to another surface. Under these rules a sphere, for example, cannot be continuously deformed into a torus. To see the relevance of this to wormholes, imagine a universe shaped like an elongated balloon. Suppose this balloon began to narrow in its middle so as to form a waist. This waist would be a wormhole. Yet its appearance would not have altered the topology of the balloon universe.

The above definition allows for the possibility of a degenerate throat. Suppose, for example, that the universes in Figure 4a were connected instead by a cylinder of constant radius. There would then be no single surface of minimal area in this cylindrical tunnel connecting the asymptotically flat regions. There would be an infinite number of them. In other words,

P2. *The geometric throat of a wormhole is not necessarily unique.*

As will become clear when we further explore the nature of wormhole throats, the thing to do in the case of such degeneracy is to identify the boundaries of the degenerate region as the locations of the active throats (Figure 4.5e).

Wormhole Categories

P3. *Wormholes are of two major types – Euclidean or Lorentzian. Both of these can be further classified as permanent or transient, periodic or continuous, stable or unstable, traversable or not traversable, macroscopic or microscopic, virtual or real, static or dynamic, thin-shell or thick-shell, single-throated or multi-throated.*

Euclidean Wormholes

A Euclidean wormhole is a wormhole in a Euclidean spacetime. A flat Euclidean spacetime is one in which the distance s between any to points (t_1, x_1, y_1, z_1) and (t_2, x_2, y_2, z_2) is given by

$$s^2 = (t_2 - t_1)^2 + (x_2 - x_1)^2 + (y_2 - y_1)^2 + (z_2 - z_1)^2 \tag{1}$$

You might recognize this as the Pythagorean Theorem in four dimensions. The spacetime distance defined in this way plays no role in the physics of our world. Over a hundred years ago Albert Einstein realized that our physics should only be formulated in terms of quantities that do not change their values when viewed from different inertial reference frames. As mentioned in our discussion of special relativity, inertial reference frames are frames of reference that differ only by a constant relative rotation in space or a constant relative velocity. That is to say, such frames differ by what is known as a Lorentz transformation. This feature of our world – that its physics retains the same form when viewed from different inertial reference frames – is called Lorentzian symmetry. In other words, spacetime of our world is Lorentzian, not Euclidean. What, then, is the point of considering Euclidean wormholes?

The reason for considering Euclidean wormholes involves 1) quantum theory and 2) a purely mathematical technique for converting a solution to Einstein's gravitational field equations in Euclidean space into a "solution" in Lorentzian space. In quantum theory it is possible for a system to perform a classically forbidden transition through an energy barrier from one classically allowed state to another. This is called tunneling.

Consider, for example, a particle the size of a shotgun pellet placed within an empty box with walls made of sponge. Classical mechanics forbids the particle from exiting the box unless it has sufficient energy to punch through the sponge. In quantum mechanics, however, the particle has a finite probability of being outside the box irrespective of the particle's energy. The quantum mechanically permitted (but classically forbidden) path of the particle through a sponge wall of the box is obtained by solving the system's equations of motion in Euclidean space. This equation of motion would feature a complicated potential function that would describe the energy barrier created by the various nooks and crannies of the sponge walls of the box. The solution to this equation would describe the most likely path that the shotgun pellet would take though the wall of sponge to reach the exterior. It is possible to transform this Euclidean solution into a Lorentzian one. In other words, there exists a technique for converting this solution to an equation of motion in

Euclidean space into a (classically forbidden) solution to the corresponding equation of motion in Lorentzian space. When this technique, which is called *analytic continuation* and is described below, is applied, a tunneling solution, also known as an *instanton*, is obtained. An instanton is characterized by imaginary momentum (meaning that the momentum is the square root of a negative number). This is the sense in which it is classically forbidden.

In order to obtain a gravitational instanton -- a solution to Einstein's gravitational field equations that describe quantum mechanical tunneling between distinct classically permitted solutions -- one must solve the field equations in Euclidean space and analytically continue them to the corresponding Lorentzian space.

P4. *A Euclidean wormhole, being a solution to Einstein's gravitational field equations in Euclidean spacetime, is a gravitational instanton, which by definition describes tunneling between distinct solutions of the Einstein's equations in the corresponding Lorentzian spacetime, if such a spacetime exists.*

The sense in which a Lorentzian spacetime corresponds to a particular Euclidean spacetime is that the Lorentzian spacetime is obtained from the Euclidean spacetime by the replacement of the time coordinate as follows,

$$t \rightarrow it$$

where i is the square root of -1. Such a replacement, which is the essence of analytic continuation, is not, however, guaranteed to result in a valid Lorentzian spacetime. The resulting metric tensor might not, for example, be real. Hence,

P5. *A Euclidean wormhole is not guaranteed through analytic continuation to have a Lorentzian counterpart.*

The chief motivation for considering Euclidean wormholes stems from quantum field theory. In quantum field theory the probability of a particular field configuration (or of any observable quantity) is expressed as a *functional integral* – a sum of the values that result when a particular expression is evaluated at each possible history of field configurations. This sum is undefined in Lorentzian space. It does not converge to any particular value. Its analytic continuation to Euclidean space is, by contrast, well defined and calculable (at least in toy models). Hence, many practitioners of quantum gravity prefer to work in the Euclidean realm, remaining optimistic that their results will correspond to some Lorentzian reality.

Euclidean wormholes can make unexpectedly significant contributions to the values of these functional integrals. This formed the crux of a famous attempt to explain what was considered the unreasonably small value of the observed cosmological constant. This fueled interest in Euclidean wormholes amongst particle physicists at the same time that Morris and Thorne were renewing interest in Lorentzian wormholes amongst relativists. These simultaneous bursts of interest were entirely coincidental. Research efforts involving Euclidean wormholes had little to do with the work that explored their Lorentzian cousins. These activities occurred for the most part within separate communities of physicists.

While one can in principle consider various subcategories of Euclidean wormholes -- e.g. macroscopic, traversable, stable, etc. -- such sub-categorization is usually applied to Lorentzian wormholes only. Euclidean wormholes are sub-categorized indirectly through the sub-categorization of their Lorentzian counterparts.

The study of Euclidean wormholes in physics focuses less on the properties of the solutions themselves than on their role as facilitators of other mechanisms, such as coupling constant renormalization, baby universe formation, or topological fluctuation. There are, for example, no Euclidean wormhole solutions with a stature comparable to that of the Schwarzschild black hole.

Lorentzian Wormholes

A Lorentzian wormhole is a wormhole in a Lorentzian spacetime. A flat Lorentzian spacetime is one in which the distance s between any to points (t_1, x_1, y_1, z_1) and (t_2, x_2, y_2, z_2) is given by

$$s = -(t_2 - t_1)^2 + (x_2 - x_1)^2 + (y_2 - y_1)^2 + (z_2 - z_1)^2 \qquad (2)$$

Unlike the corresponding distance formula in Euclidean space, this distance retains its value when it is evaluated in other inertial reference frames. In other words, it is consistent with the Lorentzian symmetry observed in nature. A Lorentzian wormhole, then, is one that appears in a spacetime of a sort known to physically exist.

Macroscopic Wormholes

We are unable to detect physical structures that are too small to be resolved by our most powerful instruments. The more powerful the instrument – the higher the energy of the particles used by the instrument for detection -- the smaller the structures visible to it. Viruses, for example, are too small be detected by optical microscopes. To electron microscopes, however, they are visible. This is because it is easy to accelerate the electrons, used by the electron microscope, to energies that well exceed that of the optical photons used by the optical microscope. According the Heisenberg's uncertainty principle, the size x of the smallest structure detectable to an instrument, whose probing particles have energy E is given by

$$x \sim hc/E$$

In other words the resolution of an instrument is inversely proportional to its probing energy. The introduction of such a length scale immediately suggests the following definitions.

> A **macroscopic wormhole** is one whose throat is sufficiently large to be detected by an instrument probing at the maximum energy that the human race can bring to bear.
>
> A **microscopic wormhole** is one that is not macroscopic.

The size of what we are calling a macroscopic wormhole, then, depends on the current level of human technology. Mid-twenty-first-century particle accelerators will be characterized by a probing energy of about 1000 TeV (Trillion electron Volts). This defines macroscopic wormholes as those whose throats are at least one percent of the size of a proton.

As we will see in the next section, the introduction of this length scale into wormhole physics is significant in that it permits a spacetime to effectively change its topology. Because we can only detect those topology-determining features of a spacetime that exceed this length scale, we can only be aware of a spacetime's macroscopic topology. Consider, for example, a manifold with the topology of sphere (S^2). Add a handle to it, and its topology becomes that of a torus (S^1 x S^1). If, however, this handle is too thin to be detected, the *macroscopic* topology of the manifold remains that of a sphere (Figure 4.3).

Consider a transition between the manifold shown in Figure 4.3a and that in Figure 4.3b. This would be a transition in which the macroscopic topology changes (from that of a torus to that of a sphere), while the actual topology does not. To clarify what we mean by macroscopic topology, let's define it explicitly.

> The **macroscopic topology** of a spacetime is the topology that can be detected using geometry-probing instruments with the highest resolution available.

This allows us to state that

P6*. The macroscopic topology of a wormhole can change, even if its actual topology does not.*

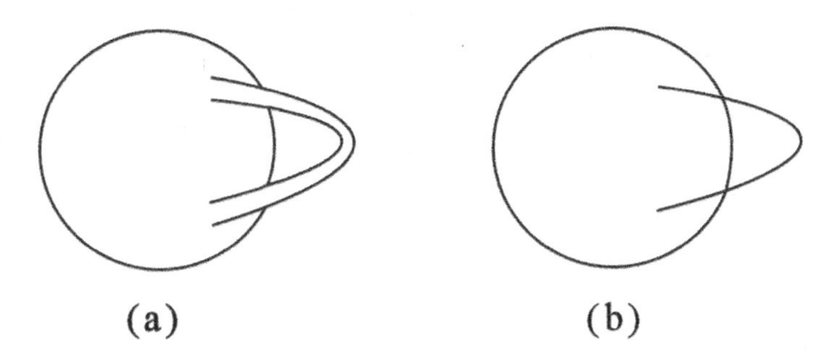

Figure 4.3. Macroscopic vs. microscopic topology. (a) A sphere with a handle added to it acquires the topology of a torus. (b) If the handle is made too thin to be detected, the macroscopic topology of the space becomes that of a sphere, even though its actual topology remains that of a torus.

Inter-Universe Wormholes

An **inter-universe wormhole** is one that does not function as a short cut. It exists only when there are no paths between its mouths other than those that pass through the wormhole itself or through other wormholes (Figure 4.1a).

An **intra-universe wormhole** is one that is not an inter-universe wormhole. It connects regions within the same universe (Figure 4.1b).

Permanent Wormholes

You might think it perfectly natural to consider the creation and destruction of wormholes. Surely, you might think, if a wormhole exists, it can be destroyed. Or if it doesn't exist, one can be created. Unfortunately, things are not so simple in wormhole physics.

The reason is that the only spacetimes that we can justifiably consider are those that are consistent with established physics – classical general relativity. By "classical general relativity" I mean classical general relativity *together with the conventional assumptions*. One of these assumptions is that spacetime is *temporally orientable* – that it is possible to globally define a future-pointing direction for every point in spacetime. Another conventional assumption is that spacetime contains *no closed timelike curves* – that the future of any observer is distinct from her past. A third assumption is that matter obeys the weak energy condition – that its density is everywhere positive and is never exceeded by its tension. With these assumptions and theorems proved by Robert Geroch in 1966 and Frank Tipler in 1976, both of which were extended by Arvind Borde in 1994, we can conclude that the topology of space cannot change.

However, as we will see in Chapter 10, the conventionally assumed absence of closed timelike curves and adherence to the weak energy condition are utterly inconsistent with the assumption of traversable wormholes in the classical theory. So we cannot use the theorems of Geroch and Tipler to conclude that the topology of a space containing wormholes is immutable. Nevertheless, we will (at least until we consider wormhole engineering) take the conventional position that the topology of space is fixed.

P7. *The exclusion of wormhole creation* ex nihilo *and other topology changes from general relativity follows from the theorems of Geroch and Tipler and the assumptions that spacetime is temporally orientable, free of closed timelike curves, and only contains matter that obeys the weak energy condition.*

By assumption, then, classical general relativity does not permit topology change. The topology of space does not enter into the dynamics of the theory. There is no way, for example, for an exceedingly high curvature or an exceedingly high rate of curvature fluctuation to force a space to change its topology. Hence,

P8. *According to the conventional view of general relativity wormholes may exist, but a wormhole cannot be created nor can an existing wormhole be destroyed.*

Again, I need to point out that there's a bit of a problem with this conventional view. Classical general relativity permits solutions, free of weak-energy-condition-violating matter, that nevertheless contain closed time like curves. The presence of these curves invalidates an assumption on which Geroch's theorem, which bans topology change, depends. The conventional proscription of topology change from general relativity, to which we will for the most part adhere, is not entirely justified under certain conditions of interest.

If we take P8 seriously, all wormholes are permanent. How, then, does it make sense to consider permanent wormholes as a particular variety of the species? Although we assume general relativity to forbid topology change, and therefore the creation and destruction of wormholes, it does permit *pseudo* topology change as emphasized by Visser.

> A **pseudo topology change** is a change in the geometry of spacetime that changes its macroscopic topology without changing its actual topology.

A wormhole, then, can *appear* to be created, if, through a pseudo topology change, an existing microscopic intra-universe wormhole inflates to macroscopic size. Similarly, a macroscopic intra-universe wormhole can *appear* to be destroyed, if a pseudo topology change shrinks it until it is microscopic. For inter-universe wormholes, such apparent creation and destruction need not even require a pseudo topology change. Hence,

P9. *The transition of a wormhole between microscopic and macroscopic sizes simulates the creation and destruction of wormholes. For intra-universe wormholes this results in a pseudo topology change. For inter-universe wormholes it need not.*

So wormholes can appear to be created or destroyed without this actually occurring. Given this, we now have a basis for defining a permanent wormhole as one that appears to persist for a sufficiently long time. The time for which a wormhole can be said to persist is the time for which the wormhole remains macroscopic. We need now only decide what we mean by a "sufficiently long time". Let's employ the concept of permanence implicit in astrophysics.

> A **permanent wormhole** is a wormhole that persists macroscopically

longer than the most short-lived astrophysical objects.

Accordingly,

A **transient wormhole** is a wormhole that is not permanent.

By these definitions a wormhole that persists macroscopically for 5 thousand years is merely transient; one that persists for 5 billion years is permanent.

Stable Wormholes

A permanent wormhole is not necessarily stable. To understand this, consider a house of cards. It can exist indefinitely, but only if it is not perturbed. So it is permanent, but unstable. What about a house of brick? If it is exposed to a sufficiently powerful perturbing force, it too will fall. How, then, can we refine our intuitive notion of stability so that it applies to a house of brick, but not to a house of cards? The answer is to define stability relative to a particular class of perturbations. We would, for example, consider a house of brick to be stable if it continues to stand while we hold wild dance parties in it, but falls when we detonate a bomb in it. Accordingly, we can define wormhole stability as follows.

A **stable wormhole** is a permanent wormhole that persists despite exposure to perturbations resulting from human activity not intended to destroy it.

As defined in the previous section, a wormhole is "destroyed" only if it *appears* to be destroyed as a result of its becoming microscopic, which for intra-universe wormholes corresponds to a pseudo topology change.

Traversable Wormholes

In order to be able to use a wormhole as a shortcut for interstellar travel, a human being must be able to pass through it. Unfortunately, such a traveler's survival is threatened by gravitational tidal forces that might crush her, radiation levels that might incinerate her, a throat that might close before she can pass through it, or a trip duration that might exceed her remaining life span. These considerations motivate the following simple definition.

> A **traversable wormhole** is a wormhole through which a human traveler can pass unharmed in either direction an arbitrarily large number of times.

P10. *A traversable wormhole is necessarily macroscopic and stable.*

These definitions will be later expanded to include convenient but inessential characteristics. One of these is a reasonably short traversal time -- one shorter than the remaining-life-span criterion used above.

Periodic Wormholes

> A **periodic wormhole** is a wormhole that, as a consequence of its repeated transitions between macroscopic and microscopic sizes, appears at regular intervals to be created and subsequently destroyed.

As for any wormhole, a mouth of a periodic wormhole possesses at any time an effective mass, charge, momentum and angular momentum. The motion of either of its mouths is the same as that of any other object for which these properties are identical. In particular,

P11. *The mouth of a periodic wormhole will in successive incarnations appear at the same spatial location, as viewed by observers stationary with respect to the mouth.*

Virtual Wormholes

As I mentioned in Chapter 2's summary of wormhole history, John Wheeler hypothesized that the vacuum of the gravitational field is characterized by what he termed "quantum foam". Just as the vacuum of an ordinary quantum field features virtual particles, Wheeler supposed, the vacuum of the quantum gravitational field features virtual geometries. Just as a virtual particle may be considered a configuration of a quantum field that briefly comes into existence, a virtual geometry is a similarly ephemeral configuration of the quantum gravitational field. Some of these virtual geometries include wormholes, in which case they are called *virtual* wormholes or *Wheeler* wormholes.

Because virtual wormholes are expected to be exceedingly short lived, in analogy to virtual particles, virtual wormholes are transient in addition to being microscopic.

P12*. Virtual wormholes are likely transient and microscopic.*

The existence of Wheeler wormholes does not imply quantum fluctuations in the topology of space. As mentioned above, the topology of space is conventionally taken to be fixed and equal to that of the initial three-dimensional spacelike hypersurface of the spacetime manifold – the stage on which the dynamics of the theory are enacted. A fixed topology requires the number of wormholes in any three-dimensional spacelike hypersurface to be fixed as well. Quantum fluctuations in geometry that involve Wheeler wormholes are those in which the wormhole transitions from being submicroscopic to microscopic. In such a transition the radius of its throat would grow from the smallest possible size of about 1 Planck length to, for example, a size of 1000 Planck lengths. It would exist at this enlarged yet microscopic size for a time restricted by the uncertainty principle, before returning to its original submicroscopic dimensions. In order for the appearance of a Wheeler wormhole to be possible at a particular location, the wormhole would already have to be there in its submicroscopic form. The number of such wormholes in any volume of space, then, would be the ratio of this volume to that of the wormhole. Assuming the latter to be a Planck volume, a cubic centimeter contains no fewer than 10^{98} fluctuating virtual wormholes.

P13. *The existence of virtual wormholes does not necessarily imply that there are quantum fluctuations in the topology of space.*

Thin-Shell Wormholes

Matter in an ordinary wormhole is distributed within a region of finite extent centered about the wormhole's throat. If the wormhole is spherically symmetric, this region is a spherical shell of a given finite thickness. Such a wormhole is an example of an ordinary *thick-shell* wormhole. If its matter is instead distributed within a shell that is infinitesimally thin, the wormhole is said to be a *thin-shell* wormhole.

Thin-shell wormholes are mathematical idealizations that simplify the analysis of wormhole stability.

Ringholes

Consider a spherically symmetric wormhole. Such a wormhole could be surrounded by a spherical mouth within which there would be another sphere that would define the wormhole's throat. If this wormhole is traversable, a traveler would, upon passing through the outer sphere (the mouth) and entering the inner sphere (the throat) soon find herself in another universe or in another part of her universe of origination.

Replace the outer sphere in the above description with an outer torus. Replace the inner sphere defining the wormhole's throat with an inner torus. A traveler would traverse this wormhole by passing through the mouth defined by the outer torus, into the throat defined by the inner torus, and finally out to another universe or another region of her universe of origin. A wormhole of this topology is called a *ringhole.*

> A **ringhole** is a wormhole whose mouth and throat both have the topology of a torus.

Dynamic Wormholes

Consider a pendulum. If the mass hanging at the end of the string is in its lowest position, there it will remain indefinitely. This is an example of a static solution to the pendulum's equations of motion. If instead we displace the mass and release it, the pendulum will swing back and forth until it is brought again to rest by the action of air resistance. This is a dynamic solution to its equations of motion. We similarly define static and dynamic wormholes as follows.

> A **static wormhole** is a wormhole whose geometry does not change with time.

> A **dynamic wormhole** is wormhole whose geometry changes with time.

Why do we need the categories of "static" and "dynamic", when we have already defined those of "permanent" and "transient"? Consider the following principle.

P14*. All static wormholes are permanent, but permanent wormholes need not be static.*

Recall that a permanent wormhole need only persist for a finite time. Moreover, during the time it persists, it need not be unchanging. Regarding transient wormholes, we can say,

P15. *All transient wormholes are dynamic, but dynamic wormholes need not be transient.*

A transient wormhole is one that briefly exists. Hence, it must be dynamic. A dynamic wormhole, by contrast, may persist indefinitely. It need not, therefore, be transient. It just needs to change with time during some part of its existence.

Multi-throated Wormholes

The static wormholes that have been the subject of the overwhelming majority of research efforts have been single-throated, as shown in Figure 4.4a. Multi-throated wormholes, however, are also possible, as shown in Figure 4.4b. While a multi-throated wormhole can in principle be static -- given an appropriate distribution of mass-energy and stress -- they are more likely to arise in the study of dynamic wormholes. In that case a multi-throated wormhole could be a temporary configuration that a perturbed (artificially stabilized and spherically asymmetric) wormhole assumes, as it settles down, through the emission of gravity waves, into a final single-throated state.

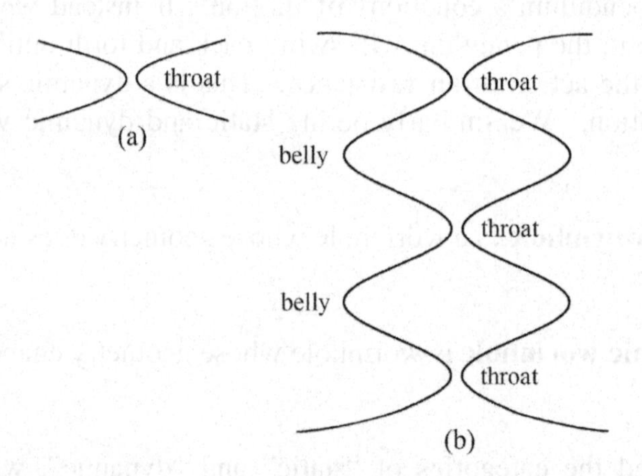

Figure 4.4. Single and multi-throated wormholes. (a) Cross-section of a standard single-throated wormhole at a given instant (hypersurface of simultaneity). (b) A multi-throated wormhole that might momentarily arise in a dynamical wormhole. It can also be interpreted as two intermediary bubble universes connected to each other and to asymptotically flat universes by single-throated wormholes. Both types of wormholes are highly unstable.

The functional Throat

In addition to defining a wormhole's throat geometrically, as we did above, we can also define it in terms of its action on beams of light.

> A **functional wormhole throat**[*] is a two-dimensional closed surface that satisfies the following.
>
> 1) Beams of light traveling perpendicular to the surface do not diverge from each other at this surface – i.e. they are parallel precisely at the surface.
>
> 2) Such beams begin to diverge upon crossing the surface.

The reason that we need this functional definition, when we already have a geometric one, is that the latter ceases to be of relevance for dynamic wormholes. The antigravitating action of the wormhole on light, which occurs at the geometric throat of a static wormhole, occurs elsewhere if the wormhole is dynamic. In other words,

P16. *The functional throats and geometric throat of a wormhole only coincide if the wormhole is static.*

Figure 4.5, which is drawn in two-dimensions instead of three for the sake of simplicity, helps to clarify the functional definition. Imagine a circle (the two-dimensional analog of a spherical surface). Ingoing light beams perpendicular to the circle converge (4.5a,b). Attach to this circle a short cylinder of equal diameter (4.5c). Observe that the beams travel parallel to each other on this cylinder. This is what happens at a wormhole's throat. Add to the short cylinder two flaring funnels, one above, the other below (4.5d). Realize that ingoing light beams traveling perpendicular to the circular base of the lower funnel converge, while they are on the lower funnel. When these beams reach the short cylinder between the upper lower funnel, they travel parallel to each other. When they travel on the surface of the upper funnel, they diverge. A typical (non-degenerate) wormhole throat results from decreasing the height of the short cylinder until it is zero.

[*] In the case of dynamic wormholes it will be useful to modify this definition by restricting it to either ingoing or outgoing beams depending on the dynamics.

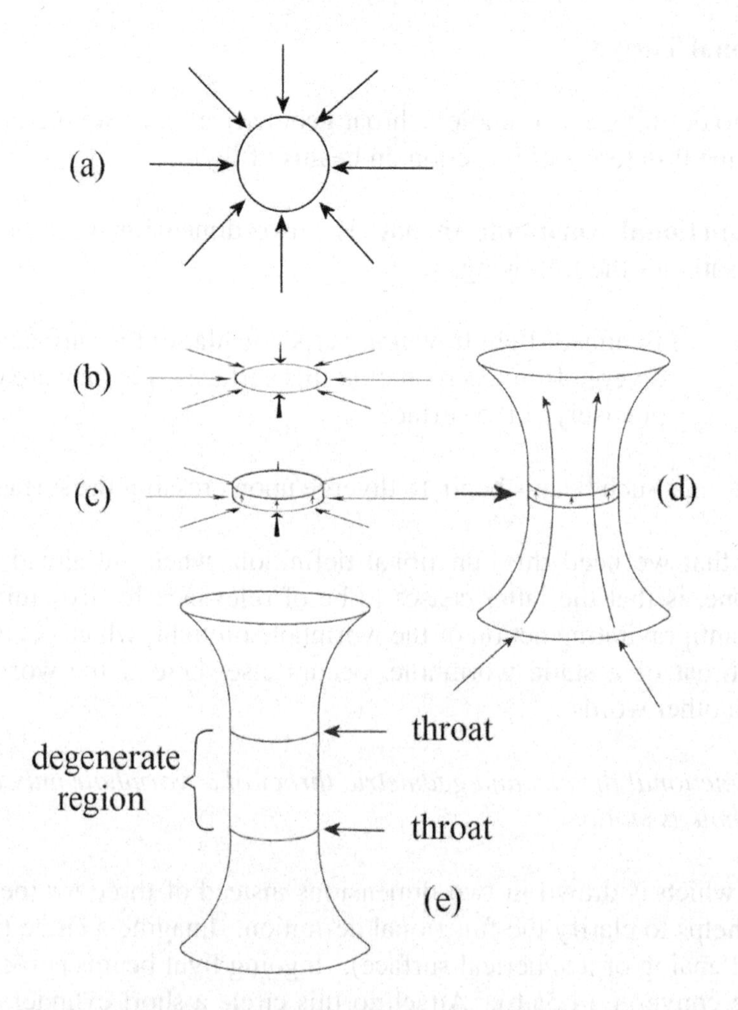

Figure 4.5. Light rays and throats. (a) Ingoing light rays converge on a circle. (b) another view. (c) Adding a short cylinder to the circle. (d) Adding funnels to the cylinder. (e) Degenerate throat region results from leaving the cylinder height finite. Active throats are at the boundaries. Unique throat results from shrinking cylinder height to zero. In either case, rays are parallel at the throat. As usual, these are 2-dimensional representations of 3-dimensional hypersurfaces.

The Belly

Multi-throated wormholes possess another interesting surface. I shall call it a "belly". Like the throats, it may be defined geometrically and functionally. Geometrically, the belly is the surface of maximal area between two throats. Functionally, it is the surface between throats at which light rays travel in parallel. They do not appear anywhere in the wormhole literature, because one normally regards any swollen region that arises between throats as a tiny independent universe. Here, I present them instead as a transient state of an arbitrarily dynamic wormhole, or as permanent state of an artificially stabilized one. One should be aware, however, that unfettered dynamics of such a region would likely result in its pinching off to form a baby universe. We have, then, two definitions

> The **geometric wormhole belly** is the surface of minimal area located between the geometric throats.
>
> A **functional wormhole belly**[‡] is a two-dimension closed surface that satisfies the following.
>
> 1) Beams of light traveling perpendicular to the surface do not diverge from each other at this surface – i.e. they are parallel precisely at the surface.
>
> 2) Such beams begin to converge upon crossing the surface.

As with throats, geometrical and functional bellies differ only if the wormhole is dynamic. Or,

P17. *The functional bellies and geometric belly of a wormhole coincide only if the wormhole is static.*

A glance at Figure 4.4 suggest that

P18. *The number of bellies possessed by a spherically symmetrical static wormhole is one less than the number of its throats.*

Figure 4.5 suggests another principle. Suppose that we had not taken the limit of an infinitesimally short cylinder. Suppose instead that we had considered a cylinder of finite height connecting the upper and lower funnels. In that case there would not have been a single surface at which the light beams were parallel. There would have been an infinite number of them. In this case there is not an infinite number of throats but precisely two of them, as shown in figure 4.5e. The upper throat is where light originating in the lower universe begins to diverge. The lower throat acts similarly for light originating in the upper universe.

[‡] In the case of dynamic wormholes it will be useful to modify this definition by restricting it to either ingoing or outgoing beams depending on the dynamics.

The Interior Region and its Boundaries

The geometric throat of a single-throated static wormhole will bifurcate into two functional throats, if the wormhole begins to expand or contract. For a general wormhole the region between these functional throats – the *interior region* -- will contain surfaces that cause parallel light beams to converge (bellies), diverge (interior throats), or do neither (degenerate regions). It turns out that the surface that functions as a throat for a bundle of light rays -- in that it defocuses them -- depends on the wormhole's dynamics and on the ray's direction of traversal. If the wormhole is contracting, traversing rays will encounter their effective functional throat on the far side of a wormhole. The effective functional throat will be the last surface that the rays cross that meets the above definition of a functional throat. If the wormhole is expanding, the effective functional throat will be the first such surface encountered. These first or last encountered throats -- the exterior throats -- are to be identified as *the* throats of a wormhole. To restate,

P19. *A dynamic wormhole will in general have two functional throats.*

P20. *The region between a dynamic wormhole's functional throats will in general contain bellies, interior throats, and degenerate regions.*

P21. *The effective functional throat for light traversing a wormhole is either the first surface encountered that meets the definition of a functional throat or the last, depending on whether the wormhole is expanding or contracting.*

For the sake of clarity, it is important to draw a distinction between two cases. The first is that of a static wormhole with two geometric throats. The second is that of a dynamic wormhole with one geometric throat that has, as a consequence of the wormhole's dynamics, bifurcated into two functional throats. Although both are cases of dual-throated wormholes (the geometric throat ceasing to act as a throat in the dynamic case), their three-dimensional geometries differ (Figure 4.6). The bifurcation of a dynamic wormhole's geometric throat does not change the qualitative character of its geometry in any spacelike hypersurface. In short, neither expansion nor contraction causes the wormhole of Figure 4.6a to look like that of Figure 4.6b. In both cases we will take the interior region to be that between the exterior throats.

By the way, bellies can bifurcate too. Just as the onset expansion or contraction will bifurcate the geometric throat of a previously static wormhole, so will it bifurcate its bellies. Just as for throats, the effective functional belly for traversing light will depend on whether the wormhole is expanding or contracting and on the light's direction of traversal.

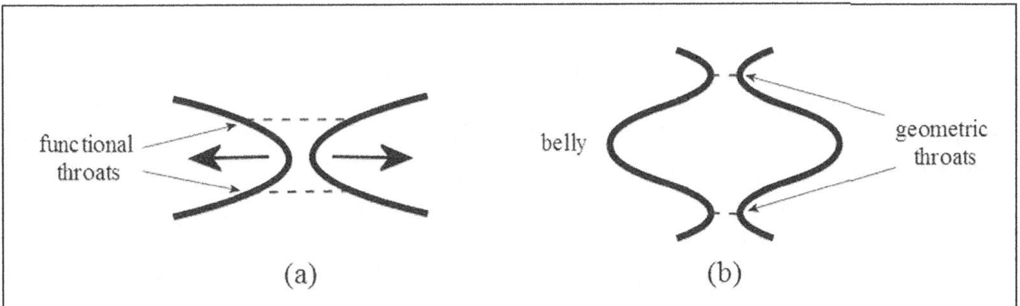

Figure 4.6. Dual-throated wormholes. (a) When a wormhole expands, its geometric throat bifurcates without drastically changing the wormhole's 3-dimensional geometry. The surface of minimal area remains, but ceases to defocus light, i.e. it stops functioning as a throat. (b) A static wormhole may have dual throats at the cost of acquiring a belly. If this wormhole begins to expand or contract, both its throats will bifurcate as will its belly.

The larger the area of a wormhole's belly in comparison to that of its throat, the less tenable is our interpretation of the geometry as that of a fat multi-throated wormhole. At some point, we would be forced to cease viewing it as a wormhole interior and, as mentioned above, regard it as its own universe. While multi-throated wormholes can be studied as static entities, they are (like all conventional wormholes) inherently unstable. Were they allowed to evolve dynamically, the most likely result would be the birth of at least one new universe. This would occur when the throats effectively pinch off. Whether these newly born universes would expand or contract would depend on the nature and state of the matter they contain.

P22. *A dual-throated static wormhole with a belly is more commonly regarded as a universe connected to two distinct asymptotically flat regions by two distinct wormholes.*

Classical and Quantum Wormholes

A classical wormhole is not a special type of wormhole. It is a wormhole of any type whose properties are elucidated using classical physics only. Similarly, a quantum wormhole is one whose description relies solely on quantum physics. A wormhole becomes "semiclassical" if both classical and quantum physics are used to describe it. Hence,

P23. *A wormhole is classical, quantum, or semiclassical depending on whether its description relies on classical physics, quantum physics, or both.*

While quantum theory can always be used to accurately (though perhaps awkwardly) describe a system, classical physics – which is an approximation to quantum theory -- is only valid under certain conditions. These are specified by the famous Heisenberg uncertainty principle. Classical physics will fail to determine with arbitrary precision both a wormhole's geometry and the magnitude of its tendency to fluctuate. If the product of precisions with which these quantities can be measured does not well exceed a small value determined by Plank's constant, quantum theory should be used to describe the wormhole.

While the classical theory of gravitation, General Relativity, permits a thorough description of a classical wormhole, there is no quantum theory of gravity of equal maturity. Fully quantum treatments of wormholes are currently limited to unrealistic toy models in which all but one degree of freedom is eliminated.

5. Classical Wormholes

Classical wormholes are solution to the purely classical Einstein equations. Three of the solutions described in this chapter are untraversable, singularity-containing wormholes better known as black holes.

It might seem strange to refer to black holes as wormholes. Keep in mind, however, that we will be considering the *maximally extended* black hole solutions, which all contain at least two exterior regions across which the black hole seems to form a bridge. It was never *a priori* obvious that these bridges were untraversable. As late as 1997, for example, there was still some question as to whether it was truly impossible to traverse any Reissner-Nordstrøm wormhole, as had been previously and is currently believed.

Each of these black holes – i.e. untraversable wormholes -- has a traversable wormhole counterpart. These counterparts could in principle each have been created by gradually irradiating the corresponding black hole with negative energy. Because of this close relationship we will consider in turn three pairs of black-hole-wormhole counterparts. The first is the traversable Morris-Thorne wormhole and its untraversable counterpart, the Schwarzschild wormhole. The traversable Kim-Lee wormhole is similarly paired with the untraversable Reissner-Nordstrøm wormhole. The traversable Teo wormhole finds its untraversable counterpart in the Kerr wormhole. These three pairs of wormhole solutions describe respectively the following three classes of wormholes: uncharged and nonrotating, charged and nonrotating, and uncharged and rotating. All of the wormholes considered in this chapter are static in the sense that they do not explicitly depend on the passage of time as experienced by distant observers.

The aforementioned untraversable wormholes, like all black holes, are characterized by two important features – *event horizons* and *singularities.*

> A **future event horizon** is a boundary between 1) a region of spacetime from which light signals can be sent that will travel arbitrarily far into the future and arbitrarily far in space, and 2) a region from which such signals cannot be sent.
>
> A **past event horizon** is a boundary between 1) a region of spacetime in which light signals can be received that have traveled from arbitrarily far in the past and arbitrarily far in space, and 2) a region in which such signals cannot be received.

The first of these definitions is somewhat more prominent in that the term "event horizon" without further modification normally denotes a future event horizon. The word "antihorizon" is often used as a synonym for a past event horizon.

Anyone unfortunate enough to find herself within a future event horizon would be totally cut off from the spacetime exterior to it. She will not be able to travel to this exterior region. Nor would she even able to send signals to it. By contrast, anyone within a past event horizon *would* be able to travel to the exterior spacetime [She would in fact be required to]. She would not, however, be able to receive signals from the exterior. In either case she would have to contend with the presence of a singularity.

To define what general relativists mean by a singularity, we must review another definition. Recall that particles moving solely under the influence of a gravitational field – under influence of the curvature of spacetime – follow certain paths called geodesics. Geodesics help to define the concept of a spacetime singularity as follows.

> A **singularity** is a location in spacetime at which geodesics end or begin and around which curvature increases arbitrarily as the location is approached.

Normally, one thinks of singularities as locations at which the curvature of spacetime is infinite. We will not define them this way, however, because we need a definition that will apply as well to *practically* singular spacetimes. By "practically singular spacetimes" I mean those from which locations of infinite curvature have been surgically excised, while their surrounding regions of arbitrarily high curvature have been retained.

The most important thing to realize about spacetime singularities is that they are locations of the total breakdown of physics. Think of spacetime as the stage upon which a play called *Physics* is enacted. At singularities this stage is destroyed. There the play cannot proceed.

Uncharged, Nonrotating Wormholes

The Schwarzschild Wormhole

Karl Schwarzschild served in the German army during World War I. Shortly before his painful death in 1916 from a disease contracted at the Russian front, Schwarzschild found, with an ease that surprised Einstein, an exact solution to the field equations of general relativity, a theory that was less than a year old. The solution describes the curvature of spacetime induced by a spherically symmetrical, nonrotating mass. In 1935 Einstein and Rosen noticed that an extended version of this solution seems to describe a sort of bridge between two universes.

Like any solution to Einstein's equations, the Schwarzschild solution is expressed in terms of a spacetime metric. Recall that the metric is a mathematical object that describes the geometry of spacetime. It is a matrix of numbers that is used to obtain the distance between any two points. When these points are close enough, the distance between them can be expressed as a generalization of the Pythagorean Theorem that involves the components of the metric.

We can visualize a metric by using what is known as an *embedding diagram*. Such a diagram relies on the use of an extra dimension to create a higher-dimensional embedding space in which the space of interest is a "surface" – a space of lower dimensionality. For example, we can embed a three-dimensional slice of the Schwarzschild spacetime in a four-dimensional embedding space. Because it is difficult to depict four dimensions on a two-dimensional page, one normally suppresses one of the coordinate dimensions of the original space. In the case of the spherically symmetrical Schwarzschild solution, we can without loss of important information suppress one of the angular coordinates along which our slice of Schwarzschild spacetime is constant. Figure 5.1 shows such an embedding.

The interesting thing about the Schwarzschild solution is that it is impossible to understand it fully in the seemingly reasonable coordinate system that Schwarzschild used to discover it. He was looking for a static, spherically

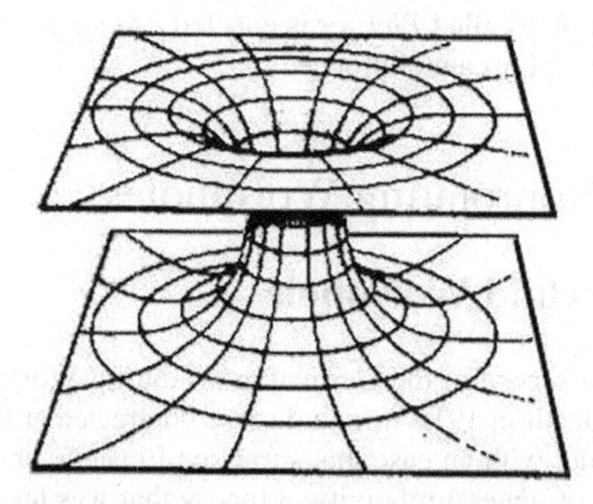

Figure 5.1. Snapshot of Schwarzschild wormhole. Embedding diagram of Schwarzschild solution at the instant at which its throat is at its maximum diameter. As usual, one degree of freedom has been suppressed.

symmetrical solution, so naturally he choose spherical coordinates – radius, azimuth, elevation – together with a time coordinate. Expressed in this coordinate system, it appears that spacetime around a spherically symmetric, nonrotating, uncharged body of mass M behaves very badly at the radial coordinate value $r = 2GM/c^2$, where G is the gravitation constant and c is the speed of light. There the radial component of the metric becomes infinite and the temporal component becomes zero.

For years it was thought that the Schwarzschild spacetime did in fact exhibit some sort of radial singularity at $r = 2GM/c^2$. Eventually physicists came to realize that it was not Schwarzschild's spacetime that was behaving badly. It was his choice of coordinate system. Not only did it imply the existence of a pathology in spacetime where one did not exist, it hid the existence of such a feature where it did exist. Moreover, it tended to obscure that part of the Schwarzschild solution where the spacetime actually was behaving badly – the true singularity at $r = 0$. By the 1950s physicists began to study the Schwarzschild solution in alternative coordinates systems. These coordinates systems revealed the true nature of the solution, such as the absence of any sort of physical singularity at the event horizon. One of these coordinates systems, that proposed by Roger Penrose of Oxford University, has become particularly popular. Within a finite diagram, it conveniently permits the display of a spacetime of infinite extent. Figure 5.2 shows two such *Penrose diagrams* – one for an empty space, the other for a static traversable wormhole. The Penrose diagram of the Schwarzschild solution appears in Figure 5.3.

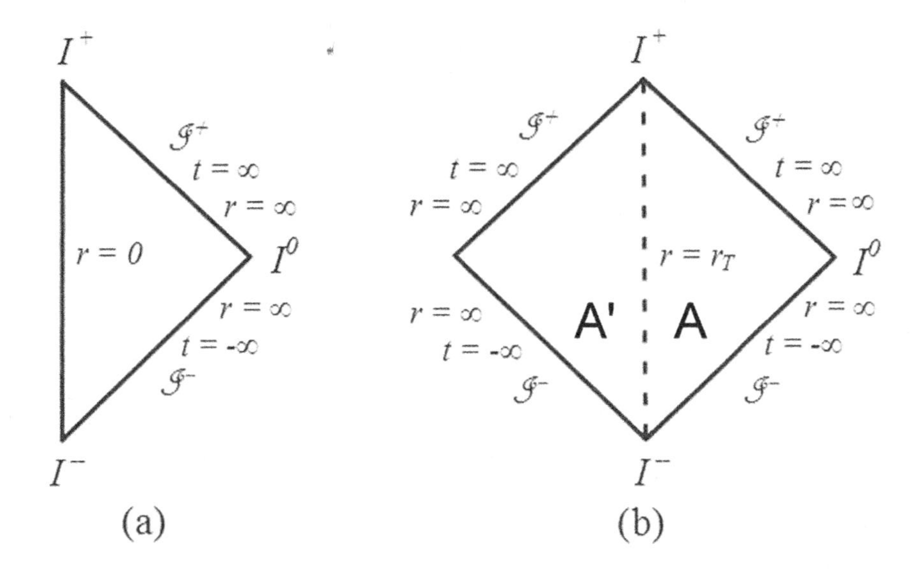

Figure 5.2. Two Penrose diagrams. (a) Empty spacetime. The world lines of all eternal objects possessing mass begin at I^- (past timelike infinity) and end at I^+ (future timelike infinity). The world lines of all eternal massless objects, such as photons, are at 45-degree angles and begin at $\mathscr{I}^-$ (past lightlike infinity) and end at $\mathscr{I}^+$ (future lightlike infinity). (b) Traversable wormhole with throat radius r_T. World lines may now begin in universe *A*, pass through the wormhole (dotted line), and enter universe *A'*. Both universes are of infinite spatial extent, i.e. I^0 represents spatial (spacelike) infinity.

Understanding Penrose Diagrams

To make sense of the Penrose diagram, it will help to review a couple of concepts from special relativity – that of *Minkowski* space and that of a *world line*. Minkowski space is just the flat spacetime of special relativity. Figure 5.4 shows Minkowski space with two of the three spatial dimensions suppressed for the sake of simplicity. Each point in Minkowski space represents an event – a time and a place. The vertical axis marks the time at which an event occurred, the horizontal axis the place. Persistent objects in Minkowski space follow paths called word lines. For example, an object that remains at the position $x = a$, will have a world line that is a vertical line that intersects the horizontal axis at $x = a$ (Figure 5.4a). An object that moves at varying speeds from $x = a$ to $x = b$ will have a world line like that shown in Figure 5.4b. Note that the speed at which an object moves at a particular instant is the inverse of the slope of its world line at that instant. By convention, the

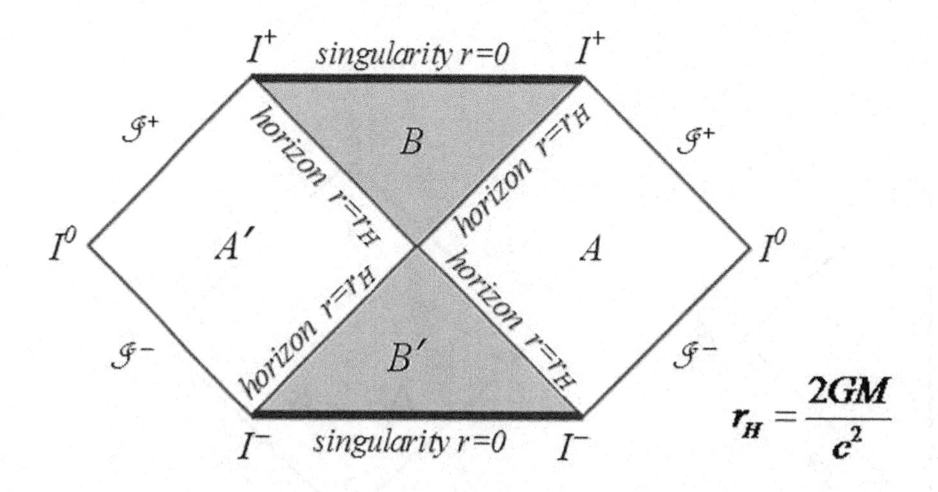

Figure 5.3. Penrose diagram of the maximally extended Schwarzschild solution.
I^+ -- timelike future infinity. I^- -- timelike past infinity. I^0 – spacelike infinity.
$\mathcal{J}^+$ -- lightlike future infinity. $\mathcal{J}^-$ -- lightlike past infinity.
A – exterior region of "upper" universe. *A'* – exterior region of "lower" universe.
B – black hole interior region. *B'* – white hole interior region.

slope associated with the speed of light is 45 degrees from the horizontal. Accordingly, Figure 5.4c shows the world line of a light pulse of light. Finally, we will need a bit of terminology. A line or surface whose slopes always exceed 45 degrees is called *timelike.* A line or surface whose slopes are all precisely 45 degrees is called *lightlike.* A line or surface whose slopes never exceed 45 degrees is called *spacelike.* According to these definitions, the world line of a particle unable to exceed the speed of light is timelike, that of a pulse of light is lightlike, and that of a (hypothetical) particle unable to travel more slowly than light is spacelike.

A Penrose diagram maps out what are known as the *global properties* of a spacetime – the locations of its singularities, the regions of spacetime from which they can be reached, those from which they cannot, whether singularities are spacelike or timelike, the locations of event horizons, the regions of spacetime interior to the horizons and those exterior to them, and which regions of spacetime are in causal contact. Vast regions of spacetime can be mapped to a single point in a Penrose diagram. For example, the destination in the arbitrarily distant future of *all* timelike curves is mapped into

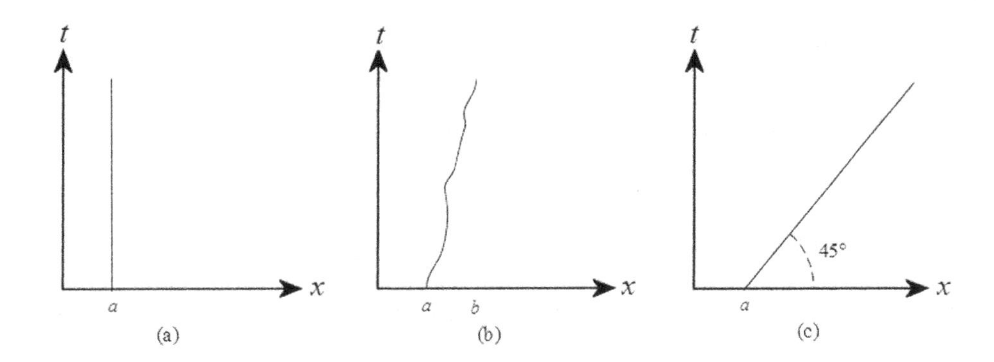

Figure 5.4. World lines. (a) World line of a particle stationary at $x = a$. (b) World line of a particle that travels from $x = a$ to $x = b$ at varying speeds. (c) World line of a pulse of light emitted at $x = a$.

a single point. The best thing about these diagrams is that they share important properties with Minkowski spacetime: Timelike lines are steeper than 45 degrees, lightlike lines are precisely 45 degrees, and spacelike lines are shallower than 45 degrees. From this one can at a glance determine whether a particular event is in causal contact with another.

Consider the Penrose diagram of a Schwarzschild black hole shown in Figure 5.3. The regions labeled *A* and *A'* are the exterior parts of the Schwarzschild wormhole. They correspond to the "upper" and "lower" universes of Figure 5.1. Region *A* was that found by Karl Schwarzschild while serving at the Russian front in 1916. Region *A'* was found by Austrian Ludwig Flamm later that year. The connection between these regions was the "bridge" discussed by Einstein and Rosen in 1935. The shaded interior regions *B* and *B'* were not known to be an important part of the Schwarzschild solution until they were understood by the Irishman John Synge in 1950. Each interior region contains a singularity, both of which are shown as thick lines. The boundaries between the interior and exterior regions are the event horizons, which appear as diagonal lines that cross in the center of the diagram.

The diagram shows three types of spacetime regions. We can best understand these by considering the world lines of indestructible particles – particles whose world lines are arbitrarily long. We have

I^+ -- future timelike infinity: where all timelike world lines go.

I^- -- past timelike infinity: where all timelike world lines come from

$\mathscr{I}^+$ -- future lightlike infinity: where all lightlike world lines go

$\mathscr{I}^-$ -- **past lightlike infinity**: where all lightlike world lines come from

I^0 -- **spacelike infinity**: where all faster-than-light world lines go (ordinary spatial infinity)

Imagine for a moment that the universe is infinitely old and will persist for an infinitely long time. Imagine also that this applies to each of its stars. Then all of these stars will end up in the region called I$^+$. Although this is a vast region, the Penrose diagram maps it to a single point. It similarly maps the region called I$^-$, which is in the infinite past -- the huge area from which these eternal stars came. It so maps spatial infinity, I^0, as well. The light from the universe's stars will in the arbitrarily distant future enter the region $\mathscr{I}^+$. Similarly, the starlight from the arbitrarily distant past comes from the region $\mathscr{I}^-$. The Penrose diagram maps $\mathscr{I}^+$ and $\mathscr{I}^-$ a bit differently. It maps each of them into a line (Figure 5.3). While our universe is not infinitely old or of infinite extent, as these mappings assume, it is very old and very large -- perhaps sufficiently so for Penrose diagrams to apply in some approximate sense.

The Penrose diagrams of Figure 5.5 show various world lines. From this it is easy to see why a traveler that has crossed the event horizon and entered into the Schwarzschild black hole – the upper interior region – is trapped there. There is no timelike world line segment that begins within this interior region and reaches an exterior region. Any world line connecting an interior point with an exterior region must be shallower than 45 degrees. In other words, such a world line requires the traveler to exceed the speed of light (Figure 5.5g). One sees, moreover, that the traveler cannot avoid the singularity at $r = 0$. A remarkable feature of the Schwarzschild solution, which the Penrose diagram makes clear, is that the singularity at $r = 0$ is effectively in the traveler's future. The same law that moves us all forward in time, drags the traveler toward the singularity and certain doom. Within the event horizons, the radial coordinate and the time coordinate exchange their functions. The Schwarzschild radial coordinate assumes the role of time coordinate, and the Schwarzschild time coordinate assumes the role of a spatial coordinate.

This reversal of roles between the Schwarzschild time and the radial coordinate can be seen in Figure 5.6. In the exterior regions the roughly horizontal curves correspond to constant values of the time coordinate, and the roughly vertical curves correspond to constant values of the radial coordinate. Within the event horizons – within the black hole and white hole – the situation is reversed. The roughly horizontal curves now correspond to constant values of the radial coordinate, the roughly vertical curves to the time coordinate. What determines whether a particular coordinate acts as a temporal coordinate is whether its curves of constant value are roughly horizontal. Hence the Penrose diagram makes clear that in the interior regions time is marked by the radial coordinate. Motion, then, toward the singularity of the black hole and away from that of the white hole is a manifestation of the universe's relentless, ongoing march from the past toward the future.

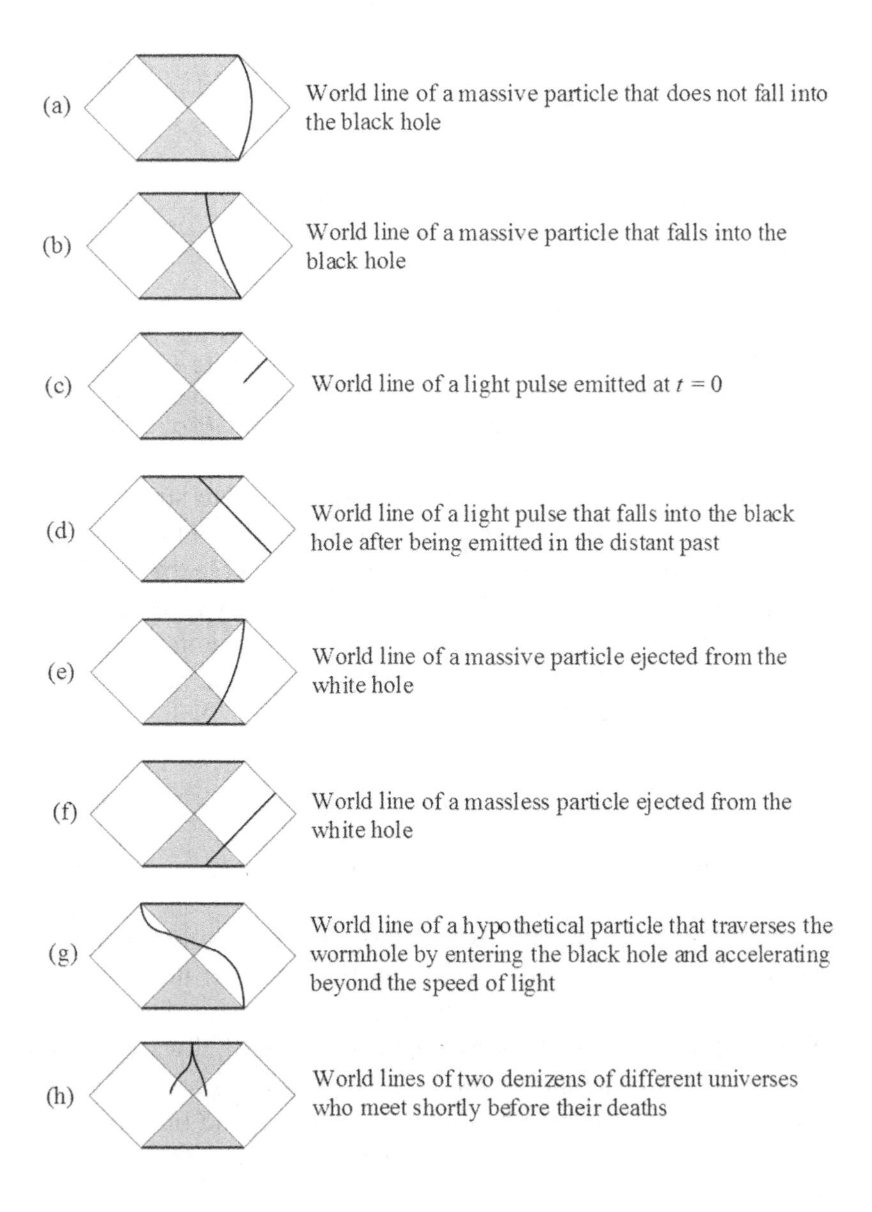

Figure 5.5. World lines in the Schwarzschild solution.

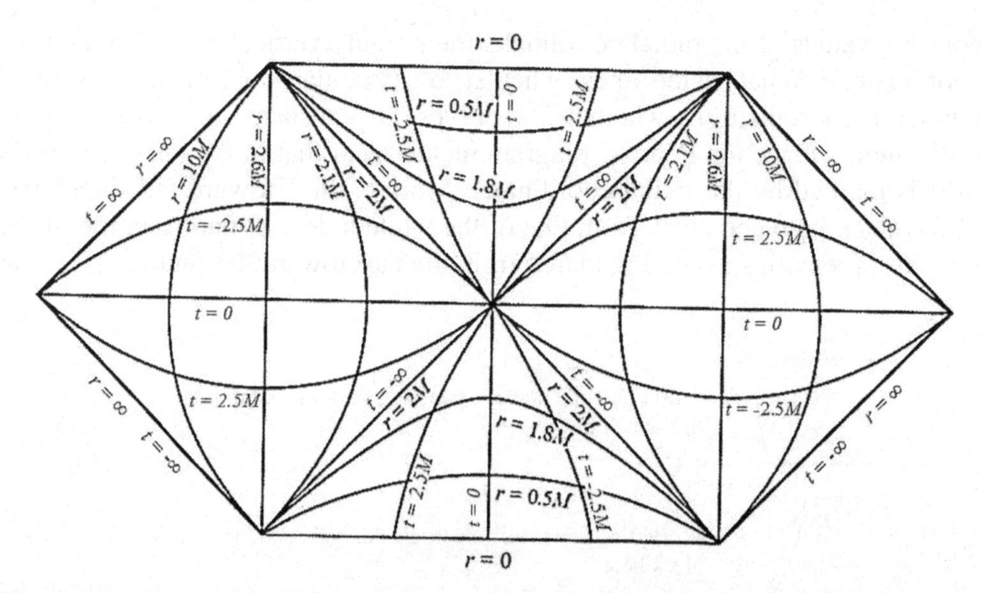

Figure 5.6. Correspondence between Schwarzschild and Penrose coordinates. Schwarzschild coordinates drawn on a Penrose diagram illustrates the relationship between Penrose and Schwarzschild coordinates. Space coordinate is in units of $GM/c^2 \rightarrow M$ (M = black hole mass, G = gravitational constant, c = speed of light. Units have been chosen so that $G = c = 1$.) Time coordinate is in units of $GM/c^3 \rightarrow M$.

We are now in a position to read the Penrose diagram in order to summarize a few basic facts about the extended Schwarzschild solution: It contains a black hole from whose interior no light can escape and a white hole from whose interior all light must escape. The black hole is surrounded by a future event horizon, the white hole by a past event horizon. Objects can enter the black hole but cannot enter the white hole. Travel from one exterior region to the other requires the speed of light to be exceeded. The Schwarzschild wormhole, therefore, cannot be traversed.

The Hidden Dynamics of the Schwarzschild Solution

Another way of understanding why the Schwarzschild wormhole cannot be traversed follows from a feature of this wormhole that escaped the notice of physicists, including Albert Einstein, for well over thirty years. Despite the fact that the Schwarzschild metric does not depend on the Schwarzschild's time coordinate, the solution is actually dynamic. How can this be? How can a solution that does not depend on time be dynamic – i.e. depend on time? To understand this and to see how this feature was missed, we need to consider a few things about coordinate systems.

Coordinate systems need not be rectilinear. Nor need they be symmetrical. They only need to be useful in uniquely specifying the location of any point within the patch of space to which the coordinate system is applied. This is accomplished in general by using intersecting families of parametrized curves. One such family is required for each dimension of the space of interest. Each curve that is a member of a family is identified by a particular value of a parameter. To see a two-dimensional example, look at Figure 5.7. The coordinates of a particular point are the parameters of the two curves that intersect at the point.

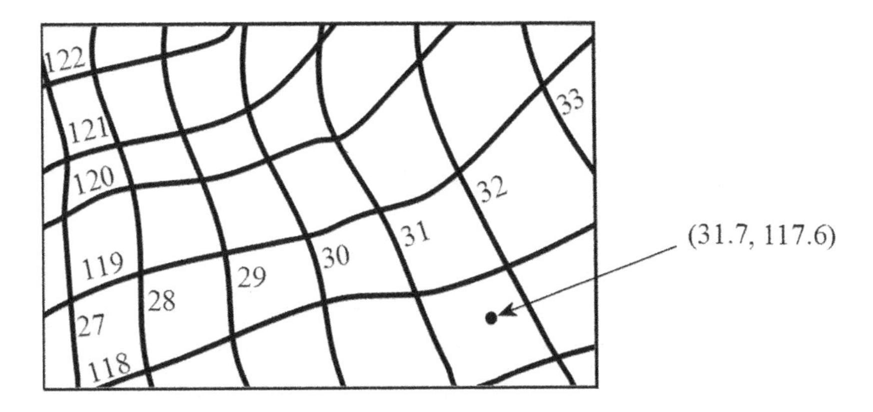

Figure 5.7. An arbitrary coordinate system. Any intersecting family of parametrized curves will serve as a local coordinate system.

The same applies to spacetime – in particular the spacetime of the extended Schwarzschild solution. Figure 5.8a shows the family of curves that define the time coordinate of the system in which Schwarzschild originally expressed his solution. This diagram helps explain why the dynamics of the Schwarzschild solution were missed for so many years. Schwarzschild's choice of time coordinate – reasonable as it seems – does not penetrate the interior regions of his solution, the location of the solution's dynamics. This could be remedied by using instead a coordinate system whose time coordinates are defined by the curves shown in Figure 5.8b. It is important to realize that the curves of constant time shown in Figure 5.8a and 5.8b are actually *hypersurfaces* --three-dimensional subspaces of four-dimensional spacetime. A hypersurface is the three-dimensional equivalent of an ordinary surface. An ordinary surface is, of course, a two-dimensional subspace of a three-dimensional space. Recall that the choice of a particular set of hypersurfaces (such as those chosen in Figures 5.8a and 5.8b) for the purpose of defining a time coordinate is called a *foliation* of the spacetime. Foliations may be chosen arbitrarily. To summarize, then, the foliation shown in Figure 5.8a is inferior to that shown in Figure 5.8b for the sole purpose of understanding the dynamics of the Schwarzschild solution. The solution seems to be static in the former foliation, but is revealed to be dynamic in the latter. However, a foliation such as that of Figure 5.8b and its associated coordinate system is exceedingly inconvenient mathematically. It is also unnatural in that it does not accurately reflect the relative value of measurements of intervals of length and time taken by separate observers.

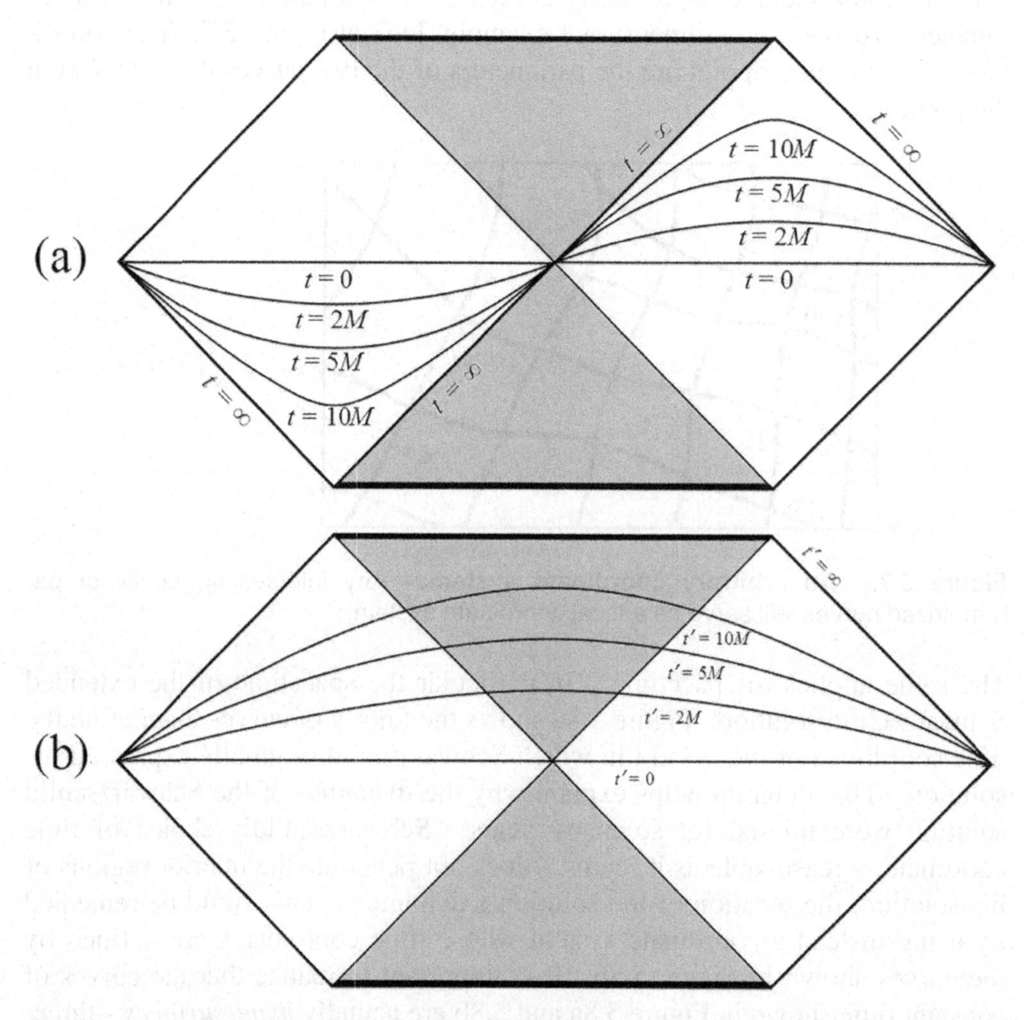

Figure 5.8. Choice of foliation reveals dynamics. (a) In the foliation created by Schwarzschild coordinates the wormhole appears static, because none of the hypersurfaces enter the region within the event horizons. [Hypersurfaces for negative values of t are obtained by reflection about the t=0 line.]. (b) In a foliation due to a certain unnatural and mathematically inconvenient coordinate system, hypersurfaces penetrate the interior regions. This can be used to reveal the expansion and contraction of the wormhole throat. Radial coordinates are not shown. As always, the angular coordinates are suppressed.

To better understand these dynamics it helps to establish a relationship between a hypersurface of the spacetime foliation shown in Figure 5.8b and an embedding diagram, such as that in Figure 5.1. Such a hypersurface corresponds to a snapshot of the geometry at the instant of time to which the hypersurface corresponds. This relationship is illustrated in Figure 5.9. Part a) of this figure shows a particular element of the foliation, the spacelike hypersurface labeled $t' = 5M$. Part b) shows a cross section of the corresponding embedding diagram. The points labeled identically in both parts are to be identified. Point *A* lies just outside the event horizon in the upper universe. Point *C* is its counterpart in the lower universe. Point *B* is the wormhole's throat. Notice that the thickness of the throat in the embedding diagram corresponds to the distance between Point B of the hypersurface of Figure 5.9a and the black hole singularity at $r = 0$.

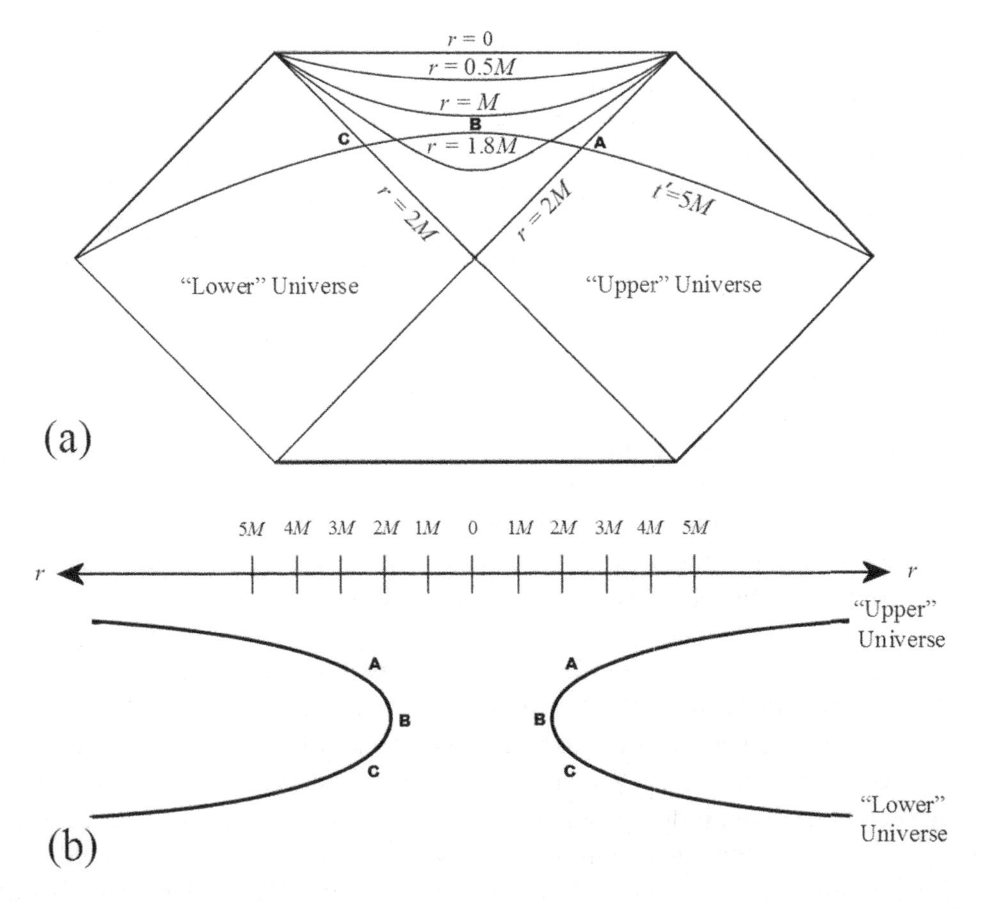

Figure 5.9. A hypersurface and its embedding diagram. (a) The spacelike hypersurface defined by $t' = 5M$. (b) A cross section of the corresponding wormhole embedding diagram.

Having established that a foliation of spacetime corresponds to a sequence of embedding diagrams, we are now able to create and understand the dynamical sequence shown in Figure 5.10. It shows that as time elapses – as successive hypersurfaces of constant t' are selected – the embedding diagram changes. Initially there is no wormhole. Then it appears, its throat expands to a maximum value at $t' = 0$, it contracts, it pinches off, and it leaves a black hole sans inter-universe connection. The time required for this sequence to occur is very brief – only a small fraction of a second for a black hole whose mass is that of a typical star. These considerations allow us to explain a traveler's inability to traverse a Schwarzschild wormhole in slightly different language:

P24. *A traveler who attempts to traverse a Schwarzschild wormhole will find that the wormhole will pinch off even before she is crushed in the singularity.*

Even if the wormhole remained open for an entire year and the singularity could somehow be avoided, a naturally occurring Schwarzschild black hole would make an unlikely inter-universe or intra-universe portal. The chances are exceedingly slim that the single year in which the black hole forms a wormhole will coincide with the particular year that a given traveler would want to traverse it. The would-be traveler would in general have missed the wormhole phase of the black hole by a few billion years, or she will have to wait a few billion years for it to occur.

Occurrence in Nature

It is exceedingly unlikely that any naturally occurring manifestations of the maximally extended Schwarzschild solution exist. There are at least three reasons. The first is that it is very unlikely that a black hole formed by natural processes would have zero angular momentum. Stars or conglomerations of stars, from which black holes could be expected to form, all spin.

The second reason is that the maximally extended solution extends backward in time to the arbitrarily distant past. This past includes existence of a white hole. However, a star destined to become a black hole was not at any time in its history a white hole. When its surface contracts sufficiently for it to become a black hole, it does not magically acquire a new history. It becomes a black hole, but does not manifest the full – i.e. maximally extended – Schwarzschild solution. The white hole and the "other universe" sections of the solution are excluded.

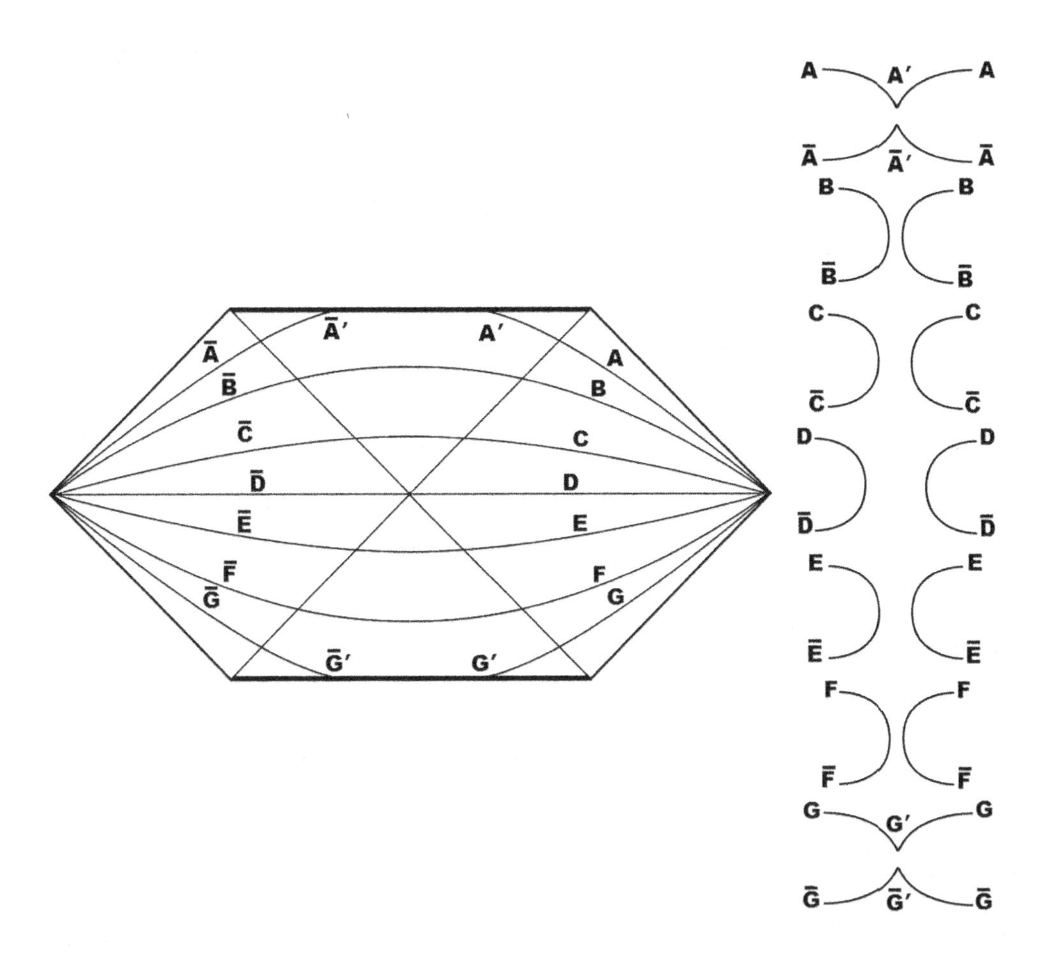

Figure 5.10. Dynamics of the Schwarzschild wormhole. Foliation of Schwarzschild solution and corresponding embedding diagrams. A wormhole appears, reaches maximum diameter at t'=0, then disappears. Sequence is G, F, E, D, C, B, A. For a black hole whose mass is that of a large star, the wormhole would appear to a local inward-falling observer to exist for no more than a second.

The third reason is that even if the white hole had somehow come into existence -- by being, for example, part of the initial conditions of the universe -- its past horizon would be unstable. To see this, consider again the Penrose diagram shown Figure 5.3. Recall that the white hole is the lower interior region *B'* of the figure. Its past horizons, or antihorizons, are the boundaries between *B'* and the "universes" – the exterior asymptotically flat regions. These regions include timelike past infinities I^-. The world lines of virtually every permanent object of each universe come from there. These include the world lines of all older stars. Light from these stars falls onto the antihorizons. Because the world line of the light and that of the antihorizons are both slanted at 45 degrees, it appears that light emitted toward the white hole from the vicinity of I^- remains the same distance from the antihorizons. In fact the light continues to approach the antihorizons. You can see this by observing that the lines of constant r emanate like spokes from I^- (Figure 5.6). These spokes are crossed – r decreases -- as the world line of a light beam parallels that of an antihorizon on the Penrose diagram. To distant observers the light appears, due to gravitational time dilation, to slow down as it approaches the antihorizon.

Star light from the entire universe, then, accumulates at the antihorizon. Here at the surface of the white hole the gravitational field is exceedingly strong – precisely as strong as that at the horizon of the corresponding black hole. Light that falls into such a deep gravitation well consequently acquires an enormous amount of energy. Because the energy of a particle of light is proportional to its frequency, the frequency of this inward falling light is shifted upward. The light is said to be "blue shifted". The arbitrarily high energy of the blue shifted light accumulating on the antihorizons warps spacetime there and thus destroys the antihorizons. They are unstable, and so is the white hole solution of which they are a part. The infinitely high energy density resulting from this accumulation of blue shifted radiation is called a *blue sheet singularity*. It is also known as a *mass inflation singularity*.

Summary

The maximally extended Schwarzschild solution features a connection between universes – a wormhole. This wormhole cannot be traversed. The inter-universe (or intra-universe) connection exists only in the sense that the solution allows denizens of the separate universes to meet each other within the black hole just before they are killed in the singularity. There can be no naturally occurring instances of the maximally extended solution. We may further summarize the contents of this section with several principles.

P25. *The dynamics of a spacetime can be obscured by a poor choice of foliating spacelike hypersurfaces, i.e. by a poor choice of coordinate system.*

P26. *The maximally extended Schwarzschild solution is dynamic. It features the brief appearance of an untraversable wormhole connection between universes.*

P27. *There is no physical singularity at the event horizons of the Schwarzschild wormhole solution.*

P28. *Penrose diagrams display the causal structure of extended wormhole spacetimes by retaining the same orientation as in the Minkowski spacetime of timelike, lightlike, and spacelike intervals.*

P29. *The event horizons of the Schwarzschild wormhole solution occur at* $r = 2GM/c^2$.

P30. *Within the event horizons of the Schwarzschild wormhole solution the role of Schwarzschild's original time coordinate and that of his radial coordinate are exchanged.*

P31. *A traveler who attempts to traverse the Schwarzschild wormhole is driven to the singularity at* $r = 0$. *She is driven there to her inexorable death by the same feature of the universe that drives all objects forward in time.*

P32. *The singularities of the Schwarzschild wormhole spacetime are spacelike.*

P33. *By definition, singularities are characterized by the termination or creation of geodesics and the presence of arbitrarily high curvature.*

P34. *The spherical coordinate system in which Schwarzschild presented his solution breaks down at the event horizons but is valid everywhere else in the maximally extended Schwarzschild solution.*

P35. *The maximally extended Schwarzschild wormhole solution becomes apparent when the original Schwarzschild wormhole solution is expressed in a coordinate system that does not breakdown at the event horizons.*

P36. *The maximally extended Schwarzschild wormhole solution contains two exterior and two interior regions separated by event horizons. The two exterior regions are identified with separate universes (or separate regions within the same universe). One of the interior regions corresponds to a black hole, the other to a white hole.*

The Morris-Thorne Wormhole

What prevents the Schwarzschild wormhole from being traversable? One answer is that it requires would-be travelers to pass through an event horizon. Event horizons only permit travel in one direction, which rules out two-way traversability. This suggests that we might be able to design a traversable wormhole by modifying the Schwarzschild solution in such a way as to eliminate its event horizons. Kip Thorne of Caltech and his student Michael Morris accomplished precisely this in 1988.

They wrote down a generalized metric of which the Schwarzschild wormhole is a special case. Their generalized metric depends on two unspecified functions. The first of these functions is called the *redshift function*. It specifies that part of the metric responsible for determining the magnitude of the gravitational redshift. The gravitational redshift is the reduction in the frequency that a photon – a particle of light -- will experience when it climbs out of a gravitational potential well in order to escape to infinity. In doing so, the photon expends energy. Its energy is proportional to its frequency. A reduction in energy, then, is equivalent to a reduction in frequency, which is also known as a redshift. If the wormhole has an event horizon, it means that a photon emitted outwardly from the horizon cannot escape to infinity. In other words, it would take an infinite amount of energy for the photon to escape. Its frequency would be infinitely reduced, i.e. its redshift would be (negatively) infinite (Figure 5.11). Hence,

P37. *A negatively infinite value of the redshift function at a particular value of the radial coordinate indicates the presences of an event horizon there.*

If a wormhole is to be two-way traversable it cannot be surrounded by an event horizon. Or,

P38. *If a wormhole solution is two-way traversable, the magnitude (absolute value) of its redshift function must be finite everywhere.*

It may be useful as well to know that the redshift function is also related to the amount by which time slows down for an observer at a given distance from the wormhole compared to an observer far away. The more negative the redshift function, the more time slows down. When the redshift function reaches negative infinity, a horizon appears, and the local elapse of time stops. Another interesting tidbit is that in the limit of a weak gravitational field, the redshift function becomes the familiar gravitational potential of Newtonian physics.

The second unspecified function is called the *shape function*. It has the same dimension, viz. distance, as the radial coordinate on which it depends. It specifies that part of the metric that determines the shape of the wormhole.

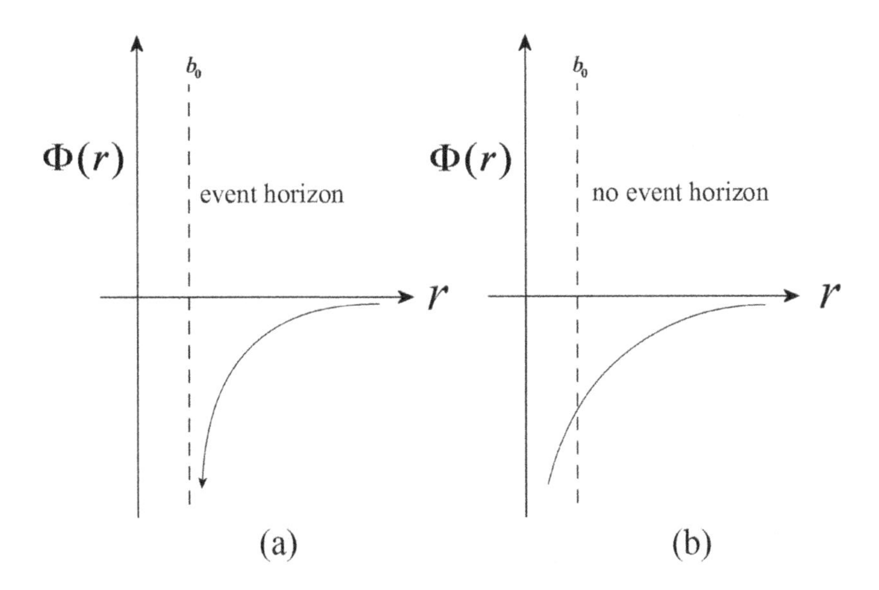

Figure 5.11. Constraint on redshift function. (a) When the redshift function $\Phi(r)$ goes to negative infinity at the wormhole throat (at $r = b_0$), it indicates the presence of an event horizon there. (b) Redshift function of a traversable wormhole must be finite at the throat to ensure the absence of an event horizon there.

This shape should be that of a spherical hole in space whose diameter widens with distance from its throat and merges smoothly with two asymptotically flat regions, one in each universe. In order for this to occur the shape function must satisfy what is known as a *flare-out condition.* This condition requires the ratio of the radial coordinate to the value of the shape function at that coordinate to increase on average as the radial coordinate increases. In addition to the flare-out condition, the value of the shape function at the wormhole's throat must be the radial coordinate of the throat. In other words, the aforementioned ratio must be 1 at the throat. This ensures that there are no kinks at the throat, as shown in Figure 5.12. The shape function, then, is constrained as follows.

P39. *In order for the wormhole to have the proper shape, the ratio of the radial coordinate to the shape function at that coordinate must be 1 at the throat and generally increase with distance from the throat.*

Like the Schwarzschild wormhole, the Morris-Thorne wormhole does not explicitly depend on a time coordinate and is spherically symmetrical. Hence, the redshift and shape functions are time-independent and depend solely on the radial coordinate.

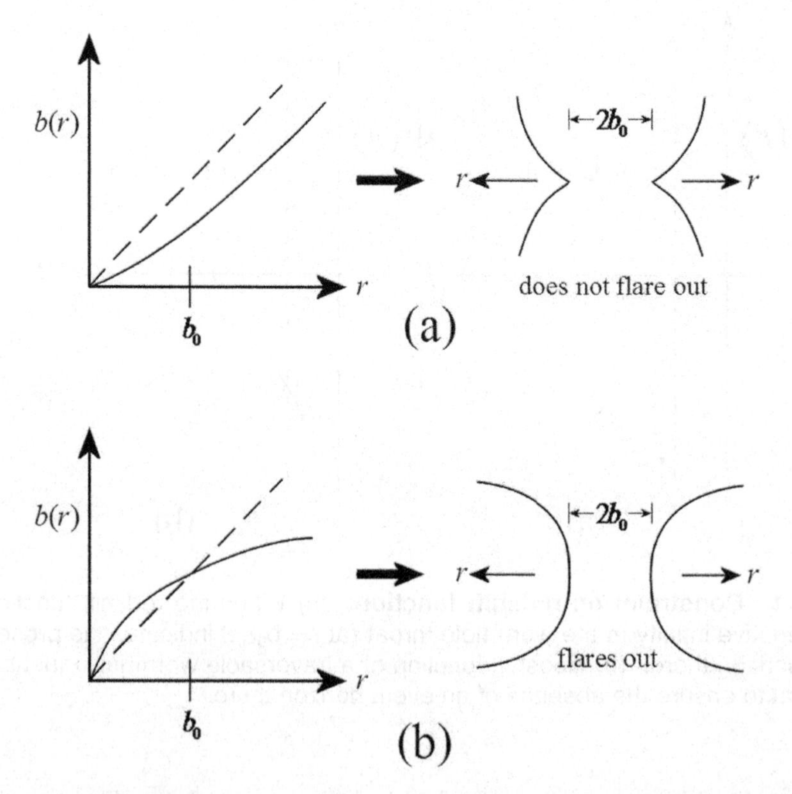

Figure 5.12. Constraints on the shape function. Shape function b(r) must satisfy 1) r/b increases with r, and 2) $b(b_0) = b_0$, where b_0 is throat radius. (a) Conditions not met. (b) Conditions met.

Desired Properties

Beyond the constraints that we have already imposed, the wormhole should also satisfy the following.

1. No event horizons
2. Connects asymptotically flat regions (universes)
3. Solves the Einstein field equations,

Morris and Thorne listed other desired properties of their wormhole:

4. Traversal time less of less than a year
5. Gravitational tidal accelerations experienced by traveler no greater than 1g
6. Motion-related accelerations experienced by traveler no greater than 1g

7. Stress-energy of supporting matter consistent with known physics
8. Stable under traversal attempts
9. Can be assembled by a civilization with a conceivable level of advancement

Properties 4, 5, and 6 impose additional constraints on the redshift and shape functions. Property 9 would be violated, according to Morris and Thorne's example, if a wormhole needs for its construction more mass than exists in the universe. Property 8 did not induce them to perform an elaborate stability analysis. Rather, they pointed out that if the wormhole is unstable, a civilization advanced enough to construct it would be advanced enough to artificially stabilize it. Property 7 fell by the wayside along with the traditional, but now obsolete, energy conditions. It imposes no additional constraint.

An arbitrarily large number of redshift and shape functions can be found that satisfy the desired constraints. Consequently, a Morris-Thorne wormhole is not a particular wormhole solution that, like the Schwarzschild wormhole, depends upon a single parameter such as the wormhole's mass. It is instead a class of solutions, all of whose members satisfy the above constraints.

Once suitable redshift and shape functions have been chosen, the geometry of the wormhole is determined. Recall that according to the Einstein equations of general relativity, a particular geometry of spacetime results from a particular distribution of the stress-energy of matter. By inserting the geometry for the chosen Morris-Thorne wormhole into the Einstein equations, we can ascertain the distribution of stress-energy responsible for it.
The result of this insertion reveals that the stress-energy of the geometry-supporting matter is *negative*. A more precise statement is this: In the vicinity of the wormhole's throat, the sum of this matter's pressure and its energy density is negative. And for observer's moving with sufficient radial velocity, this matter's energy density itself is negative. In other words, the matter violates the Weak Energy Condition.

Such matter was dubbed *exotic* by Morris and Thorne. A generation ago its existence was assumed impossible. However, as noted by Morris and Thorne and described in subsequent chapters, quantum theory provides a means of creating it in principle. There is, moreover, reason to believe – as we will later see -- that exotic matter might be produced through purely classically means.

Four Examples of MT Wormholes

In the appendix of their first paper Morris and Thorne provided four examples of their new class of wormholes. These are each characterized by a particular division of space into concentric regions together with a specification of the redshift and shape functions in each region. They are further characterized by the resultant "exoticity" of each region. Exoticity is a dimensionless measure of the degree to which the Weak Energy Condition is violated within the region. Values of exoticity have the following meanings.

Exoticity	Meaning	Example
negative	normal matter	steel
zero	quasi-exotic matter	electromagnetic field
positive	exotic matter	Casimir vacuum

The word "matter" here is used in the sense of general relativity, viz. any source of stress-energy. Chapter 7 discusses the Casimir vacuum, a means of exploiting quantum field theory to create a field of negative energy.

Each of the following examples refer to the size of the wormhole's mouth. The value of the radial coordinate defining the mouth is that at which Morris and Thorne imagined it safe to place space stations. In each universe a space station sits at the edge of the wormhole where gravity has weakened sufficiently to comfortably accommodate the human staff and visiting human travelers. Morris and Thorne's definition of sufficiently weakened gravity specifies three conditions:

1) Space is nearly flat. This is equivalent to requiring the ratio $b(r)/r$ of the shape function $b(r)$ to radial coordinate r to be small. Here "small" is defined as 1% or less. [This corresponds to a slope in the embedding function of Figure 5.13 being 10% or less.]

2) The redshift function is small. Specifically a beam of light sent outward from the space station is redshifted by no more than 1%.

3) Gravity is no stronger than on Earth. Imagine surrounding the wormhole with a spherical superstructure whose surface reaches the position of one of the stations. The weight of man standing on this surface should not exceed his weight on Earth.

Let us elevate this definition to the status of a principle.

P40. *The mouth of a Morris-Thorne wormhole is a sphere surrounding it, at whose surface the following is true 1) the ratio of the shape function to the radial coordinate is no greater than 1%, 2) the redshift function is no greater than 1%, and 3) the acceleration of gravity is no stronger than on the surface of Earth.*

I discuss briefly each of these examples below. I have selected the smallest possible value for the throat radius (10m) in order to make the examples concrete. A short summary of the features of the example precedes each discussion. In what follows r is the radial coordinate.

1. Infinite-Exotic-Region Wormhole

(Exotic matter distributed throughout space)

Number of Regions:	1
Redshift Function:	0
Shape Function:	increases as $\sqrt{r}$
Minimum Throat Radius:	size of traveler's spacecraft ~ 10 m
Minimum Mouth Radius:	100 km
Exoticity:	1.0
Normal Matter Distribution:	no normal matter
Exotic Matter Distribution:	from throat to infinity
Traversal Time:	$\geq$ ~1 hour

In this example, which Morris and Thorne called the "zero tidal force solution", the redshift is zero everywhere. This eliminates the radial tidal force. This force is due to the traveler's body parts accelerating downward at different rates. Were the traveler falling feet first toward the wormhole, for example, his feet would be accelerating faster than his head. The radial tidal force would, if present, seem to him to be pulling the lower and upper halves of his body apart. As the traveler falls further into the wormhole, the radial tidal force could increase sufficiently to exceed the tensile strength of his neck or other part of his body. Ouch.

Although the constancy of the redshift function eliminates the radial tidal force, the tangential tidal force remains. This force, which results from the various parts of the traveler's body all attempting to fall along converging radial lines, tends to squeeze his body inward from all directions perpendicular to his path. Unlike the radial tidal force, which only depends on the traveler's position in the gravitational field, the tangential tidal force depends on the traveler's velocity. The slower he descends, the smaller the tangential tidal force. The capacity of the traveler's body to withstand such squeezing, which Morris and Thorne assumed to be its capacity to withstand its terrestrial weight, limits the speed with which a traveler's spaceship can journey through the wormhole. This in turn places a lower limit on the time required to traverse the wormhole from space station to space station, i.e. from mouth to mouth. The minimum value of this traversal time works out to be just under 1 hour.

The shape function in this example increases with the square root of the radial coordinate in such a way as to ensure the desired wormhole shape.

To obtain the matter distribution, we insert the spacetime that corresponds to the above choices for the redshift and shape functions into the Einstein gravitational field equations. The resulting matter is purely exotic and extends from the throat to infinity. Its density decreases with the radial coordinate r as $1/r^{5/2}$, ensuring that at any particular point far from the wormhole the traces of its matter are negligibly small.

Needless to say, exotic matter is hard to come by, perhaps even for an exceedingly advanced civilization. And to require that the entire universe be covered in exotic matter that falls off precisely as $1/r^{5/2}$ seems to be a bit much to ask. The only realistic wormhole solutions, then, would seem to be those whose matter is confined within some finite radius. We consider three such examples next.

2. Large-Exotic-Region Wormhole
(Exotic Matter Restricted to Finite Radius)

Number of Regions:	3 interior $10\text{ m} \leq r < 10^{11}\text{m}$ boundary $10^{11}\text{m} \leq r < 1.00001 \times 10^{11}\text{m}$ exterior $1.00001 \times 10^{11}\text{m} \leq r$
Redshift Function:	interior region: constant boundary region: increases as r^2 exterior region: Schwarzschild value
Shape Function:	interior region: increases as $\sqrt{r}$ boundary region: increases as r exterior region: Schwarzschild value
Minimum Throat Radius:	size of traveler's spacecraft ~ 10 m
Minimum Mouth Radius:	10^8 km
Exoticity:	interior region: 1.0 boundary region: < -1 exterior region: 0
Normal Matter Distribution:	confined to boundary
Exotic Matter Distribution:	confined to interior
Traversal Time:	≥ 7 days

The strategy here is to divide space into three concentric regions: an interior, a thin boundary region, and an exterior. The interior region contains exotic matter. The boundary layer contains normal matter, whose purpose is to provide a positive radial pressure to cancel out the negative radial pressure of the exotic matter of the interior. Once the boundary layer has thus brought the matter pressure to zero, the solution in the boundary layer is ready to be matched to the standard empty space Schwarzschild solution.

The reason that we cannot eliminate the boundary layer and match the interior solution directly to the Schwarzschild exterior has to do with the behavior of the Einstein equations in the presence of discontinuities. This behavior is captured in so-called *junction conditions* that are summarized in this case as follows.

P41. *The energy density and tangential pressure of matter may vary discontinuously with the radial coordinate, but the radial pressure, shape function, and redshift function may not.*

In other words, Einstein's equations allow the matter density in the interior to discontinuously drop from some finite value in the interior region to zero in the exterior region. The radial pressure, however, may not. It needs to gradually (i.e. continuously) drop from its interior value to its exterior value of zero. The boundary region is where this transition occurs.

The redshift function similarly transitions from a constant value in the interior to the Schwarzschild value in the exterior in a manner imposed through the Einstein equations by the change within the boundary region of the radial pressure and the shape function. The constancy of the redshift function in the interior eliminates radial tidal force, but not tangential tidal force, as in the previous example. As before, the traveler's ability to withstand the velocity-dependent tangential tidal force limits the speed with which he can traverse the wormhole.

Having confined the exotic matter to an interior region, we now require a finite quantity of it. This is a great improvement over the zero-tidal-forces example, which, if taken literally, requires an infinite amount of exotic matter dispersed in a particular manner over the entire universe. The price to be paid for this improvement is a stronger gravitational field. This significantly widens the wormhole's mouths, which means that the space stations are much farther apart, and station-to-station traversal time is correspondingly increased. Instead of traversing the wormhole in an hour, it will now take a week.

A week-long traversal is certainly acceptable, especially given that our alternative is a trip that might take years according to shipboard clocks and even longer to stationary observers awaiting our return. What we cannot yet accept is the huge quantity of exotic matter required to sustain this wormhole. Although the exotic matter density falls off as $1/r^{5/2}$, it must nevertheless fill a sphere whose radius is 10^8 km – about 70% of the distance between the earth and the sun.

The next example attempts to economize on exotic matter by confining it to the immediate vicinity of the throat.

3. Medium-Exotic-Region Wormhole
(Exotic Matter Loosely Restricted to Throat)

Number of Regions:	4	
	throat	$10\text{ m} \leq r < 10^5\text{m}$
	interior	$10^5\text{m} \leq r < 10^{14}\text{m}$
	boundary	$10^{14}\text{m} \leq r < 1.01 \times 10^{14}\text{m}$
	exterior	$1.01 \times 10^{14}\text{m} \leq r$

Redshift Function: throat region: constant
interior region: constant
boundary region: constant
exterior region: Schwarzschild value

Shape Function: throat region: increase as $\sqrt{r}$
interior region: increases as r
boundary region: increases as r^3
exterior region: Schwarzschild value

Minimum Throat Radius: size of traveler's spacecraft ~ 10 m

Minimum Mouth Radius: 10^{11} km

Exoticity: throat region: 1.0
interior region: 0.0
boundary region: -0.99
exterior region: 0

Normal Matter Distribution: quasi-exotic confined to interior region
non-exotic confined to boundary region

Exotic Matter Distribution: confined to throat region

Traversal Time: ~200 days

Here Morris and Throne attempted to improve upon the previous example by confining the exotic matter to a region more immediately within the vicinity of the throat. They replaced the interior region of the previous example, which was entirely composed of exotic matter, with an interior region that consists of a concentration of exotic matter near the throat supplemented by quasi-exotic matter everywhere else in the interior. As in the previous example, they followed this interior with a thin boundary layer in which the radial pressure was reduced to zero so that it would match the pressure of the Schwarzschild exterior of empty space.

Their division of the interior into two regions (throat, and interior proper) raises the number of regions used to define this solution to four: throat, interior, boundary, and exterior.

In the throat region the redshift and shape functions are as they were in the previous example. As before, all of the matter within this region is exotic. Because they chose not to use highly exotic matter – matter with a negative energy density as measured by stationary observers, the throat region still needs to be large. For a throat size of 10 meters this region's radius must be 100 km.

Because the region interior to the boundary now consists mostly of quasi-exotic matter, such as an electromagnetic field, instead of purely exotic matter, the wormholes shape does not flare out as rapidly. Or put another way, the matter density does not fall off as rapidly. As a result the locations at which the gravitational field is weak enough to allow the stations to be positioned there are much farther out from the throat. About 1000 times farther out. This necessarily lengthens the station-to-station traversal time. For this example it is about 200 days. Though 30 times longer than in the previous example, 200 days lies well within the traversal time constraint of 1 year.

Still, a 100 km ball of exotic matter is a tall order. Surrounding it with a 10^{11} km sphere of quasi-exotic matter around which there is a thin shell of suitably dense matter is no picnic either. Do the laws of physics allow a simpler solution? They do.

4. Small-Exotic-Region Wormhole

(Exotic Matter Closely Restricted to Throat)

Number of Regions:	2	
	throat	$10 \text{ m} \le r < 15 \text{ m}$
	exterior	$15 \text{ m} \le r$
Redshift Function:	throat region: 0	
	exterior region: 0	

Shape Function:	throat region: decreases as $1/r^2$ to 0 exterior region: 0
Minimum Throat Radius:	size of traveler's spacecraft ~ 10 m
Minimum Mouth Radius:	15 m
Exoticity:	throat region: 1.25 decreasing to 1.0 exterior region: 0
Normal Matter Distribution:	no normal matter
Exotic Matter Distribution:	confined to throat region
Traversal Time:	$\geq$ 0.7 seconds

Morris and Thorne called this "an absurdly benign" wormhole. It's not hard to see why. This wormhole does little more than poke a spherical hole in flat space. Unlike the other examples, it extends no gravitational field out beyond its immediate vicinity. Its secret is that it combines the throat region with the boundary layer. Instead of using a shape function that increases with r and thus slowly flattens the embedding diagram of the wormhole, it uses a shape function that goes to zero at the outer edge of the throat region. This causes the embedding to flatten at the edge of the throat region. Because space is flat beyond the throat, and the redshift function was chosen to be zero, there is no gravitational field beyond the edge of the throat region. The space stations – which would in this case be no more than space toll booths – could be situated right at the edges of the throat region. Moreover, the spherical shell that constitutes the throat region may be chosen to be as thin as one likes. As a result, traversal times can be made arbitrarily small – easily under one second.

Surely, there must be a catch! Indeed there is. This wormhole requires, unlike any of the previous examples, highly exotic matter – matter whose energy density is negative as measured by stationary observers. The thinner the throat region, the more highly exotic will such matter need to be. Because quantum theory has rendered the assumptions used to prove the Positive Mass Theorem inapplicable, the existence of negative mass can no longer be said to be prohibited by general relativity. It can perhaps exist. There just does not seem to be any in our part of the universe. If it can exist, an exceedingly advanced civilization presumably will have mastered its creation, acquisition, and manipulation for wormhole construction. They would, of course, still be faced with the problem of topology change. The hole around which the highly exotic matter would be suitably distributed must first be created.

In the limit of an infinitesimally thin throat region, this wormhole becomes a special case of a *thin-shell* wormhole. The properties of such wormholes in the absence of Morris' and Thorne's restriction to spherical symmetry were explored by Matt Visser in the late 1980s.

Charged, Nonrotating Wormholes

The Reissner-Nordstrøm Wormhole

In 1916, after Karl Schwarzschild's results had been announced, German Heinrich Reissner worked out the spherically symmetric solution to Einstein's equations in the presence of a charged, massive, spinless object. Gunnar Nordstrøm's work on the same subject did not appear until 1918. Unlike Reissner, Nordstrøm, a Finn, had published his paper in English (in a Dutch journal). It thus may have been more widely read by the postwar British and American physicists, who gave this the solution the label by which it is currently known.

The Reissner-Nordstrøm metric differs from the Schwarzschild metric in possessing the following.

- Two event horizons – outer and inner
- An additional type of interior region
- Connections to multiple universes
- Timelike singularities – i.e. they can be avoided by travelers
- Endlessly repeated structure

Three Types of Regions

These are best understood by examining the Penrose diagram of the Reissner-Nordstrøm solution, shown in Figure 5.13. It displays regions of three different types, which I shall call *A*, *B*, and *C* (which are further explained in Figure 5.14). The *A* regions are the familiar asymptotically flat regions that we have identified with universes. One of these is our own universe. The others are identified with other universes or with distant regions within our own universe. The *B* regions are similar to the black hole and white hole interior regions of the Schwarzschild solution but with an important difference – they contain no singularities. The *C* regions are new. They each contain singularities. Unlike the Schwarzschild singularities, these are timelike. Accordingly, they are represented by thick *vertical* lines instead of the thick horizontal lines used to depict the Schwarzschild singularities. This means that a traveler's world line need not intersect them. This opens new possibilities for traversing the wormhole. Instead of attempting to traverse the wormhole by moving in a generally horizontal direction across its Penrose diagram – as in the Schwarzschild case, one can traverse it by moving vertically. The former, requiring superluminal travel, was the reason that the Schwarzschild wormhole could not be traversed. Vertical, i.e. timelike, world lines, however, are not analogously restricted. They permit inter-universe travel. A traveler can *in principle* begin in the *A* region of one universe, pass through an outer horizon to enter a *B* region, pass through an inter horizon to enter a *C* region, avoid the *C* region's singularity, and finally pass through an outer antihorizon to emerge in the *A* region of another universe.

Two Horizons

In order to find the location of event horizons for a particular solution of Einstein's equations, one seeks the value of the radial coordinate that causes the purely temporal component of the solution's metric to be zero. Setting this component to zero, defines an equation for the radial coordinate r. In the Schwarzschild cases, this equation is linear in r. In the Reissner-Nordstrøm case it is quadratic. As you will recall, quadratic equations have two roots. These roots, which we will call r_+ and r_-, represent respectively the outer an inner event horizons of the black hole.

The outer event horizon at r_+ is like that of the Schwarzschild black hole. This also holds true of the outer antihorizon at r_+ that separate the interiors of Reissner-Nordstrøm white holes and their exteriors. As in the Schwarzschild case this antihorizon is unstable. Blue shifted radiation of arbitrarily high energy accumulates there. The energy density becomes sufficiently great to create a singularity and thus destroy the antihorizon.

Precisely the same reasoning applies to the inner event horizon at r_-. All of the light emitted from the world lines of stars at the I^+ points of Figure 5.13 will fall toward the nearest inner horizon at r_-. Another way to say this is that the entire future history of a universe will be observed by a traveler as she crosses this inner horizon. As in the case of the white hole antihorizon, the light emitted during this history comes arbitrarily close to the horizon but never

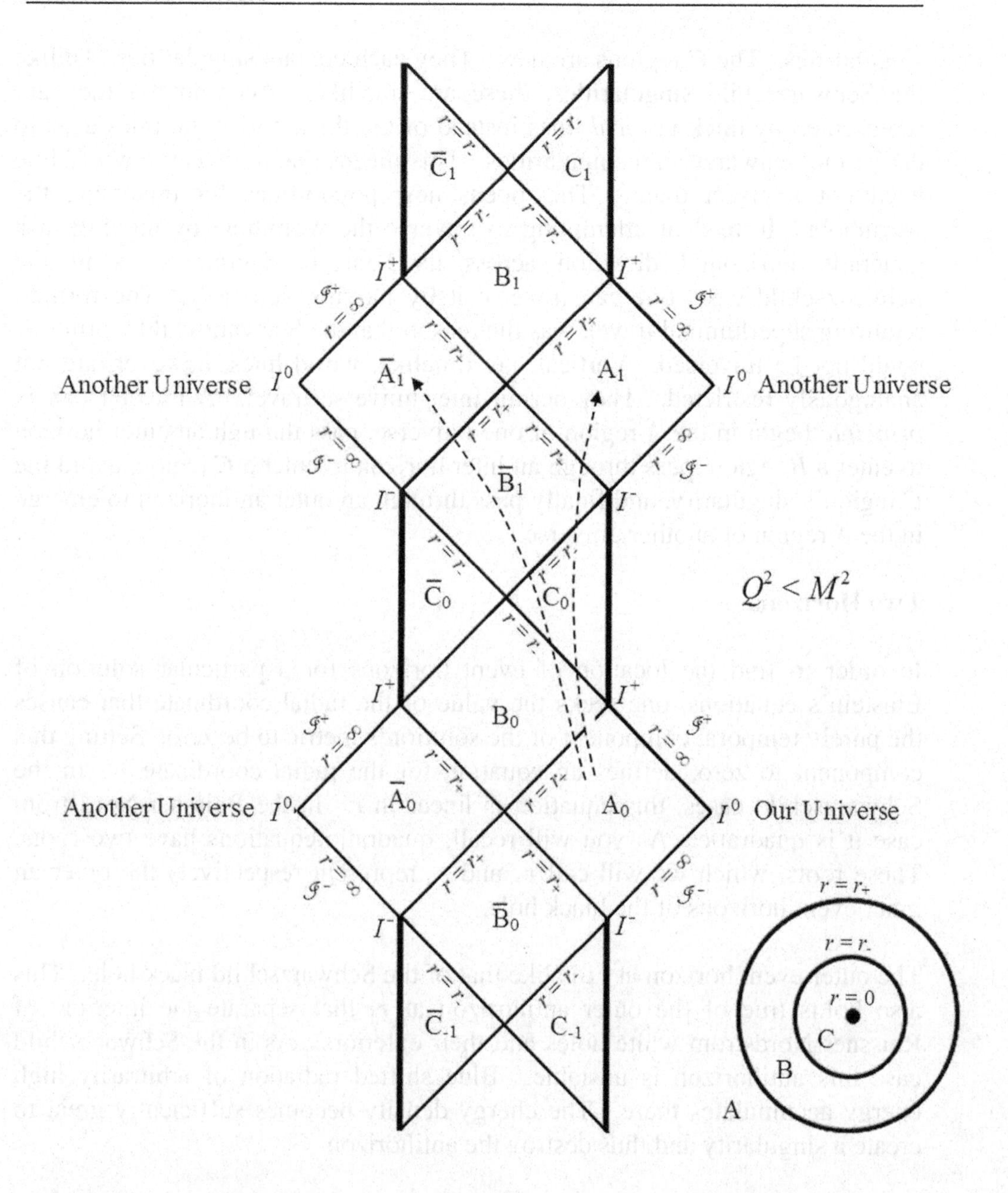

Figure 5.13. Penrose diagram for the Reissner-Nordstrøm solution. A traveler from universe A_0 can in principle travel to universes A_1 and $\overline{A}_1$ but not to $\overline{A}_0$. Singularities are shown as thick vertical lines. The magnitude of the wormhole's charge Q is less than its mass M ($Q^2 < M^2$).

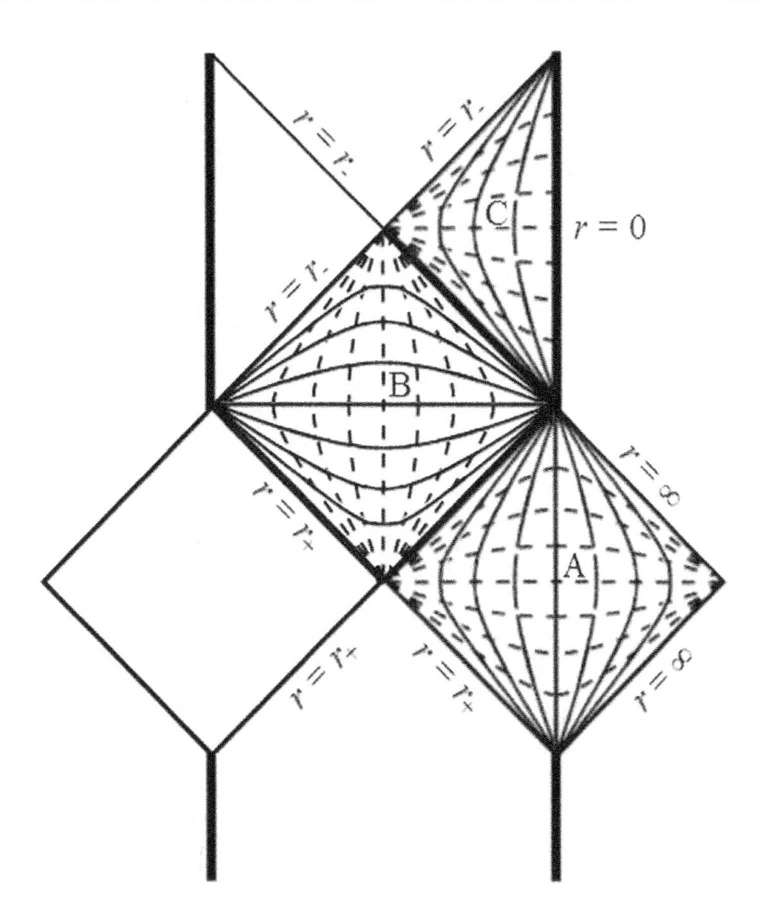

Figure 5.14. Types of regions. Lines of constant r (solid) and line of constant t (dashed) are shown in each of the three types of regions. Region A: r goes from infinity to r_+. Region B: r goes from r_+ to r_-. Region C: r goes from r_- to 0. Note that, as in the Schwarzschild interior, r becomes temporal in Region B (surfaces of constant r are spacelike). [Moving toward $r = r_-$ within B is the same as moving forward in time.] In region C, r returns to being a spatial coordinate.

crosses it. In effect, it piles up on the inner horizon. Because the gravitational field is precisely as strong at the inner horizons at the outer horizons, the same arbitrarily high blue shifting occurs as at the white hole antihorizon. As a result the inner horizons are similarly unstable.

The inner horizons forming the past boundary of the type C regions bear another important distinction. They are called *Cauchy horizons* after the 19th-century French mathematician Augustin-Louis Cauchy. These horizons are defined with reference to so-called *Cauchy surfaces* in the regions in the horizons' past. To understand what this means, consider Figure 5.15. It shows spacelike hypersurface *S*. On S we will specify the position and velocity of a particle. We will have no problem working out the future motion of the particle until it crosses the inner horizon at r_-. Upon crossing this horizon the particle has entered a region in which it is exposed to emissions from a singularity.

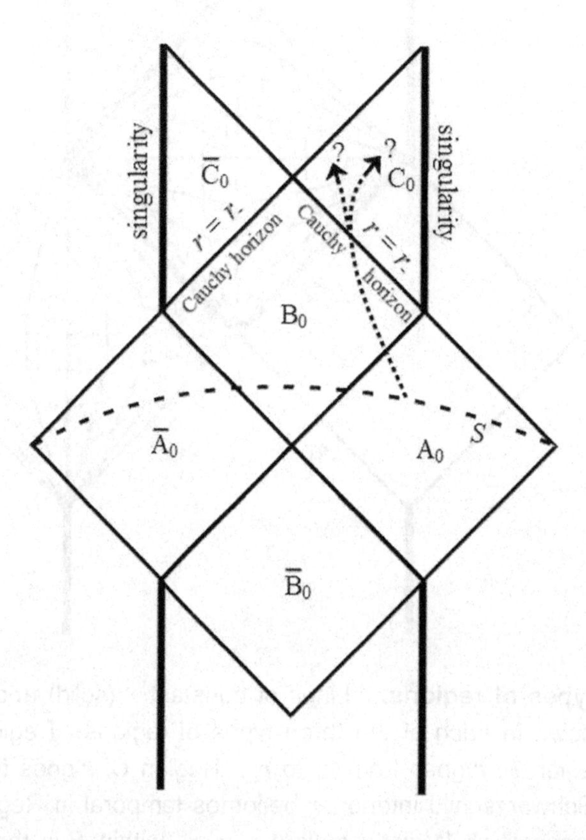

Figure 5.15. Cauchy Horizon. The motion of a particle whose position and velocity are known on the spacelike hypersurface *S* can be predicted until the particle crosses the inner horizon at $r = r_-$. Its motion cannot be predicted in the regions C_0 or $\overline{C}_0$, because these regions are exposed to unpredictable emissions from their singularities.

Physics breaks down at the singularity. There spacetime, the very stage upon which all physics is performed, becomes infinitely distorted. This means that there are no rules restricting what the singularity might emit. It could in principle emit a toaster, a sports utility vehicle, or a sperm whale. Were such objects emitted, however, they would be reduced to elementary particles by the arbitrarily powerful tidal forces in the vicinity of the singularity. In practice, then, a singularity emits radiation or elementary particles of unknown type or intensity.

These unpredictable emissions will interact with our particle, thus making its motion unpredictable. Because the inner horizon is a boundary between unpredictability and predictability based on initial data – also known as *Cauchy data* -- specified at *S*, it is called a *Cauchy horizon* for *S*. Because S could have been any spacelike hypersurface in the region in the horizon's immediate past, the horizon is said to be a Cauchy horizon for this entire region (*A* and *B*).

These spacelike hypersurfaces must, however, satisfy a particular condition, before we can use them to specify Cauchy data. Every timelike world line that begins at past timelike infinity I^- and ends at future timelike infinity I^+ must pass through the hypersurface exactly once. The same must hold true for any lightlike world line that begins at past lightlike infinity $\mathscr{I}^-$ and ends at future lightlike infinity $\mathscr{I}^+$. Such a hypersurface is called a *Cauchy surface*.

The unpredictability in the region within the inner horizon -- the Cauchy horizon -- pertains, not just to any particular particle, but to any physical system. We cannot predict beyond the Cauchy horizon the behavior of any system of particles or fields. This includes the gravitational field itself. Any discussion of the goings on within the type C regions of the Reissner-Nordstrøm metric, then, must be understood to be qualified by the phrase, *"assuming that emissions from the singularity do not interfere."*

Endlessly Repeating Global Structure

The most striking difference between the Penrose diagrams of the Schwarzschild and Reissner-Nordstrøm solutions is that the latter consists of an endlessly repeating pattern. What does this mean physically? Graves and Brill considered this question in 1960. They concluded that the wormhole is executing a cycle of expansion to a maximum diameter, followed by contraction to a minimum diameter. This minimum, they discovered, is imposed by the *gravitational* repulsion induced by the presence of an electric charge. Not only does the radius of the wormhole's throat expand and contract, it effectively turns itself inside out. In the first part of the cycle, the wormhole has its expected shape. Travel away from the throat leads to spatial infinity. As the cycle proceeds, the wormhole turns itself inside out so that travel away from its throat leads not toward spatial infinity but to a singularity. Figure 5.16 illustrates this. At negative temporal infinity as experienced by distant observers in Universe I, there is no wormhole (1). A wormhole appears with a throat radius of r_- (not shown). The wormhole's throat enlarges (2), reaches a maximum value of r_+ (3), contracts (4), and vanishes at positive temporal infinity as measured by distant observers (5). As this foliation of spacetime continues, the wormhole turns itself inside out (6), although "outside in" might be a more accurate description. Motion on this "disk hole" away from the singularity in one coordinate patch leads to the singularity in another. The disk hole contracts to a minimum radius of r_- (7), re-expands (8), and disappears at negative temporal infinity of Universe II (9), which completes the cycle that repeats endlessly.

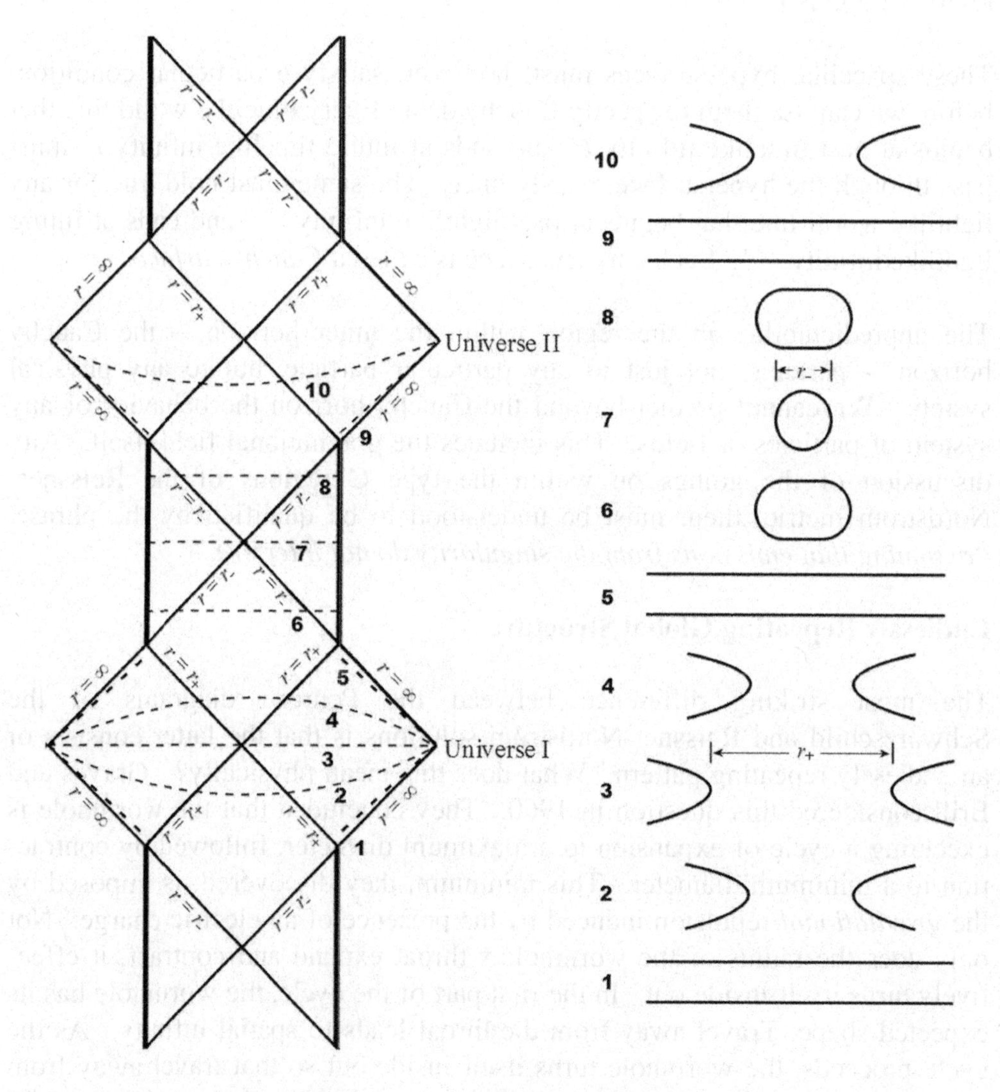

Figure 5.16. Cyclic oscillation of Reissner-Nordstrøm wormhole. Numbered hypersurfaces of the spacetime are on the left. The cross sections of the corresponding embedding diagrams on the right. Note that cases 1, 5, and 9 occur respectively at negative, positive, and negative temporal infinity.

Timelike Singularities

Unlike the case of the Schwarzschild wormhole, travelers falling freely into a Reissner-Nordstrøm wormhole need not be concerned about being crushed by a singularity. Rather than being drawn toward the singularities, they experience instead a gravitational repulsion away from them. The closer they approach a singularity, the stronger the repulsion. Were this repulsion not present, they still need not fear. The singularities are timelike. This means that in the type C regions containing these singularities, the radial coordinate retains its meaning as a spatial locator. Unlike the Schwarzschild case, the position $r = 0$ remains spatial and therefore avoidable. A traveler would prevent her death in the singularity by simply steering clear of it.

Connections to Multiple Universes

The Schwarzschild wormhole provided an untraversable connection to a single universe or distant region within our own universe. In addition to this the Reissner-Nordstrøm wormhole provides connections, which are traversable in principle, to an infinite number of other universes or distant regions. Traditionally, this connection to an infinity of universes is dismissed as a curiosity of no significance, because 1) maximally extended Reissner-Nordstrøm wormhole solutions are exceedingly unlikely to exist in nature, and 2) even if one did exist the blue sheet singularities that form at its inner event horizons and at its white hole antihorizons prevent the wormhole from being traversed. However, if the solution to the field equations exists, it should be possible for sufficiently advanced civilizations to create an instance of it, irrespective of the likelihood of such an instance arising naturally. Regarding the blue sheet singularities, calculations first performed by Amos Ori of Caltech indicate that the inner horizon might nevertheless be traversed. Moreover, it is not unreasonable to suppose that one of the consequences of a complete theory of quantum gravity – once one has been developed – would be the mitigation of singularities. Regions of infinite curvature would in the context of such a theory become regions of exceedingly large but finite curvature. In short, dismissing as unimportant the unsolicited appearance of an infinite number of universes might not be warranted.

I can't help but wonder whether there's something deeply significant about this. Why would such an elementary solution to the field equations force us to deal with an infinite number of universes? We could, of course, choose to identify them all with our own universe. But this would require us to make an unjustified, ad hoc assumption. Moreover, it would turn the Reissner-Nordstrøm wormhole into a faulty time machine. Travelers entering the black hole would emerge from a white hole before they left. As we will see in Chapter 10, that is not in itself a problem. The problem in this case is that the theory must be consistent. The traveler's world line that emerges from the white hole must seamlessly merge with her world line that enters the black hole. However, there is no way to ensure that this occurs. Once she travels past the Cauchy inner horizon, her world line can no longer be predicted. We are forced, then, to drop our ad hoc, one-universe assumption and take the infinite universes seriously.

One might be tempted to ask whether there might be a connection between these infinite universes and the multiverse of the Many Worlds Interpretation. The conventional answer to this question is "no". It is nonetheless interesting that in order to achieve self consistency both quantum theory and general relativity must invoke the concept of a multiplicity of universes. Another possible connection results from a traversable wormhole's capacity to serve as a time machine. This generates the usual time travel paradoxes and pseudo paradoxes. The only means of resolving them completely is to apply the Many Worlds Interpretation. When a wormhole-traversing time traveler reaches the past, he does not enter the past of his world – in the Many Worlds sense -- of origin, he enters the past of another world. In other words, he used a wormhole to travel between different worlds (universes) of the multiverse of quantum theory. I would not be astounded, then, were developments in quantum gravity to somehow show that the universes into which those traversing Reissner-Nordstrøm-like wormholes emerge are in some way related to universes of the quantum theoretic multiverse.

Traversability

The Reissner-Nordstrøm wormhole is not two-way traversable. In principle, however, it can be traversed in one direction to reach a universe other than our own, as shown in Figure 5.13. In practice, such traversals are unlikely. Blue sheet singularities form at the inner horizon as well as at the antihorizon of the wormhole's white hole sector. Moreover, these horizon instabilities appear to persist in attempts to apply quantum corrections to the physics (Poisson 1997). The only hope that this wormhole could enable a one-way trip to another universe is Ori's result indicating that a traveler could survive passage through a blue sheet singularity. If this is true, and if the blue sheet does not destroy the white hole sector of the solution, such a trip might be possible. With the familiar disclaimer that the matter will be settled definitely by an as yet undiscovered quantum theory of gravity, it is safe to state that

P42. *The Reissner-Nordstrøm wormhole is not two-way traversable. It is one-way traversable in principle. In practice it is one-way traversable, if blue sheet singularities at its inner horizons and antihorizons do not destroy its global structure, and if human travelers can survive passage through such singularities.*

This statement bears on the truth of an important idea in physics due to Roger Penrose. Cosmic censorship is the idea that nature conspires to prevent the observation of singularities. The weak form of this *Cosmic Censorship Hypothesis* is

> *The laws of physics prevent the existence of singularities exterior to an event horizon.*

Singularities not cloaked or "censored" by an enveloping event horizon are called *naked singularities.* The strong from of the Cosmic Censorship Hypothesis may be stated as follows.

> *The laws of physics prevent anyone from observing a singularity under any circumstances.*

If Ori is right, a traveler should be able to enter the Reissner-Nordstrøm wormhole, survive passage through the blue sheet singularity at the inner horizon, and gaze upon a timelike singularity. Even if he fails to escape the wormhole, his having viewed the singularity will cause a problem for believers in strong cosmic censorship. According to the strong version of the Principle of Equivalence, the laws of physics are the same for any observer, anywhere, anytime, moving anyhow. If the traveler can view the singularity within an event horizon, someone else should be able to view one exterior to an event horizon -- in total violation of cosmic censorship.

Occurrence in Nature

Reissner-Nordstrøm black holes do not occur in nature. As with the Schwarzschild black hole, the known processes of black hole formation result in a spinning object. There is no natural way, moreover, for a substantial charge to accumulate. The presence of a charge of a given sign repels additional charge of the same sign, while counterproductively attracting charge of the opposite sign. In other words, the marginal costs of adding charge of the same sign rises with the charge of the black hole. These costs, for a black hole sufficiently charged to display the global structure discussed above, would easily be afforded by a civilization sufficiently advanced to build Morris-Thorne wormholes.

In 1998 Shahar Hod and Tsvi Piran of The Hebrew University in Israel calculated the result of a natural means that such a civilization might have employed (if they were sufficiently patient) to construct a charged wormhole. They worked out the future evolution of a self-gravitating, charged, massless scalar field. The causal structure of the resulting collapsed object turned out *not* to be that of the Reissner-Nordstrøm solution. Rather, this object's Penrose diagram (Figure 5.17) is a sort of hybrid between that of the Reissner-Nordstrøm and Schwarzschild solutions. In its interior, its weakly singular Cauchy horizon is shortened and joins together with a (strongly singular) *spacelike* singularity. This object's exterior geometry evolves toward that of the Reissner-Nordstrøm metric. Upon passing through this object's outer horizon, an intrepid traveler would have a brief time during which she must decide her fate. If she steers toward the Cauchy horizon, she will survive – assuming of course that traversing a weak mass-inflation singularity is not terminally traumatic. If she fails to do so in time, her fate will be sealed. She will be doomed to a crushing death in the spacelike singularity. The smaller the total charge of the object, the shorter its Cauchy horizon, the less time will there be for a traveler to act to avoid this fate.

This numerical calculation of Hod and Piran's was not evolved beyond the Cauchy horizon. Whether a traveler that crosses this horizon can yet arrive in another universe is currently unknown.

Case of Extremely High Charge

A wormhole's charge is considered large, if the electrical force between two such identical wormholes is comparable to the gravitational force between them. This turns out to be true if the square of the wormhole's charge approaches GM^2/k, where G is Newton's gravitational constant, k is Coulomb's constant, and M is the wormhole's mass. Physicists working with black holes or wormholes normally use a system of units in which G and k are both equal to 1. In this system charge and mass have the same unit, which allows us to compare them directly.

When the magnitude of a wormhole's charge Q, expressed in these units, is large enough to equal its mass M, the causal structure of the wormhole differs drastically from the ($Q^2 < M^2$) case considered above. When $Q^2 = M^2$, the type B regions of the $Q^2 < M^2$ case (Figure 5.13) vanish as shown in Figure 5.18a.

The pairs of inner and outer event horizons coalesce into single event horizons separating asymptotically flat, type A universes and type C regions. Gone,

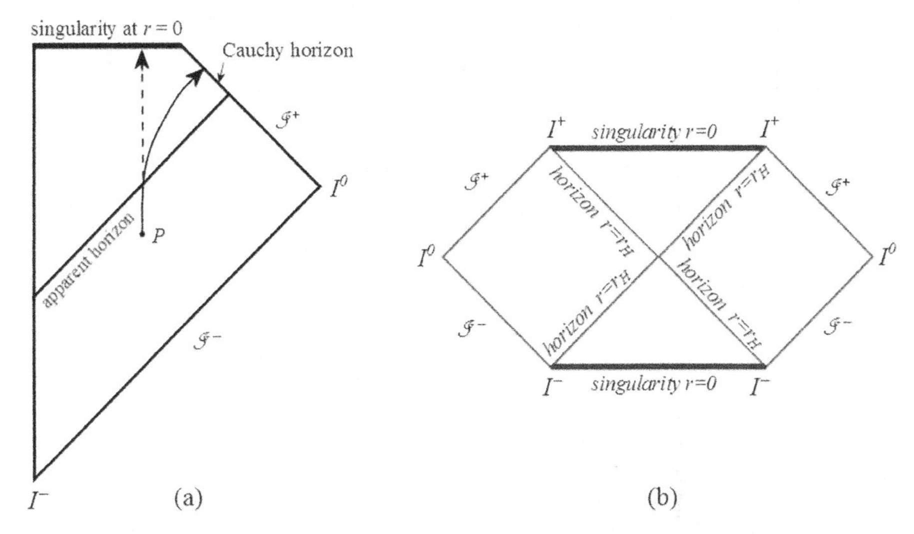

Figure 5.17. Causal structure of numerically calculated collapse of charged scalar field compared to that of the Schwarzschild solution. (a) Penrose diagram corresponding to the calculation of Hod and Piran (1998). [The "apparent" horizon is essentially a "predicted" event horizon – a boundary between locally trapped and untrapped surfaces drawn before information from future infinity is available to definitively locate the event horizons.] The traveler whose world line begins at *P* will be crushed in the singularity, unless she veers toward the Cauchy horizon soon enough to be able to reach it. (b) Schwarzschild solution for comparison.

too, are the type-A counterparts for each type A region. These pairs of type A regions (universes) were the ones connected by spacelike intervals across the Penrose diagram for the $Q^2 < M^2$ case. These were the universes that we would have sought to connect with timelike paths by using the technique of Morris and Thorne to make wormholes traversable. This is not to say that travelers cannot in principle move from one type A region to another. If they are able to survive any mass-inflation (blue sheet) singularity at the intervening inner horizon, they can.

It is of course exceedingly unlikely that the wormhole's charge would precisely equal its mass, even if there were astrophysical processes that permit a body to accumulate charge to this degree. The $Q^2 = M^2$ case, then, could only be expected to occur if it were maintained artificially by a wormhole-building civilization, or as a transitional case briefly encountered as such a civilization increased the magnitude of a wormhole's charge beyond its mass. When this latter condition ($Q^2 > M^2$) is achieved, the wormhole's global structure undergoes another drastic change (Figure 5.18b). All event horizons vanish. As does all access to other regions of any type. What remains is a solitary type A universe with a naked point singularity at $r = 0$. Like the case in which $Q^2 = M^2$, the techniques of Morris and Thorne cannot be applied. A traversable wormhole cannot be constructed in this case due to the absence of distinct

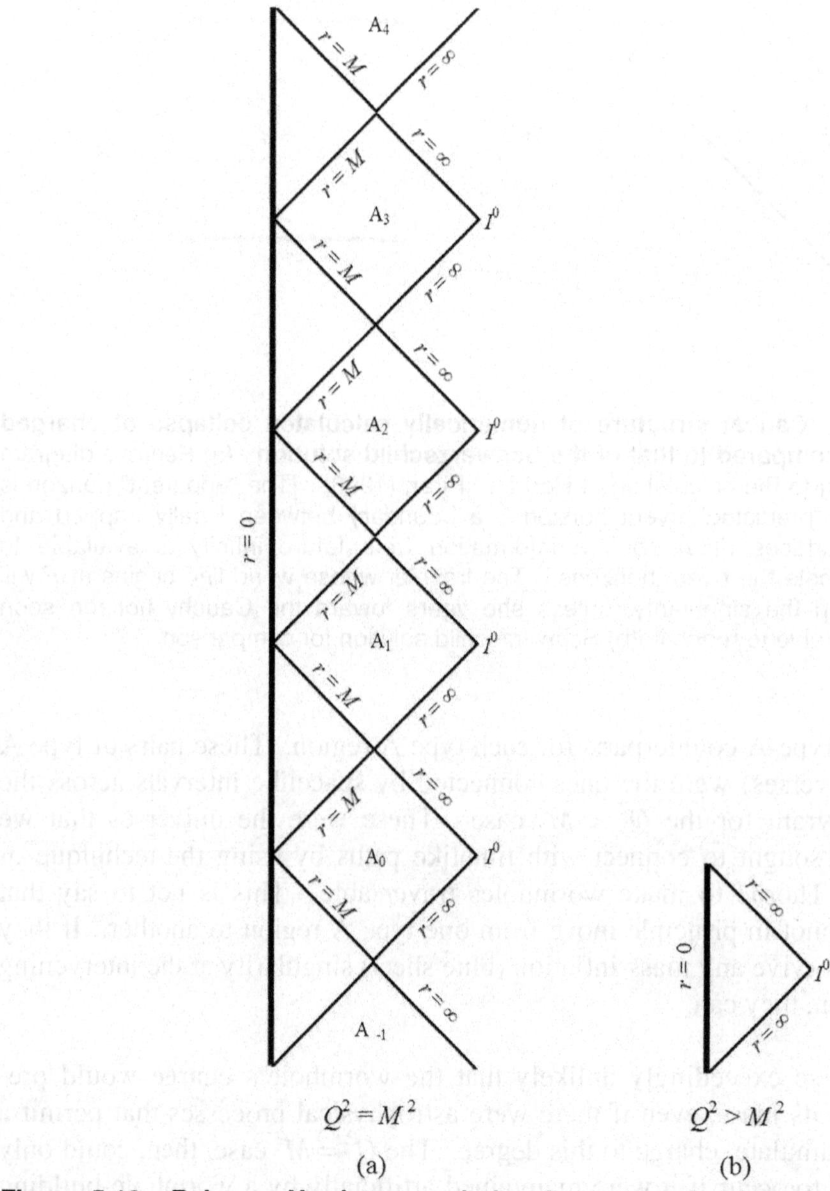

Figure 5.18. Reissner-Nordstrøm solution for extremely high charge Q compared to mass M. (a) $Q^2 = M^2$. Singularity at $r=0$ is surrounded by an event horizon at $r=M$. (b) $Q^2 > M^2$. Singularity at $r=0$ is naked, i.e. it is not surrounded by an event horizon.

asymptotically flat regions to connect. In short, the maximally extended Reissner-Nordstrøm solution is not a wormhole in this case.

The Kim-Lee Wormhole

It would be natural to suppose at this point that we could repeat Morris and Thorne's generalization of the Schwarzschild solution by similarly generalizing the Reissner-Nordstrøm solution in order to create a charged traversable wormhole. Unfortunately, things are not quite that simple. Unlike the Schwarzschild case, the Reissner-Nordstrøm solution involves not only a gravitational field but also the electromagnetic field generated by the wormhole's charge. This means that a solution must not merely satisfy the Einstein equations. It must instead satisfy a couple set of equations known as the Einstein-Maxwell equations. They may be translated roughly as follows. For every point in spacetime

essential curvature of spacetime =
stress-energy of matter (1)

essential rate of change of electromagnetic field =
charge-current of matter (2)

The first equation is the same caricature of the Einstein equations that appeared in Chapter 3. The second equation is a rough translation of Maxwell's equations. It says that a particular way in which the electromagnetic field changes in space and time – what I've called the "essential" rate of change – is generated by the charges and currents of matter.

Solving this pair of (sets of) equations poses a challenge, because they are not independent. The electromagnetic field counts as matter. As such it possesses stress-energy that appears on the right side of equation (1). So the solution to equation (1) depends on the solution to equation (2). Moreover, the "rate of change" of the electromagnetic field – how it changes in space and time – cannot be calculated until the curvature of spacetime is known. So the solution to equation (2) depends on the solution to equation (1). Lastly, the density of charges and currents must be consistent with the stress-energy of matter. It would not do, for example, for a solution to locate electric charge at a point in space devoid of matter.

When the Einstein equations are coupled to the equations that govern matter, the solution to this joint system of equations is called a *self-consistent solution.* Such a solution simultaneously reflects the dependence of matter on spacetime and the dependence of spacetime on matter.

In generalizing the Reissner-Nordstrøm solution we are driven to find a self-consistent solution only because Reissner had in 1916 specified the type of the matter involved, viz. electromagnetism. This specification of the nature of the matter field informed us that our solution would have to solve the Einstein-Maxwell equations. If, instead, Reissner had not so specified the type of matter, if the Reissner-Nordstrøm solution was not known to be associated with the electromagnetic field, we could guiltlessly proceed in the manner of Morris and Thorne. We could generalize the Reissner-Nordstrøm metric to a suitable wormhole shape, use Einstein's equations to determine the corresponding stress-energy of matter, and leave to an "advanced civilization" the details of finding a matter field that generates this stress-energy. As soon as we specify the matter to be of the type governed by X's equations – e.g. Maxwell, Yang-Mills, Klein-Gordon, Dirac – we are obliged to find a self-consistent solution to the combined Einstein-X equations.

P43. *A self-consistent wormhole solution is required only when the type of matter field contained in the wormhole is specified.*

Needless to say, it is always nicer to have a self-consistent solution than not. Even so, it was not until 1997 that David Hochberg, Arkadiy Popov, and Sergey Sushkov obtained a self-consistent solution for an uncharged Morris Thorne wormhole. By 2001 Sung-Won Kim and Hyunjoo Lee of Ewha Womens University in South Korea had found a (partially) self-consistent charged wormhole solution.

Properties of the Kim-Lee Wormhole

Kim and Lee obtained their solution by beginning with a particular (uncharged) Morris-Thorne wormhole. This wormhole did not follow from a self-consistent solution; its matter field – the base matter -- was not specified. To this solution they added a charge-dependent term, identical to that of the Reissner-Nordstrøm solution. They added this term to the temporal and radial components of the uncharged Morris-Thorne metric tensor. This in turn resulted in a charge-dependent addition to spacetime curvature and a corresponding charge-dependent addition to the stress-energy. The latter was assumed to be due to a radial electric field, for which they were able to solve. Their solution has the following properties.

- No event horizons
- Becomes zero-tidal-force MT wormhole when charge is set to zero
- Becomes zero-mass Reissner-Nordstrøm when shape function is set to zero
- Cannot become massive Reissner-Nordstrøm solution as $r \rightarrow \infty$
- Self-consistency limited to electromagnetic matter
- Ignores interaction between electromagnetic matter and base matter
- Charge cannot exceed the minimum value of the shape function

In order to make the Kim-Lee wormhole fully self-consistent it will be necessary to complement the Einstein-Maxwell equations with another set of equations that govern the behavior of some form of exotic matter. It might be possible, for example, to use the Hockberg-Popov-Sushkov wormhole as a self-consistent base onto which one could add the charged-dependent modifications of Kim and Lee.

Abandoning Self-Consistency

If we decide that it's simply too hard to work out a self-consistent (meaning self-contained) solution for a charged wormhole, we may revert to the original philosophy of Morris and Thorne: Leave the matter specification to an advanced civilization. Upon reaching this decision, we soon find that our work is all but done. The general Morris-Thorne traversable wormhole metric, expressed in terms of the redshift function $\Phi(r)$ and the shape function $b(r)$, already includes the case of a charged wormhole. In order to make their examples of uncharged traversable wormholes apply to the charged case, we need only make the following change to express the solution in terms of the shape function $b_{ch}(r)$ of the charged wormhole.

$$b(r) \rightarrow b_{ch}(r) - Q^2/r$$

Instead of matching the exterior to a solution to the Schwarzschild black hole geometry, we would match it to the Reissner-Nordstrøm geometry. The condition on $b_{ch}(r)$ that ensures the presence of exotic matter turns out to be the same as that on $b(r)$ (viz. $b_{ch} - rb_{ch}' > 0$). Also, the minimum value of $b_{ch}(r)$ determines the maximum value of the wormhole's charge. In essence,

P44. *The charged Morris-Thorne wormhole is a special case of the general Morris-Thorne wormhole in which the solution in the exterior region is matched to the Reissner-Nordstrøm solution.*

Lost connection to other universes

One of the unfortunate (or fortunate, depending on your point of view) features of the charged traversable wormhole is that the price of being able to traverse its Penrose diagram horizontally appears to be the loss of the ability to do so vertically. The elimination of the outer event horizon seems to seal off the access to the plethora of asymptotically flat regions (universes) that characterize the Reissner-Nordstrøm solution. Whether it is possible to retain access to

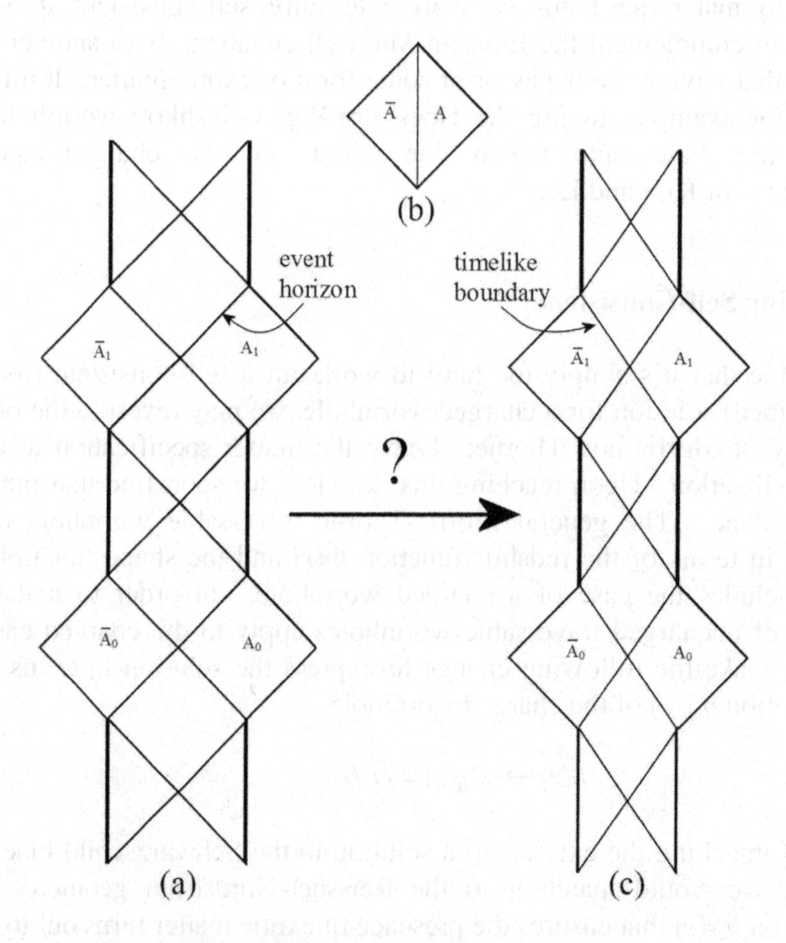

Figure 5.19. Traversable charged wormhole. (a) Reissner-Nordstrøm wormhole. (b) Kim-Lee wormhole. (c) Hypothetical charged traversable wormhole. Traversals between A and $\bar{A}$ universes is ensured by converting the lightlike horizons of the RN solution into timelike boundaries.

these universes while being able to traverse the Penrose diagram horizontally, is currently unknown. Were it possible, the event horizons of the Reissner-Nordstrom wormhole would be replaced by softer timelike boundaries (Figure 5.19). Just as an event horizon marks the points of no return for an object unable to exceed the speed of light, these timelike boundaries would be the point of no return for particles unable to exceed a lesser speed. In other words, all objects able to exceed some minimum, subluminal speed relative to the center of this hypothetical wormhole could traverse it.

Uncharged, Rotating Wormholes

The Kerr Wormhole

It was not until 1963, forty-seven years after Reissner's paper, that New Zealander Roy Kerr solved the Einstein equations for the case of a dense rotating body. The maximally extended Kerr wormhole solution is similar to that of the maximally extended Reissner-Nordstrøm wormhole. These wormholes share the following features.

- Two event horizons – outer and inner
- A new type of interior region (beyond those of Schwarzschild case)
- Connections to multiple universes
- Timelike singularities – i.e. they can be avoided by travelers
- Endlessly repeating structure

In addition to these the Kerr wormhole is characterized by

- Absence of spherical symmetry
- Ring singularities
- More universes (beyond Reissner-Nordstrøm case)
- Closed timelike curves
- Frame dragging
- Utility as an energy source

Absence of Spherical Symmetry

Figure 5.20 is a diagram of the Kerr black hole. When we view it by looking "downward" along its axis of rotation (shown as the z axis in Figure 5.20b), we see that it is circularly symmetrical. It does not change with the azimuthal angle φ. When we view it by looking along the x axis (Figure 5.20a), we see that its cross section in the z-y plane is not circular. It is elliptical. The outer surface's distance from the center varies with the declination angle θ. This outer surface — which is called the *static limit* for reasons explained below — forms an oblate spheroid. The Kerr black hole, then, bulges at its equator much as the earth does.

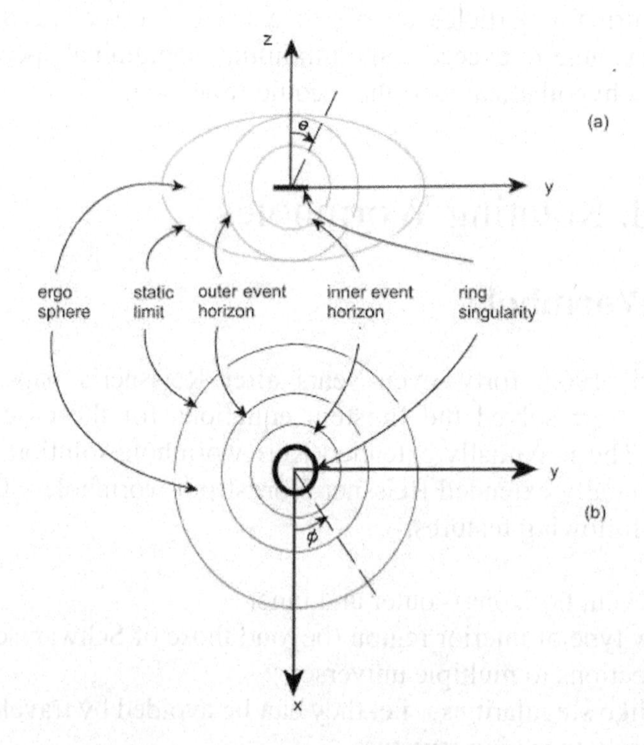

Figure 5.20. Kerr black hole. (a) View perpendicular to axis of rotation (in x-direction). (b) View along axis of rotation (in z-direction).

Ring Singularities

One of the big surprises of the Kerr black hole is that it does not have a singularity located at its central point, as do the Schwarzschild and Reissner-Nordstrøm black holes. Its singularity is instead a circular ring centered about this point. As can be seen in Figure 5.20, the plane of the ring is perpendicular to the black hole's axis of rotation. As with the singularities of the Reissner-Nordstrøm wormhole, the Kerr wormhole's singularities are endlessly repeated in every type C region of its Penrose diagram, as shown in Figure 5.21.

To understand this diagram it helps to know something about its relation to the coordinate system normally used to analyze the Kerr wormhole. This coordinate system -- called Boyer-Lindquist coordinates -- is a generalization of the

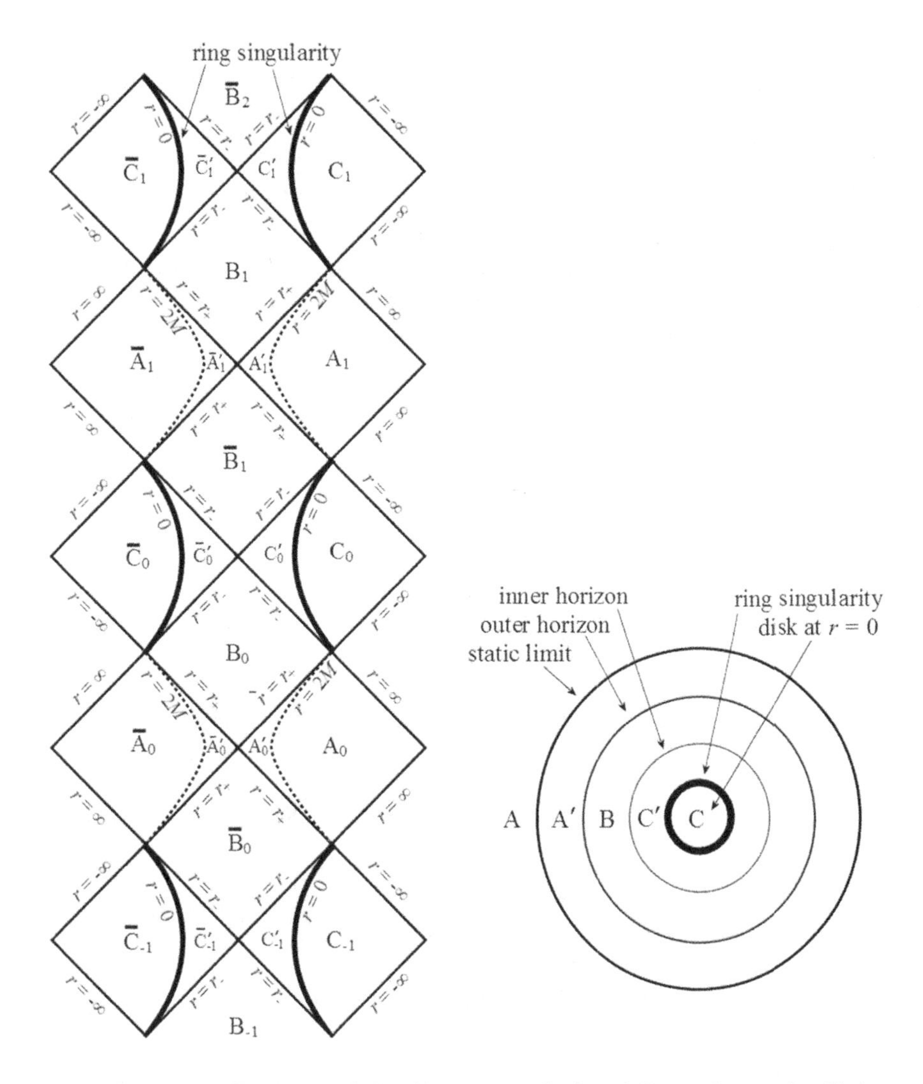

Figure 5.21. Penrose diagram of the Kerr wormhole. It is similar to the Reissner-Nordstrøm case except that in the type C regions $r = 0$ corresponds to a *disk*, whose edge is a ring singularity, instead of a point. Therefore, travelers can in principle pass through this disk to enter type C universes. Static limit (dashed lines) shown at equatorial ($\theta = \pi/2$) value of $r = 2M$. The magnitude of the measure of angular momentum, a, is less than the mass, M ($a^2 < M^2$).

spherical coordinates used to describe the Schwarzschild wormhole. As in the Penrose diagrams of the Schwarzschild (Figure 5.3) and Reissner-Nordstrøm (Figure 5.13) wormholes, Figure 5.21 refers to values of the Boyer-Lindquist "radial" coordinate r. In the Schwarzschild and Reissner-Nordstrøm cases r is the radial coordinate of the familiar spherical coordinate system. Surfaces of constant r are concentric spheres. In the case of the Kerr wormhole, however, r has a different meaning. Surfaces of constant r are not spheres. They are ellipsoids as shown in Figure 5.22. If we slice these ellipsoids with a plane containing the wormhole's axis of rotation, we obtain the family of ellipses shown in the figure. These ellipses all share the same two foci. These foci are separated by a coordinate distance of $2a$, where the parameter $\underline{a}$ is a measure of the spin of the wormhole. It is the ratio of the wormhole's angular momentum to its mass. The larger the value of r^2 in comparison to a^2, the more closely does the corresponding ellipsoid resemble a sphere of radius r. The smaller the value of r^2 in comparison to a^2, the more oblate does the corresponding ellipsoid become. In the limit of $r = 0$, the ellipsoid becomes so oblate that it flattens to a disk of radius a. On this disk the curvature of spacetime is finite everywhere except at its edge. This circular edge is the location of the ring singularity.

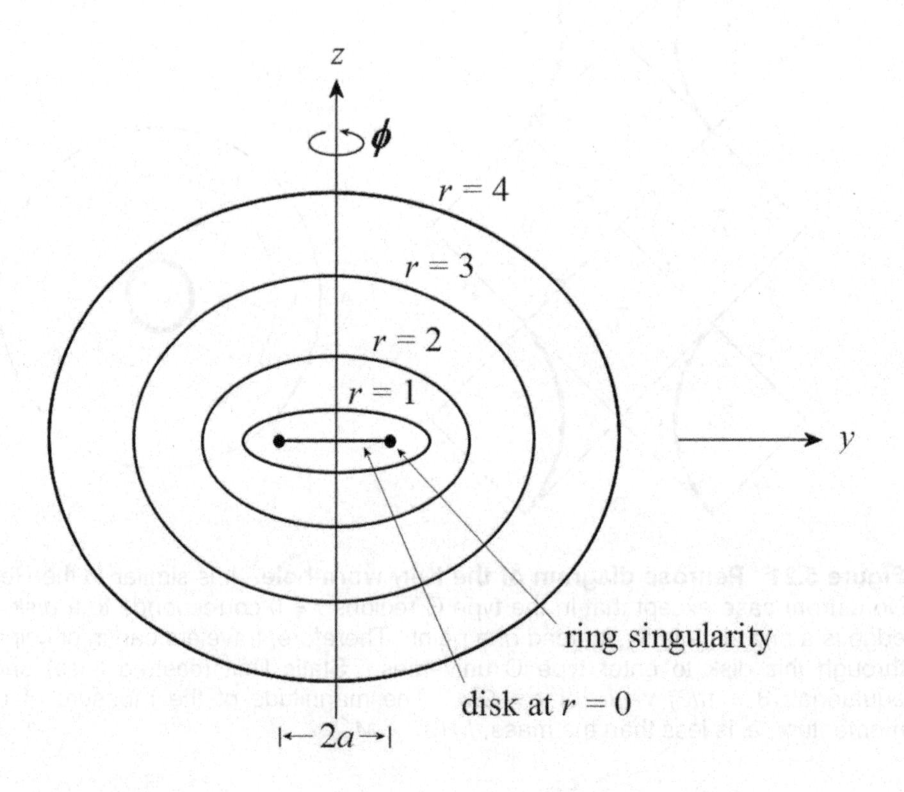

Figure 5.22. Surfaces of constant *r* in coordinate system (Boyer-Linquist) often used to describe Kerr wormhole. These surfaces are oblate ellipses that flatten to a disk at $r = 0$. $r = 0$, then, corresponds not to single point but to this entire disk. Black hole rotates about z axis with angular momentum Ma (M = black hole mass). Diameter of ring singularity is $2a$, which is also the separation of the foci of the ellipses formed by intersecting the ellipsoids with a plane containing the z axis.

For a nonrotating wormhole, $a = 0$. In this event the ellipsoids become spheres as their foci contract to a single point. The coordinate system reduces to spherical coordinates, and the Kerr wormhole reduces to the Schwarzschild case. In summary,

P45. *The singularities of Kerr wormholes are ring shaped, occurring at the edge of a circular disk whose radius is the ratio of the wormhole's angular momentum to its mass. This disk is the surface specified by setting $r = 0$ in the Boyer-Lindquist coordinate system.*

More Universes than Reissner-Nordstrøm Wormhole

The interesting thing about the disk at $r = 0$ is that a traveler can in principle pass through it. As long as he avoids the singularity at its edge, he should be fine. Of course, if the wormhole is rotating too slowly, its ring diameter might be too small to prevent him from being incinerated by radiation or crushed by tidal forces. Let us suppose, then, that the wormhole's angular momentum is sufficiently large to spare the traveler from such travails. As he passes through the disk, his radial coordinate goes from being positive, to being zero at the disk, to being negative once he has passed through. How, you might ask, can a radial coordinate be *negative*? It can be negative as a means of distinguishing one side of an $r = 0$ disk from the other side. Denizens of the negative-r side are free to interpret their distance from the $r = 0$ disk as a positive number. The consequence of this interpretation would be their having to regard the mass of the wormhole as negative. This is because the wormhole's gravitational field – as expressed in the metric of its spacetime geometry – depends on r in only two ways. It depends on r^2, which is insensitive to a change in the sign of r. It also depends on Mr, where M is the mass of the wormhole. The change in the sign of this term may be interpreted as being due to a negative M instead of a negative r.

Our traveler, then, upon passing through the ring singularity, would notice that he is being accelerated away from it by the antigravity generated by the wormhole's negative mass. He would notice, moreover, that the ring singularity is naked and that it can be viewed from arbitrarily great spatial distances. In short,

P46. *In the asymptotically flat regions entered by passing through the ring singularity, the singularity appears to be naked and to have a negative mass that generates a repulsive gravitational field.*

These asymptotically flat regions or universes are type C regions, as shown in Figure 5.21. Recall that the type C regions of the Reissner-Nordstrøm wormhole reached a dead end at the point singularities at $r = 0$. In the Kerr wormhole, these point singularities are replaced with rings, which serve as gateways to universes that are inaccessible to travelers who chose to fly Reissner-Nordstrøm.

Closed Timelike Curves

The negativity of the Boyer-Lindquist pseudo radial coordinate r on the far side of the ring singularity yields another interesting consequence. It negates the diagonal metric tensor component associated with the azimuthal angle φ – the angle that changes as one circles the axis of rotation. As you might recall from PP3 (Preliminary Principle 3), the negativity of this metric tensor component implies that its associated coordinate φ becomes the local time coordinate.

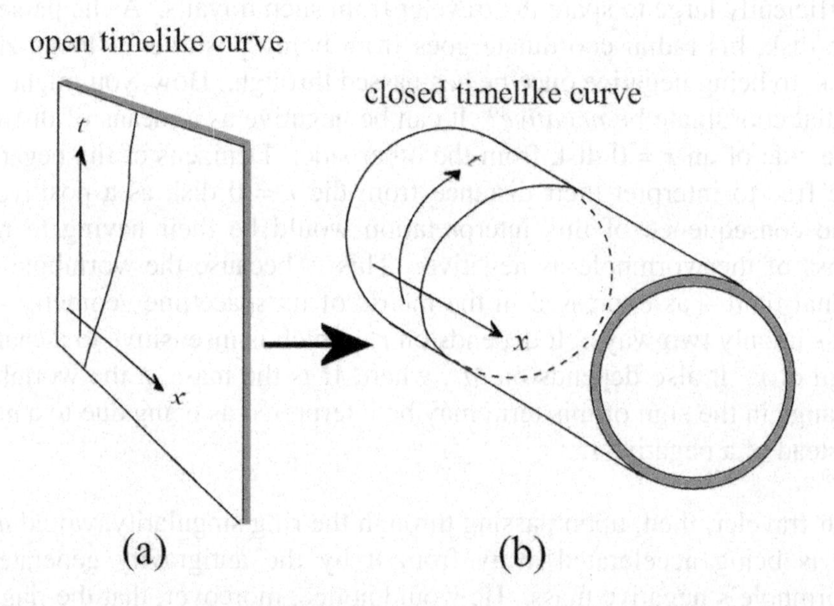

Figure 5.23. Open and closed timelike curves. (a) A world line in ordinary Minkowski space. (b) A world line in spacetime with a cyclic time coordinate is a closed timelike curve.

We have seen this before. As a traveler crosses the event horizon of a Schwarzschild wormhole, she finds that the radial coordinate, which had been spatial exterior to the horizon, assumes the role of a local time coordinate. What is different in the current case is that the new local time coordinate φ is *cyclic*. Just as a child on a carousel repeatedly returns to the same point in space, a traveler experiencing such a time coordinate must repeatedly revisit the same point in time. The world line of such a traveler would be a *closed timelike curve* – a timelike curve that moves forward in time until it joins itself in the past. To visualize such a curve, imagine ordinary Minkowski space in which the time coordinate is cyclical as shown in Figure 5.23.

We see, then, that a traveler on the far side of the singularity, who chooses to remain within a coordinate distance of about $2GM/c^2$ of it, will be in for strange experiences. If she holds her distance to the singularity, she will be stuck in a time loop – endlessly repeating the events of the last minute, hour, or day depending on the wormhole's angular momentum and her distance from the axis of rotation. If instead she had programmed her spaceship to continue moving away from the singularity, she would escape this fate. However, she would be treated to the interesting phenomenon of being able to observe future versions of her spaceship heading away from the singularity (Figure 5.24). These unusual effects cease once she has traveled sufficiently far from the singularity.

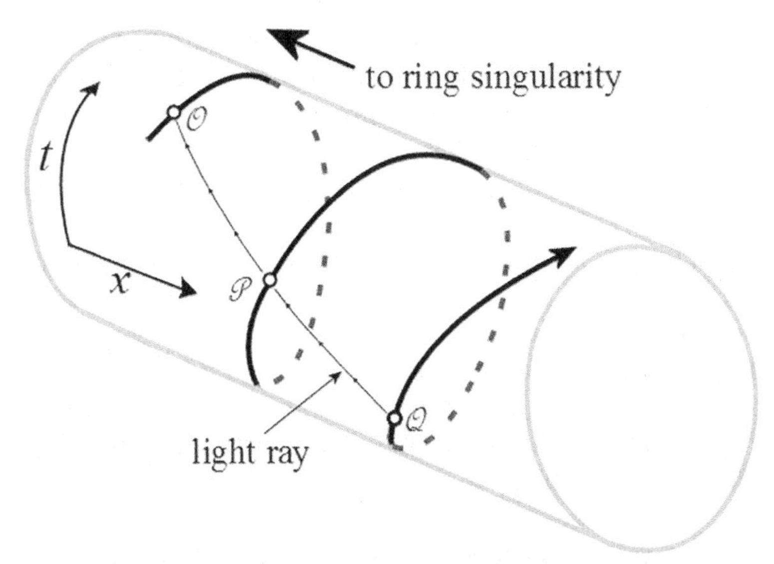

Figure 5.24. Close timelike curves in the vicinity of Kerr ring singularity. Traveler at point 𝒪 will observe future images of herself at points 𝒫 and 𝒬 as she moves away from the ring singularity. [$t \equiv \varphi$, $x \equiv -r$]

Physicists have been aware of the presence of closed timelike curves (CTC) in general relativity since the late 1940s. They arise in the presence of a rapidly rotating object – an infinitely long cylinder, a finitely long cylinder, a black hole, a cosmic string, or the universe itself. These curves have been largely ignored on the grounds that the situations that generate them are "unphysical": "You can't have an *infinitely* long cylinder." "You can't spin a finitely long cylinder that fast without its atoms flying apart." "The CTC's only exist within the event horizon of black holes, so who cares?" "There *are* no cosmic strings." "The universe isn't rotating."

In short, physicists have most commonly held that general relativity's predictions of CTCs require extreme circumstances that cannot exist in nature. CTCs, therefore, need not be taken seriously. Perhaps. It might also be that this prediction is the theory's way of informing us that it is incomplete or that there is a problem with our understanding of time. We will revisit CTCs in the context of traversable wormholes in Chapter 10.

Frame Dragging

In the days before the creation of the Global Positioning System the navigation of aircraft and the trajectories of missiles relied on what was known as an Inertial Guidance System (IGS). Such a system contained two gyroscopes at right angles. These would each point in particular directions. These directions would not change irrespective of the motion of the missile or aircraft in which the IGS was installed. Either of the gyroscopes of the IGS that pointed to a particular star before a flight, would still do so after. The two fixed directions picked out by the gyroscopes together with a third direction perpendicular to both of these and a clock define the axes of a frame of reference of the sort introduced in our discussion of special relativity -- an inertial reference frame.

If you were to put your spaceship in orbit around a Schwarzschild or Reissner-Nordstrøm black hole, the gyroscopes of your ship's IGS would continue to point in fixed directions – each toward a particular star. Were you to travel to a Kerr black hole, things would be different. As you orbit the black hole you would notice that your gyroscopes are *precessing*. Instead of pointing in fixed directions they would be steadily changing these directions, changing the stars toward which they are pointing.

This precession of your gyroscopes would be a manifestation of a feature of general relativity discovered in 1918 by Joseph Lense and Hans Thirring. They realized that a massive object rotating in a particular direction induces the spacetime in the object's vicinity to rotate in the same direction. To better understand this *Lense-Thirring effect*, imagine that you have parked your spaceship at fixed distance above a Schwarzschild black hole. You are not in orbit. You are using your ship's engines to hold the same spatial position -- as indicated by the strength of the local gravitational field, and the same orientation -- as indicated by your sighting distant stars and consulting your gyroscopes. Suppose now that we spin up the black hole. We can do this by tossing into it a stream of objects that all possess the same angular momentum. We will also inject sufficient angular-momentum-free, negative mass to ensure that the total mass of the black hole remains constant. When we are done, we will have converted the Schwarzschild black hole into a Kerr black hole of the same mass. Suppose for the sake of simplicity that we have injected spinning matter in such a way as to ensure that your spaceship is now in the equatorial plane – the orbital plane perpendicular to the axis of rotation -- of the newly formed Kerr black hole.

You continue to use your engines in exactly the manner that kept you stationary above the Schwarzschild black hole relative to the distant stars. You soon discover, however, that your ship is no longer stationary. Your ship and the

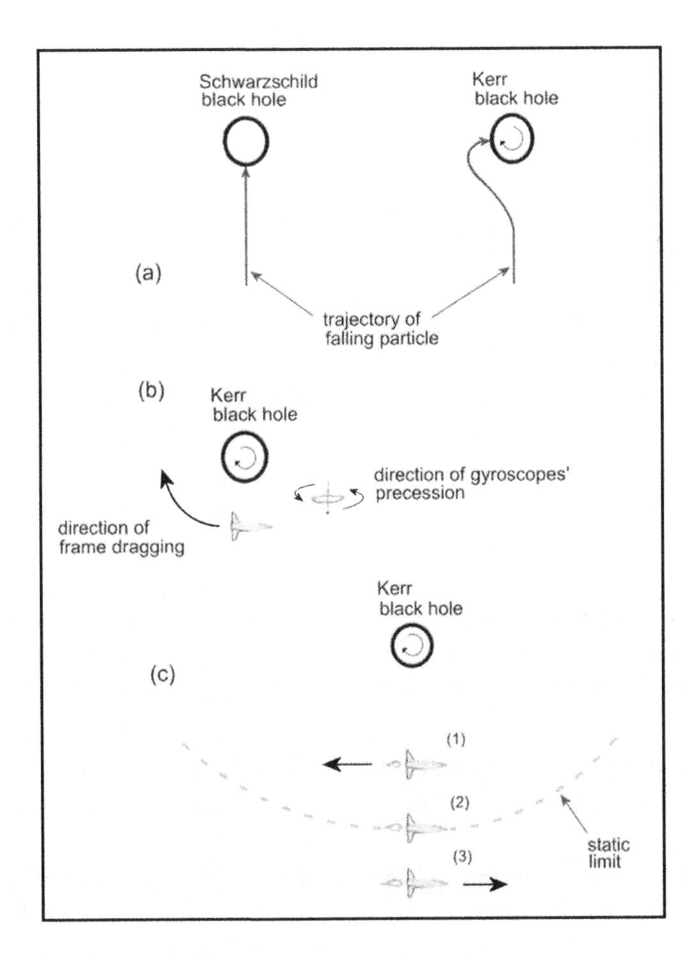

Figure 5.25. Frame dragging. (a) Effect on in-falling particle. (b) Effect on spaceship and gyroscope. (c) Spaceship firing engines at full power: (1) retrograde motion inside static limit. (2) no motion at static limit. (3) forward motion outside static limit.

inertial reference frame defined within it are being dragged around the rotating black hole in the same direction in which it rotates. The magnitude of this *frame dragging* increases with decreasing distance to the center of the Kerr black hole. While your ship is dragged in the direction in which the back hole rotates, your gyroscopes precess in the opposite direction (Figure 5.25b).

A Kerr black hole, then, causes spacetime to swirl about it in a manner reminiscent of the whirlpool currents created by a bathtub drain. The action of these swirling currents on the motion of a toy boat in your bath tub is similar to that caused by frame dragging on your spaceship. The toy boat's motor gives it a certain top speed in open water. When it is near the drain, it can remain stationary as long as its motor can compensate for the swirling drain currents. The speed of these currents increases with decreasing distance to the center of the drain. If the toy boat approaches the drain too closely, the speed of the drain currents will exceed the top speed that boat's motor can deliver. The toy boat will no longer be able to remain stationary and will begin to move backwards in the direction of the drain currents.

So it would be for your spaceship trying to remain stationary near a Kerr black hole. The closer you approach the black hole, the more you'd have to blast your engines to remain stationary. At a particular distance from the black hole, your engines would be overwhelmed. In order to remain stationary, your engines would need to be able to accelerate your ship beyond the speed of light. Because they cannot, your ship would begin to move backwards in the direction of the black hole's rotation. The surface defined by the points at which this first occurs is called the *static limit*. This surface reflects the symmetries of the Kerr wormhole. It is not spherical. It is an oblate ellipsoid – a spheroid with an equatorial bulge.

The static limit is a timelike surface (except at the black hole's poles, where it coincides with the outer event horizon). This means that after your ship has entered the region enclosed by the static limit, you are free to leave this region by crossing the static limit in the opposite (outward) direction. Of course, if you were to continue toward the center of the black hole and cross its outer event horizon, you would not be able to recross it in the opposite direction.

The difference between the type A' region (Figure 5.21) bounded by the static limit and the outer event horizon and the type B region bounded by the outer and inner event horizons is this. They forbid different sets of sets of particle motions. The type A' region's ban is the least restrictive. It only bans those particle motions that don't "go with the flow" – that do not correspond to movement around the black hole in its direction of rotation. The type B region bans in addition all particle motions that are not directed inward. This is just another way of saying that the event horizons are one-way boundaries, while the static limit is not.

Utility as an Energy Source – The Ergosphere

In 1971 Remo Ruffuni and John Wheeler came up with a name for the region between the outer event horizon and the static limit. They called it the "*ergosphere*". They chose this name, because Roger Penrose had earlier discovered something particularly interesting about this region. Work can be extracted from it. "Ergo" is the Greek word for "work". The key to extracting energy from a rotating black hole is its ergosphere's ability to allow particles within it to acquire *negative* energy. This reduces the total energy (mass) of a black hole by the magnitude of this acquisition. Because energy must be conserved, a reduction in a black hole's total energy can only occur if it emits a compensating quantity of positive energy. This emitted energy can be harnessed to do work.

To understand this feature of the ergosphere one must realize that a particle's energy is merely its momentum in a temporal direction. This is a definition from special relativity. If a particle's energy is positive, it is moving forward in time. If its energy is negative, it is moving backwards in time. Particles do not in general have negative energy, because we do not know how to cause them to move backward in time. Within an event horizon or ergosphere this changes. Recall that within such regions the temporal coordinate, which is of course timelike exterior to these regions, becomes spacelike. This means that within an event horizon or ergosphere, it is as easy to give a particle a negative temporal momentum, as it is to give it a negative spatial momentum exterior to these regions. All an experimenter within such a region need do to impart negative temporal momentum – negative energy – is to push the particle in the negative t direction.

Were an experimenter, working within an event horizon, to cause a particle to move in this direction, it would be of no consequence to external observers. Nothing can escape the event horizon including evidence of his actions. If by contrast, the experimenter were to impart negative energy to a particle within an ergosphere, he could affect the exterior universe. This is because a static limit separating the ergosphere for the external universe can -- unlike an event horizon -- be freely traversed in both directions.

An observer may impart negative energy to a particle within the ergosphere, which remains there after he returns to the exterior universe. The law of the conservation of energy – momentum conservation in the temporal direction -- must continue to hold. His loss of negative energy to a particle in the ergosphere becomes a gain of positive energy upon his emergence from that region.

This energy extraction process need not involve an experimenter. An automated cargo vessel could be programmed to extract energy by 1) falling to the ergosphere, 2) ejecting its cargo by imparting it with sufficient negative energy to ensure that the vessel itself is thrown from the ergosphere with more kinetic energy than it took down, 3) converting this surplus kinetic energy into work by, for example, using it to turn a flywheel.

Case of Extremely High Angular Momentum

As in the case of a highly charged Reissner-Nordstrøm solution, the Kerr solution changes drastically when its measure of angular momentum $\underline{a}$ equals its mass M ($a^2 = M^2$). The type B regions vanish. The inner and outer event horizons coalesce. The type $\bar{A}$ counterpart regions to each type A region is also gone (Figure 5.26). An ergosphere continues to exist. As in the Reissner-Nordstrøm case, the Morris-Thorne technique for creating timelike paths between a type A region and its type A counterpart would be inapplicable.

When the magnitude of the measure of the solution's angular momentum exceeds its mass ($a^2 > M^2$), the solution becomes a traversable wormhole. This is in marked contrast to the $Q^2 > M^2$ case of the Reissner-Nordstorm solution, which ceased to be a wormhole of any sort. The Kerr solution positions the disk specified by $r = 0$ (in Boyer-Lindquist coordinates) as a gateway between our "positive" universe and a "negative" one. This negative universe possesses the unusual features described above – closed timelike curves and a gravitationally repulsive singularity that appears to have negative mass. As in the corresponding Reissner-Nordstrøm case, the singularity is naked, and the infinity of universes -- accessible in principle by moving upward in the Penrose diagram – are gone. Unlike that case, the singularity is a traversable ring. Travelers passing through this ring singularity are likely protected from lethal effects of radiation and tidal forces by the high value of the Kerr solution's angular momentum in this ($a^2 > M^2$) case. Another interesting feature of this case is that it retains a remnant of the ergosphere as a narrow wedge centered in the equatorial plane. This wedge, which extends a small angular distance above and below the wormhole's equator, narrows with increasing angular momentum, becoming a flat disk in the limit of infinite a^2/M^2.

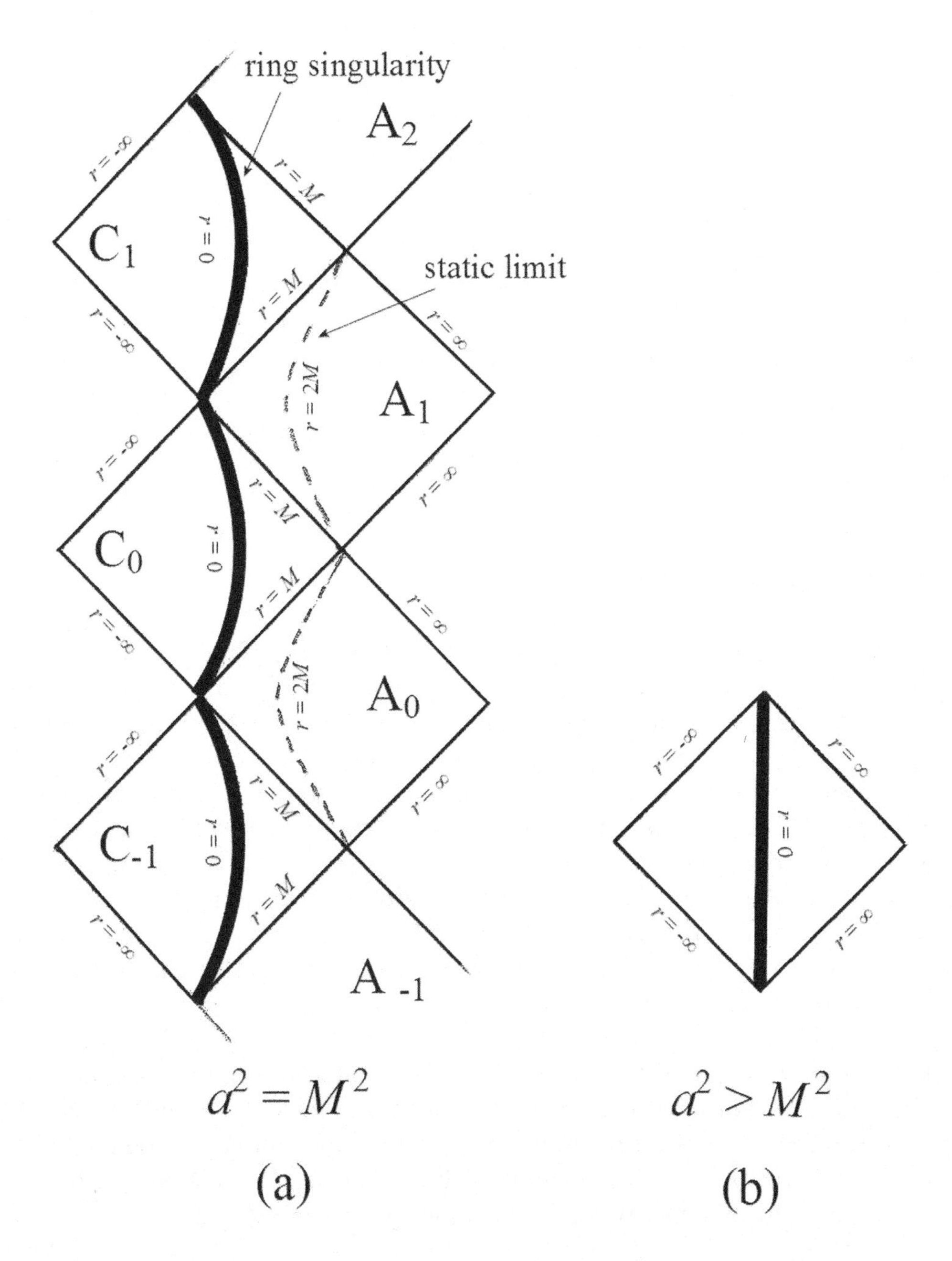

Figure 5.26. Kerr solution for extremely high angular momentum measure, *a*, compared to mass, *M*. (a) $a^2 = M^2$. Static limit (dashed lines) shown for its equatorial ($\theta = \pi/2$) value of $r = 2M$. (b) $a^2 > M^2$. Ring singularity is naked. Ergosphere reduces to a small equatorial wedge.

There remains a widespread bias amongst physicists against the possibility of this or any other case involving naked singularities. Despite the current absence of theoretical justification, the *Cosmic Censorship Hypothesis* continues to hold sway as a sort of unofficial physical axiom.

The Charged Kerr Wormhole

In 1965, two years after Roy Kerr found his solution, Ezra "Ted" Newman of the University of Pittsburgh extended it to the case of charged matter. The properties of his solution, known as the Kerr-Newman solution, are qualitatively similar to the Reissner-Nordstrøm and Kerr solutions. Its maximally extended version also describes possibly traversable wormhole connections to numerous other universes.

The Teo Wormhole

In 1998 Edward Teo, then at the University of Cambridge, generalized the results of Morris and Thorne in order to create a traversable rotating wormhole. He began with the most general expression possible for a time-independent geometry symmetric about an axis of rotation. This expression featured four arbitrary functions of the radial coordinate r and the azimuthal angle θ. A particular choice for these functions would completely specify the geometry. Two of these four functions were generalizations of the shape and redshift functions of the Morris-Thorne case. Another of them controlled the warping of space in the radial direction. The last of them specified the degree to which an in-falling particle would be dragged in the direction of the wormhole's location due to the action of the Lense-Thirring effect.

Following Morris and Thorne, Teo imposed a flare-out condition on the shape function, which ensured that the wormhole would have the desired shape. He also required the redshift function be to finite, thus eliminating the possibility of event horizons. He was then able show that these constraints result in a violation of the null energy condition at the wormhole's throat.

As an example of his new class of traversable rotating wormholes, Teo chose values for the four geometry-specifying functions that reduce to MT's zero-tidal-force wormhole in the limit of zero angular wormhole momentum. This wormhole's throat is not spherical. It is shaped like a dumbbell, as shown in Figure 5.27. As in the case of the Kerr solution for extremely high ($a^2 > M^2$), a remnant of the ergosphere straddles the equatorial plane. In this case it is a belt whose thickness varies inversely with the wormhole's angular momentum.

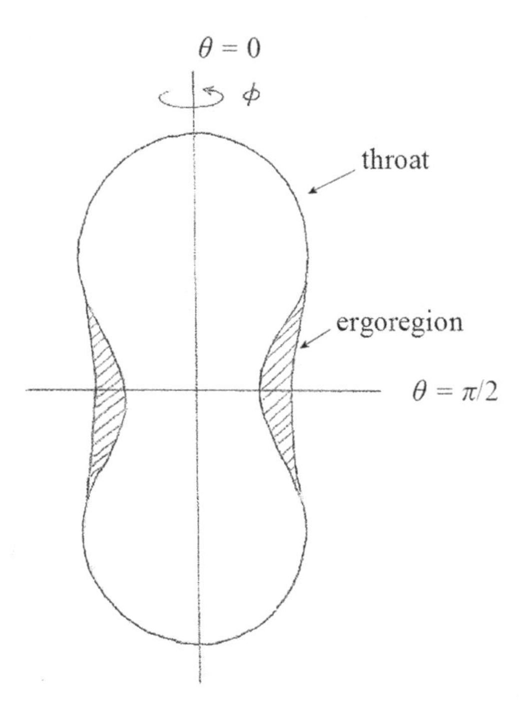

Figure 5.27. Throat of Teo's rotating wormhole. The throat of Teo's example of a traversable rotating wormhole is not spherical as it is for the Morris-Thorne wormhole.

A few years after Teo published these results S. E. Perez Bergliaffa and K. E. Hibberd of the Centro Brasileiro de Pesquisas F´ısicas noticed a possible problem. The matter required to generate Teo's wormhole cannot be modeled as a perfect fluid or even a fluid whose pressures depend upon direction. This is troubling only in that such models are the most common means of describing stellar matter. Matter that cannot be so described would be unusual. However, given that the wormhole's matter must after all be exotic, requiring an additional form of exoticity might not be a problem.

Nevertheless, this point was addressed by Peter Kuhfittig of the Milwaukee School of Engineering. He showed that for a particular choice of the redshift and shape functions some of the conditions for a fluid matter source can be met, at least approximately. His observation was incidental to his thorough analysis of the Teo wormhole, which he extended to the case of a time-varying angular momentum. Kuhfittig discovered that a Teo wormhole violates the weak energy condition to a lesser degree than its nonrotating counterpart. In other words, the construction of a rotating traversable wormhole requires less exotic matter than that of an equivalent nonrotating wormhole. He found, moreover, that some of the tidal constraints are more easily satisfied. Lastly, he found that for a *spherically* symmetrical Teo wormhole, it is possible to reduce violations of the weak energy condition by increasing the wormhole's angular momentum. The faster it spins the less exotic matter it requires. This does not, however, seem to hold true for a Teo wormhole whose sole axis of symmetry is its axis of rotation.

***P47.** A rapidly spinning Teo wormhole has an " ergobelt" -- a narrow region straddling the equatorial plane from which work can be extracted.*

***P48.** A Teo wormhole violates the weak energy condition to a lesser degree than does a Morris-Thorne wormhole containing the same matter distribution.*

6. Quantum Wormholes

General relativity, as it is normally practiced, assumes a fixed topology of spacetime. This means that wormholes cannot be created or destroyed. This is why Morris and Thorne suggested that the first step in constructing a macroscopic wormhole is to obtain a preexisting submicroscopic one. They imagined an advanced civilization dredging up such a wormhole from Wheeler's spacetime foam. Recall that spacetime foam is the conjectured quantum vacuum state of the gravitational field. The foam consists of all field configurations – i.e. spacetime geometries – not ruled out by macroscopic observations. These geometries include that of quantum wormholes.

I should point out that Wheeler's spacetime foam, containing geometries of every conceivable topology, could never result from the quantization of general relativity. The quantization of a classical system does not enlarge the set of the system's possible configurations. Consider, for example, a system consisting of a single particle whose configuration is specified by the particle's three spatial coordinates. Quantization will not result in a system for which there are configurations whose specification requires *four* coordinates. We could, nevertheless, imagine with Wheeler a classical theory whose configurations include all possible geometries and whose state of "lowest energy" is general relativity for a fixed topology. The quantization of such a theory would continue to contain foam-like geometries of every topology -- including quantum wormholes.

Unfortunately, we have no comprehensive theory of quantum gravity that we can use to understand these objects. We are reduced instead to discerning their properties through the application of incomplete quantizations of general relativity. The most popular of these are semiclassical gravity and minisuperspace quantization.

Semiclassical Wormholes

Semiclassical gravity is also known appropriately as quantum field theory in curved spacetime. The gravitational field in the form of a spacetime metric serves as a fixed background. On this background one applies quantum field theory to the particular matter fields of interest. This amounts to expressing the matter's stress-energy tensor as an operator that acts on the quantum states of the matter field.

Examples of these states are: "one particle with momentum k", "two particles – one with momentum k, the other with momentum k′", "no particles – i.e. the vacuum". In the classical theory the stress-energy at a point in spacetime is just a number. In quantum field theory such a number is obtained by the action of the stress-energy operator on such quantum states of the matter field. In other words, in order to specify the stress-energy due to a particular value of the matter field intensity, one does not, as in classical field theory, simply plug this field intensity into the classical expression for the stress-energy. Instead one chooses the quantum state of interest (usually the vacuum) of the matter field. This is equivalent to choosing the matter field intensity in the classical theory. Then one applies the stress-energy operator to this state to obtain a number called *the expectation value* of the stress-energy.

The geometry of spacetime – the gravitational field – enters, because it affects the expression for the stress-energy operator. This is because matter interacts with gravity, and this interaction contributes to the matter's stress-energy. As a result the expectation value of the matter's stress-energy depends on the geometry of spacetime.

One then uses this expectation value as a replacement for the classical stress-energy of matter in the Einstein field equations. The plain-language caricature of the field equations go from this

essential curvature = stress-energy of matter (1)

to this

essential curvature = <stress-energy of matter>, (2)

where the angle brackets denote expectation value, i.e. <X> is read "the expectation value of X". Because we are only concerned with the vacuum value, we will take <X> to mean "the vacuum expectation value X".

Both sides of equation caricatured by (2) depend on the geometry of spacetime. It is possible in principle to find a geometry that satisfies this equation. In practice this is difficult. Each side of the equation depends nonlinearly on the geometry. Moreover, each side depends on the geometry's rate of change in each dimension. This means that equation (2) is very complicated. One reduces its complexity by restricting the spacetime solutions to be of a particular class. In particular, one could require the spacetime to be a wormhole. The equation then becomes an equation for the parameter functions that specify particular elements of this class of wormhole geometries.

In order for a solution to exist it must be the case that near the wormhole's throat the vacuum expectation value of the stress-energy simulates that of exotic matter. There the vacuum of the matter fields must at least violate the weakest of the energy conditions.

In 1998 Hochberg, Popov, and Shushkov were the first to solve the equation. They used a scalar field as the matter field. For their choice of the interaction strength between the scalar field and gravity, they found quantum wormhole solutions with submicroscopic throats. These were no larger than a few hundred Planck lengths. A Planck length is a unit of distance invented by John Wheeler that represents the smallest distance – about 10^{-35} meters -- meaningful to physics.

Two years later S. Krasnikov found another wormhole solution by augmenting the matter field. To the massless scalar field he added fields corresponding to the neutrino and to electromagnetism. In this way he was able to demonstrate the existence of a class of quantum wormholes whose throats were not restricted in size.

P49. *The application of semiclassical gravity indicates that violations of energy conditions by the vacuum state of quantum matter fields are sufficient to sustain a wormhole in special cases.*

Minisuperspace Wormholes

In a way, we already possess a quantum theory of gravity. We can write down a master equation whose solution would tell us as much about the quantum states of the gravitational field as the Schrödinger equation tells us about the quantum states of atoms. The problem is that we cannot solve this master equation. This is frustrating. It would be as if we had the Schrödinger equation but lacked the ability to solve partial differential equations.

This master equation arises in what is known as the *canonical* approach to the quantization of gravity. It was first proposed by Wheeler in 1962 and thoroughly analyzed by DeWitt four years later. Hence it has come to be known – as mentioned in Chapter 3 -- as the *Wheeler-DeWitt equation.* While there has been progress toward solving this equation and an analogously formulated equation (through the use of what are called *Ashtekar variables*), the Wheeler-DeWitt equation remains intractable. It is not without controversy. It describes the quantum state of a fundamentally four-dimensional entity, space-time, in the language of three-dimensional geometries being stacked in time. It provides no clear definition for what this time variable should be. The proper choice for its boundary conditions is unclear. It suffers from the *factor ordering problem* – ambiguity in the ordering of equation's noncommuting operators. Many believe it to be inferior to its integral counterpart -- the path integral approach favored especially by workers in Euclidean quantum gravity.

Nevertheless, it is safe to say that if it were as easy to solve the Wheeler-DeWitt equation – exactly or approximately – as it is to solve the Schrödinger equation, we would be in possession of a quantum theory of gravity.

P50. *The Wheeler-DeWitt equation and its underlying assumptions would constitute a full theory*[§] *of quantum gravity were it not for the virtual impossibility of solving the equation in the domains of interest.*

The Wheeler-DeWitt equation is very much like the Schrödinger equation (actually, it more closely resembles the relativistic Schrödinger known as the Klein-Gordon equation). Recall that a solution to the Schrödinger equation tells you the proportion of universes (of the multiverse) in which a particle is located at a particular position -- or, in Copenhagen language, the probability of a particle being at a particular location. Solutions to the Wheeler-DeWitt equation tell you the proportion of universes that have a particular spatial

[§] The theory, however, would preclude the birth of new universes and not include matter.

geometry. The solutions to the Schrödinger equation are functions of position. The solutions to the Wheeler-DeWitt equation are functions (actually, functionals) of spatial geometry. The set of all possible spatial geometries is called *superspace.*

The principle that underlies the Schrödinger equation is the conservation of energy. The corresponding principle for the Wheeler-DeWitt equation is *reparameterization* invariance – the idea that physics does not depend how one chooses to slice spacetime into sequential spacelike hypersurfaces for the purpose of analysis.

The reason for bringing up the Wheeler-DeWitt equation is that we can use it to glimpse at a fully quantum description of a wormhole. Were we able to actually solve the Wheeler-DeWitt equation we would be treated to more than a glimpse. We would gaze upon a detailed artifact of a theory of quantum gravity. We are not afforded this view, however, because we are only able to solve the Wheeler-DeWitt equation for an idealized system with only the weakest resemblance to the wormhole that we wish to understand.

To understand the relationship between a system, whose Wheeler-DeWitt equation we know how to solve, and the system in which we are interested, it helps to recall an old physics joke: A farmer's chicken has stopped laying eggs. He consults with a psychologist, a biologist, a chemist, and physicist about the problem. The psychologist begins his analysis of the problem as follows, "Assuming an untraumatized chicken ...". The biologist's analysis starts with "Assuming a disease-free chicken ...". The chemist begins with, "Assuming a pH-balanced chicken ..." Finally, the farmer asks the physicist for his analysis. The physicist begins by saying, "Assuming a spherically symmetrical chicken ..."

In order to solve the Wheeler-DeWitt equation we are driven to an assumption even more absurd than that of a spherically symmetrical chicken. Instead of assuming in accordance with reality that there are an infinite number of ways in which space can be distorted, we must assume that there is only one way. We will assume that the only way in which the space constituting our wormhole can vary is in the size of the radius of its spherical throat. The Wheeler-DeWitt equation reduces in this case to an ordinary differential equation of the sort that we know how to solve arbitrarily well.

This restriction on the domain of the Wheeler-DeWitt equation -- to a wormhole whose geometry is fixed except for its radius – means that the equation now describes a tiny sliver of superspace, the infinite-dimensional space of all three-dimensional spatial geometries. This one-dimensional sliver of the infinite-dimensional superspace is called *minisuperspace.* This name applies as well to any finite-dimensional subspace of superspace.

In the late 1980s Matt Visser performed the first minisuperspace analysis of a quantum wormhole. He found an exact ground state wave function and was able to approximate the wave functions for excited wormhole states. He discovered that

> The expectation value for the wormhole radius is on the order of a Planck
>
> length
>
> The exotic matter at the wormhole's throat is quantized and on the order of the Planck mass.
>
> The value of the wave function tends to zero for wormholes whose radii are zero or much larger than a Planck length.

In short, the quantum mechanics of a wormhole with one degree of freedom is not entirely different from that of a hydrogen atom (Figure 6.1). This atom's

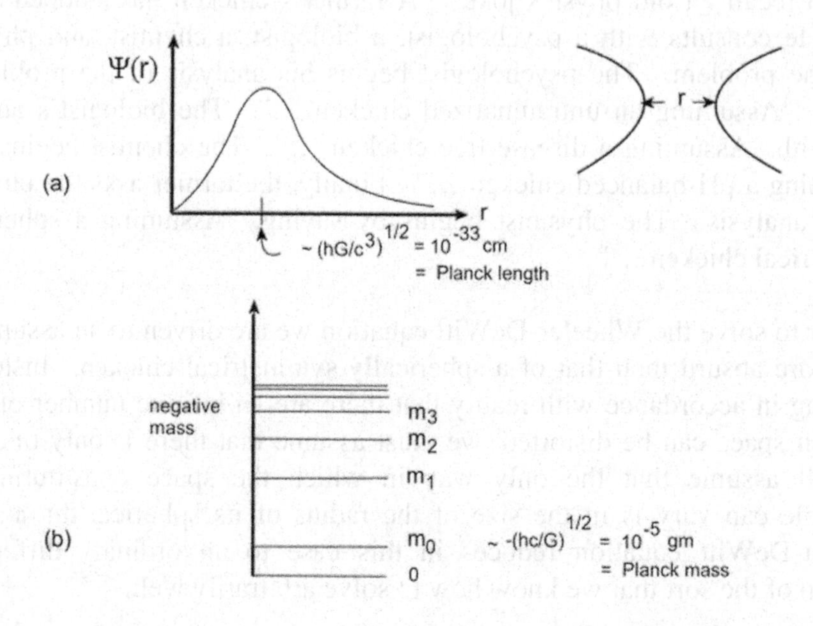

Figure 6.1. Minisuperspace quantum wormhole. (a) Domain of the wave function is reduced from superspace (all spatial geometries) to minisuperspace (a set of geometries that are fixed except for the size of their radii). Wave function is well behaved at r = 0, r = infinity, and peaks near 1 Planck length. (b) Spectrum of allowed masses are on the order of the Planck mass.

electron does not loiter at the center of the atom, nor does the electron flee to infinity. It assumes intermediate values. So it is with the radius of the wormhole. Visser believed that this stable character of the wormhole's wave function argues against the possibility of topology change.**

However, a similar analysis by Ian Redmount of the University of Wisconsin and Wai-Mo Suen of Washington State University reached a different conclusion. Beginning with a classical wormhole geometry different from Visser's choice, they were unable to find well behaved wave functions. They concluded that quantum wormholes are unstable in that they are prone to expand to an arbitrarily great extent – i.e. inflate. It appears, however, that this instability can be traced back to a difference from Visser in their model of the thin-shell wormhole and its equation of state. If one's classical model permits the accelerating expansion known as inflation, its quantization will demand it. What were impenetrable barriers to motion in the classical system become penetrable, once the system has been quantized. When this occurs, the quantum system is said to have "tunneled" through the barrier. The presences or absence of such barriers turns on the choice of the equation of state for the classical system.

Another complication in solving the Wheeler-DeWitt equation is the "nonlocality of the system's Hamiltonian". What this means is that the time evolution operator for the thin-shell wormhole system (for whatever arbitrary choice one has made for a time variable) cannot in general be expressed in terms of operators and functions that are defined at a single point. This makes the Wheeler-DeWitt equation far messier and more difficult to solve than its counterpart in elementary quantum mechanics. Nevertheless, familiar approximation methods of great use in quantum mechanics may be profitably brought to bear. They allow us to conclude that

P51. *The minisuperspace quantization of thin-shell models results in quantum wormholes characterized by radii on the order of a Planck length and quantized exotic matter on the order of a Planck mass.*

P52. *The likelihood that a quantum wormhole will tunnel into a state in which it undergoes inflation depends upon the equation of state of exotic matter.*

** Although other approaches to quantum gravity such as string theory permit topology change.

Minisuperspace Wormholes as Particle Models

Wheeler reintroduced wormholes into physics with the intention of using them as a model for elementary particles. Einstein and Rosen before him considered their "bridge" – a Schwarzschild wormhole – with a similar idea in mind. However, the minisuperspace treatment of quantum wormholes dashes any hope of such a model succeeding.

The reason is not hard to see. We have noted that the quantum states of minisuperspace wormholes have masses that are roughly that of a Planck mass. One Planck mass weighs about as much as a couple of dozen fleas. This does not sound like much, but, compared to the masses of other systems requiring a quantum treatment, it is enormous. It is, for example, over 10^{19} times the mass of a hydrogen atom. This vastly exceeds the mass of the heaviest known elementary particle.

Physicists today do not seriously entertain the idea that elementary particles are manifestations of quantum wormholes. However, a few of them have discovered a means of addressing the greatest drawback of the idea – the mass discrepancy noted above. In so doing, they have protected a flicker of hope for wormhole particle models.

In 1992 Lloyd Alty and Peter D'Eath of the University of Cambridge working with H.F. Dowker of Fermilab showed how the mass discrepancy can be eliminated. Drawing on earlier work of D'Eath with D.I. Hughes, they demonstrated that the key technique was to go beyond the consideration of quantum wormholes in mere gravity and to consider them in the context of *supergravity*. Supergravity is an example of what is known as a *gauge theory*.

A gauge theory is a way of describing the force between particles as being a consequence of a symmetry of nature. Imagine that you are in a sealed box floating in space. You perform an experiment. Your box is rotated by some angle. You perform the experiment again. You get the same result. You also notice that the results of your experiments are unchanged between states of uniform motion of your box. These are examples of symmetries of nature – transformations under which the laws of physics (the results of experiments) are unchanged. Because these transformations are fixed, not depending on your box' position in spacetime, they are called *global* symmetry transformations.

Suppose now that instead of merely rotating your box once by a fixed angle, you have the rotation angle of your box change with time and space. That is, as your box moves through spacetime, its rotation angle is a function of your box' position in spacetime. Suppose, moreover, that your motion in space is no longer uniform, that your velocity is also a function of your position in spacetime. These position-dependent transformations are called *local* symmetry transformations. You would find that your repeated experiments will no longer have the same result. The reason is clear. Your box is now rotating and changing its velocity – it is accelerating in some arbitrary way. That acceleration induces forces that are disrupting your experiments. Suppose that you would like to formulate a physical law that predicts your experimental results. You would need a law that is not only invariant under global symmetry transformations but local transformations as well. That law would have to include a description of the acceleration-induced forces. This generalized law that includes the new force in order to remain valid under local symmetry transformations is an example of a gauge theory. It is in particular a gauge theory of gravity.

So a gauge theory results when a physical law known to be invariant under certain global symmetry transformations is generalized so that it will be invariant as well under the corresponding local symmetry transformations. These symmetry transformations may involve more than mere rotations and changes in velocity. Rotations and velocity changes only act on one set of a particle's properties – those related to its existence in spacetime. But particles may also be said to have an additional existence, an existence in various *internal* spaces. An internal space is just a way of describing additional characteristics of a particle. Consider a human being. She has physical characteristics and personality characteristics, the latter being internal. Those internal characteristics could be described graphically in a sort of personality space. Suppose that this personality space consists of an x-axis that indicates a person's degree of optimism and a y-axis that marks the strength of her sense of humor. Then any point in this internal space specifies a particular personality. Although elementary particles do not have personalities, they do have characteristics that can be similarly specified through the use of an internal space. These characteristics typically involve various types of charge, of which the familiar electric charge is the simplest example. We saw above that a gauge theory of gravity results from demanding the invariance of physical law under local transformations that occur in our ordinary external space. Gauge theories of other forces result when we demand the invariance of physics under local transformations that occur in various types of internal space. These other forces include electromagnetism and the color force that binds quarks together within subnuclear particles.

In the early 1970s physicists working independently in the Soviet Union and in the West devised a symmetry transformation under which fermions and bosons appear to be different aspects of the same particle. Recall that fermions are particles with half integral spin (1/2, 3/2, ...), and bosons have integral spin (0, 1, ...). Bosons tend to be carriers of force, while fermions tend to be the particles on which these forces act. This new symmetry is what came to be known as supersymmetry. It can be described as a set of transformations in an extension of spacetime that includes an internal sector. This extended spacetime is called, unfortunately, *super-space*[††]. It should not be confused with the set of all spatial geometries, which goes by the same name. The gauge theory that results from requiring physics to be invariant under local supersymmetric transformations is called supergravity.

Supergravity, then, is a generalization of gravity. As such it possesses a generalized Wheeler-DeWitt equation. As in the pure gravity case, this equation can be specialized to consider wormholes with a single degree of freedom – the radius of its throat. When Alty, D'Eath, *et al* did this, they found quantum wormhole states of zero mass. By treating wormholes as distortions in super-space, as opposed to ordinary spacetime, they were able to avoid quantum wormholes with gigantic Planck-scale masses. Zero is, of course, the wrong answer for the mass of most elementary particles. It is, however, vastly closer to the right answer than is the Planck mass, indicating that supersymmtry might be a promising approach.

Unfortunately, supersymmetry has a problem. It has not been observed in nature. It predicts that for every boson with spin n ($n > 0$), there exists a so-called *superpartner* particle with spin n - ½. For example, there should exist for the photon with spin 1 a superpartner with spin ½. It is called a *photino*. No such particle has ever been observed. Physicists suppose, then, that supersymmetry is "broken". What they mean by this is that the symmetry holds approximately. The higher the energy used to probe the particles exhibiting the symmetry the more perfectly will it be evident. Suppose, for example, that when you face north, all of your characteristics are unchanged. But when you rotate to face south, one of your characteristics changes – your mass increases by 1%. You would describe yourself as being approximately symmetrical under rotations. Your rotation symmetry is broken, but not by much. As a consequence of this broken symmetry you would find that your freedom of rotation would be restricted. You would not be able to rotate from

[††] I have used hyphenation to distinguish this super-space from that previously defined.

facing north to facing south unless you were imparted with energy equal to 1% of your mass. Energy is conserved. You could not gain mass without acquiring it from somewhere. In order to rotate, you must be zapped with an energy beam from which you would absorb the needed mass. If your energy beam generator was insufficiently powerful to provide you with 1% of your mass, we would never see the southward-facing version of you. This southward-facing version is analogous to the photino, the northward-facing version to the photon. Physicists conjecture that we do not observe the photino, because it is exceedingly massive. We do not, they assume, currently possess the means of imparting the energy required to induce a supersymmetric transformation of a photon into a photino. The Large Hadron Collider -- a new particle accelerator near Geneva, Switzerland -- might change this.

There is another possibility that explains our failure to observe the photino and other superpartners. They might not exist. Nature might not in fact respect supersymmetry.

P53. *Unlike quantum wormholes in pure gravity, supersymmetric quantum wormholes need not have Planck-scale masses.*

7. Wormhole Traversability

Defining Traversability

Wormhole traversability has come to mean that the wormhole is repeatedly traversable as defined in Chapter 3, that it can be traversed within a year, and that it can be constructed by a civilization with a conceivable level of technological advancement. We will exclude the constructability requirement. We want our definition to include any naturally occurring wormholes whose construction would exceed the capacity of any conceivable civilization. We will say, then, that a wormhole is traversable, if it meets the following criteria:

1. It is not surrounded by an event horizon.

2. It allows travelers to pass through it without being harmed by gravitational tidal forces.

3. It allows travelers to pass through it without being harmed by radiation.

4. It allows travelers to pass through it without being harmed by interaction with wormhole matter.

5. It is stable against perturbations caused by the act of traveling through it.

6. It is possible to pass through it in a time that seems short (a small fraction of the human life span) to both the human traveler and to distant stationary human observers.

7. It endures for a time very much longer than the time required to pass through it.

P54. *Traversable wormholes cannot have event horizons, as these prevent two-way travel.*

The Fundamental Traversability Condition on Wormhole Matter

Imagine light rays converging on a traversable wormhole from all directions. They will enter the wormhole and reach its throat. As they cross the throat, the rays will be traveling parallel to each other. As they continue on their way out of the wormhole, they will diverge from each other. In general relativity light rays will only diverge, if they are traveling through a spacetime that has been curved by negative stress-energy. Just as positive stress-energy causes light to bend toward it, negative stress-energy causes light to bend away from it. For simple matter -- matter free of internal energy fluxes or stresses and whose properties are independent of direction (a so called "perfect fluid") – it is possible to quantify its associated stress-energy. It is given by

$$\text{stress-energy} = \text{pressure} + \text{density}. \qquad (1)$$

For ordinary matter -- such as iron, water, or air – this quantity is positive. For matter in the form of electromagnetic radiation it is zero. Matter is defined to be *exotic* with respect to the Null Energy Condition (NEC), if its stress-energy defined above is negative. Notice that the density of this exotic matter need not be negative. Its stress-energy can be negative as a result of sufficiently negative pressure. Equation (1) indicates, then, that it is possible to have negative stress-energy as a consequence of a negative pressure that more than cancels a *positive* mass-energy density.

We are of course supposing that any wormhole traversable by light can also be traversed by massive particles. Although this is likely true for static wormholes, it is possible for a dynamic wormhole to contract too rapidly for it to be traversed by such a particle. To ensure that beams of massive particles experience the same anti-gravitating repulsion at the throat of the wormhole as that described above for rays of light, we might insist on a negative mass-energy density at the throat. This would violate the Weak Energy Condition (WEC) without necessarily violating the NEC. Hence, violation of the WEC was initially put forward as the essential condition for wormhole traversability. WEC-violating matter with a negative mass-energy density is often referred to simply as "negative energy". However, because we are normally interested in static traversable wormholes, and because any matter that violates the NEC necessarily violates the WEC, we define the traversability of wormholes in terms of the NEC.

To summarize,

P55. *Exotic matter has anti-gravitating effect that causes a bundle of parallel light rays to diverge.*

P56. *That parallel light rays diverge as they cross a wormhole's throat implies the existence there of exotic matter.*

Another way of saying that light rays passing through a traversable wormhole must encounter negative stress-energy is to say the wormhole's matter must violate the NEC. We saw in Chapter 3 that the NEC requires that the sum of the energy density and pressure of matter in the wormhole to be nonnegative. This sum, which we have used in equation (1), is actually the particular combination of components of the stress-energy tensor that determines the path of light rays. We have

P57. *The Null Energy Condition must be violated by matter at or near a wormhole's throat in order for it to be traversable by massless particles (e.g. light).*

P58. *Wormhole traversability is defined in terms of the condition for the traversability of light rays (violation of the NEC) instead of that for massive particles (violation of the WEC), because NEC violations ensure WEC violations and the distinction is unimportant for static wormholes.*

Although NEC violations are required for a wormhole to be traversable by light, it must be traversable by light if it is to be traversable by humans. Hence, an NEC violation is only a necessary but not a sufficient condition for a human-traversable dynamic wormhole.

We see, then, that exotic matter must be found at a wormhole's throat. But how does this matter ensure that wormhole can be traversed? It holds the wormhole open.

P59. *It is the gravitational repulsion due to exotic matter that (in the absence of perturbations) prevents static traversable wormholes from collapsing.*

Suppose that the matter near the throat of a wormhole is not pure. Some of it is exotic, some of it is not. Would this wormhole be traversable? It depends. Yes, if the effect of the exotic matter exceeds that of the normal matter. No, otherwise. In other words, it would be traversable by light if on *average* the NEC is violated along the light's path. Or, more precisely, the wormhole cannot be traversable unless the average value of the stress-energy encountered by light rays is negative. For traversability by massive objects as well, the WEC must on average be violated along the object's path. This, then, is the fundamental traversability condition:

P60. *The Averaged Null Energy Condition must be violated by matter at or near the throat of a static wormhole in order for it to be traversable by both massless and massive particles..*

Achieving the Fundamental Traversability Condition

To create a traversable wormhole, we need matter that violates the Null Energy Condition. In essence we need matter with a negative energy density. It was once thought that such matter could only be obtained by exploiting quantum effects known to produce negative energy. Exploiting these effects was further complicated by constraints that seemed to limit severely the way in which negative energy could be extracted from the gravitationally altered vacuum of quantum matter fields.

It has recently come to light that NEC-violating matter is not as hard to come by as was once thought. There are straight forward ways of obtaining the required negative energy densities from ordinary classical matter fields. We will in turn describe both quantum and classical methods for creating NEC-violating (exotic) matter.

Quantum Methods

The Casimir Effect

The vacuum state of the electromagnetic field consists of an infinite number of virtual photons briefly coming into existence. Each of these photons can be described as a wave with a particular wavelength.

Like any electromagnetic wave, these waves may be subject to boundary conditions – in particular, they vanish at the surface of conductors. Consider two parallel conducting plates separated by a distance L, as shown in Figure 7.1. The only virtual photons that can exist between the plates are those that satisfy the boundary condition imposed by the presence of the plates. Because these waves are sinusoidal, this boundary condition requires half-multiples of their wavelengths to equal to L. Virtual photon waves not meeting this condition cannot exist between the parallel plates. Such photons are not excluded from the regions on the outer sides of the plates.

Each virtual photon carries energy. The vacuum energy is obtained by summing the energy of all of the virtual photons. Fewer photons between the plates mean less vacuum energy between the plates relative to the vacuum

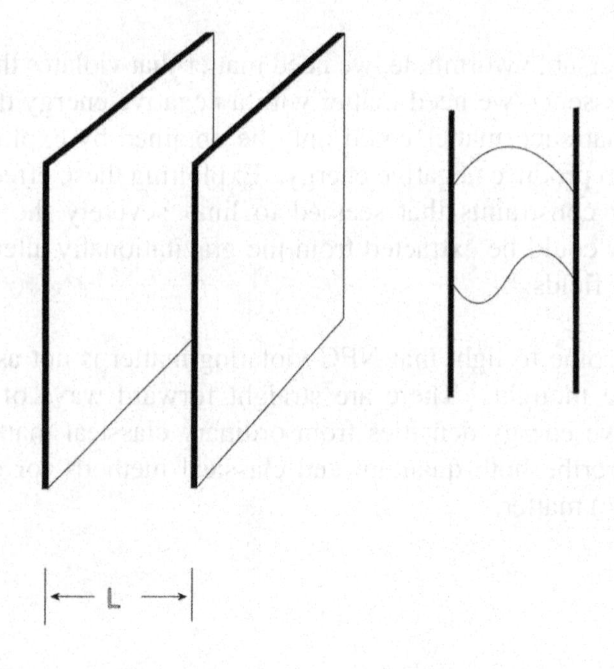

Figure 7.1. Casimir plates. Parallel conducting plates attract. Moving the plates toward each other lowers the vacuum energy between the plates by excluding longer wavelength modes of the virtual photons of the electromagnetic field.

energy in the outer regions. Because the vacuum energy in the outer regions is (an infinite) constant, we can subtract it away. This makes the effective vacuum energy of the outer regions zero and that between the plates negative.

Virtual photons whose wavelengths are longer that 2L cannot fit between the parallel plates – i.e. they do not meet the boundary conditions. The shorter the separation L between plates, the greater the number of virtual photons that are unable to fit between them. The difference, then, between the vacuum energy between the plates and that of the outer regions increases with decreasing L. The effective energy between the plates becomes more negative as the plates are brought closer together. This creates an attractive force, which was measured in 1996 to vary inversely with the fourth power of plate separation, precisely as the Dutch physicist Hendrick Casimir had predicted in 1948.

In 1988, Morris, Thorne, and Yurtsever proposed using the Casimir effect to generate the negative energy required to hold open a traversable wormhole. They conceived of two concentric spherical conduction shells. In order to create sufficient negative energy, they found that the separation between the shells needed to very much smaller than a hydrogen atom but larger than its nucleus.

P61. *Negative energy densities generally occur whenever variable boundary conditions are imposed that restrict the presence of virtual particles within a region.*

The Squeezed Vacuum

It is possible to use the phenomenon of quantum optics known as *squeezed light* to suppress fluctuations in the electromagnetic vacuum in one region at the cost of increasing them elsewhere. The averaged energy of the electromagnetic vacuum within the entire volume to which this procedure is applied would be unchanged. However, the energy density of regions of this volume within which vacuum fluctuations are suppressed would be lower than that of the normal vacuum. As the energy density of the normal vacuum is defined to be zero, the energy density of these regions would be negative. The energy removed from these regions of suppressed fluctuations would be found in adjacent regions whose energy density would be positive – exceeding that of the normal vacuum.

P62. *Regions of negative energy density occur in a squeezed vacuum.*

Hawking Evaporation

In 1974 Stephen Hawking of Cambridge University realized that quantum theory requires black holes to evaporate. The rate of evaporation progressively increases until the black hole vanishes in an explosive burst. This evaporation occurs because the quantum vacuum at the event horizon of a black hole contains ephemeral pairs of particle and anti-particles. One member of the pair can fall into the event horizon, while the other escapes to infinity. The apparent violation of energy conservation is resolved by ascribing negative energy to the in-falling particle that compensates for the positive energy with which its erstwhile partner escaped to infinity. The ultimate source of the negative energy is the black hole's strong gravitational field.

This process, then, results in negative energy being sent into the black hole. If the process continues, the black hole evaporates into nothing, as its mass becomes zero. We might speculate, however, that something else might happen. Sean Hayward of Ewha Women's University in South Korea has shown recently that a wormhole can be created from a maximally extended Schwarzschild black hole by bombarding it with negative energy pulses. While actual negative energy pulses are currently difficult to produce, the Hawking process can be used instead to send negative energy into the black hole. We can speculate that positive energy pulses could be used to damp the explosive terminal phase of the black hole's evaporation in such a way as to convert the black hole into a wormhole. The greatest obstacle to realizing this scheme is that it requires the manifestation of a maximally extended black hole solution. Such a solution would not pertain to naturally occurring black holes resulting from stellar collapse. Moreover, the solution is known to be unstable. It would appear, then, that the first step in realizing this scheme would be to create a suitable black hole, which is essentially impossible.

P63. *Negative energy densities occur in the vacuum of quantum fields in curved spacetime and in particular near the event horizons of black holes.*

Constraints on Quantum Methods -- Quantum Inequalities

The appearance of negative energy in general relativity has been regarded traditionally as wholly unphysical. Attempts to expel negative energy from physics have taken the form of various "energy conditions". These were constraints on the stress-energy tensor that were intended to ensure that the only physically reasonable matter would be considered. They were used as axioms from which important theorems of general relativity were derived, including the singularity theorems, the positive mass theorem, and the topological censorship theorem. The problem with energy conditions is that they are flagrantly violated by the most modest attempts to add quantum matter fields to general relativity. The energy conditions needed to be replaced by new rules that hold for quantum fields. In 1978 Lawrence Ford suggested such rules, which have since been refined and have come to be known as quantum inequalities.

The quantum inequalities effectively restrict the amount of negative energy to which an observer can be exposed. They require an exposure to negative energy to be followed by a compensating exposure to positive energy, whose magnitude and duration exceeds that of the negative energy exposure. Moreover, the greater the delay between the negative- energy exposure and that of the compensating positive energy, the greater must be the magnitude of the latter. This increase in the required magnitude resulting from a delay in compensation is called *quantum interest.*

In 1996 Ford and Thomas Roman applied the quantum inequalities to wormholes. They discovered that the inequalities effectively forbid macroscopic wormholes. An attempt to create a macroscopic wormhole would, according to the inequalities, require negative energy to be somehow concentrated in a shell far thinner than the effective diameter of the smallest elementary particle. The inequalities also set a lower bound for the absolute value of the negative energy density. This forces the minimum total negative energy required to hold open the traversable wormhole – the product of this lower bound and the volume of the thin shell – to be astronomically great.

In 2000 Serguei Krasnikov showed how the effective ban due to the Ford-Roman constraints on the existence of traversable wormholes can be circumvented. By relaxing the assumed conditions on the wormhole's spacetime – replacing the requirement for asymptotic flatness with one for merely increasing flatness – he was able to find a traversable wormhole solution whose negative energy was sourced by the quantum vacuum of three matter fields. Three years later, by abandoning the assumption of spherical symmetry, he was able to drastically reduce (by 34 orders of magnitude!) the negative energy required to sustain a traversable wormhole.

Six years later, Douglas Urban and Ken Olum, both of Tufts University, showed that --given the assumption of a fundamental scalar field suitably coupled to the curvature of spacetime -- the averaged null energy condition can be violated to any degree desired. Such ANEC violations, which occur in spite of the quantum inequalities, would permit the formation of traversable wormholes. A minor catch is that this would only be possible on conformally flat spacetimes – those that are obtained from flat spacetime by multiplying it by a position-dependent scaling function.

P64. *The restrictions imposed by the Quantum Inequalities can be circumvented. They do not prevent existence of traversable wormholes sustained by the negative energy from the vacuum of quantum matter fields.*

Classical Methods

In 1996 it was common wisdom that negative energy density resulted solely from quantum effects. An early indication that this was not the case was noticed in 1997 by Dan Vollick of Okanagan University College. He considered ordinary matter interacting via an ordinary scalar field. Even though the matter field and the scalar fields separately satisfied the NEC, the interaction between them did not. That is to say, positive energy matter coupled through a positive energy field produced a *negative* energy interaction. He showed that this negative energy could be used to sustain a wormhole.

More recently, Carlos Barceló and Matt Visser at Washington University realized that a scalar field that interacts directly with spacetime curvature (called a non-minimally coupled scalar field) can violate the NEC. They found a class of wormhole solutions that are sustained by the associated negative energy. They found, moreover, that each of these wormhole solutions can be made – by adjusting the parameter determining the strength of the scalar field's coupling to spacetime curvature -- to violate even the less stringent Averaged Null Energy Condition (ANEC). This condition requires the NEC to hold only on average over any path taken by a light ray (i.e. over any null geodesic).

The work of Barceló and Visser also made it clear that the scalar field of the sort assumed to drive cosmological inflation could also generate negative energy. It would do so by suitably interacting with itself. Hence there are three closely related classical methods for violating the NEC:

P65. *Certain classical scalar fields can be used to create negative energy densities through particular types of interactions with other fields, with spacetime curvature, or with themselves.*

Significance of the Classical Methods

We see, then, that an ordinary classical scalar field can create regions of indefinitely sustained negative energy density. So what? This development matters, because it results in an upheaval of known physics in three ways:

1. The assumptions underlying important theorems of General Relativity cease to be true, thus invalidating the theorems themselves.

2. Time machines that can be created from traversable wormholes are no longer prevented from existing due to the practical impossibility of exotic matter.

3. The second law of thermodynamics is violated, thus allowing perpetual motion machines to exist.

We will examine each of these in turn.

The Loss of the Singularity Theorems and their Brethren

Between 1965 and 1970 Roger Penrose and Steven Hawking proved several theorems that showed that if a spacetime contains a trapped surface and certain other reasonable conditions hold, then the spacetime contains a singularity.

A *trapped surface* is a closed surface for which outgoing light rays traveling perpendicular to it converge. And incoming light rays traveling perpendicular to it also converge. All of the closed surfaces, for example, within the event horizon of Schwarzschild black hole are trapped surfaces.

A singularity is a region of space time – a point, curve, or surface -- where physics breaks down. More specifically, it is a place at which geodesics (paths of free-falling particles) terminate after the elapse of a finite amount of proper time. It is also a place from which geodesics can begin; *anything* may emerge from a singularity.

The "other reasonable conditions" assumed by the theorems include a condition to the effect that all matter must not be "unphysical". Matter was characterized as having an energy density and three spatial pressures, one for each spatial direction. Matter was further assumed to meet the following requirements: 1) The sum of the energy density and that of the spatial pressures must be non-negative. 2) The sum of the energy density and that of each spatial pressure must be non-negative. These requirements taken together constitute the poorly named Strong Energy Condition or SEC.

The Strong Energy Condition implies the Null Energy Condition. If the NEC is violated – as it is for certain classical scalar fields – then the SEC is violated. If SEC is violated, the singularity theorems no longer hold.

We can no longer assume, then, that a singularity lurks within every trapped surface or that the universe itself will collapse into a big crunch of a singularity or that it began with a big bang singularity, as has been supposed since at least 1970.

In addition to the loss of the Singularity Theorems, the violations by classical matter of the NEC and the Averaged NEC imply the loss of other cherished theorems of general relativity including:

The Positive Mass Theorem – The energy of any asymptotically flat spacetime must be positive.

The Topological Censorship Theorem – All topologically non-simple regions of spacetime are cloaked beneath an event horizon.

The Second Law of Black Hole Mechanics – The surface area of a black hole's event horizon can never decrease.

The violation of the NEC, the weakest of all pointwise energy conditions, implies a violation of all other pointwise conditions. In addition to the SEC, the Dominant Energy Condition (DEC) and Weak Energy Condition (WEC) are violated. The violation of the ANEC similarly implies a violation of the Averaged Strong Energy Condition (ASEC), the Averaged Dominant Energy Condition (ADEC), and the Averaged Weak Energy Condition (AWEC). The proofs of virtually all major theorems of general relativity depend upon the assumption that at least one of these energy conditions is never violated.

P66. *Physically reasonable classical matter fields can be found to violate all known energy conditions. Hence, all theorems of general relativity whose proofs assume these conditions – including the Singularity Theorems – are inconsistent with physical reality.*

P67. *The assumptions of the energy-condition-dependent theorems of general relativity were already known to be violated by quantum effects. Their viola-*

tion by classical matter merely eliminates the possibility that the theorems apply when quantum effects are negligible.

The Appearance of Time Machines

If a classical matter field can be used to create effectively exotic matter without the restriction imposed on quantum effects, the existence of traversable wormholes ceases to be quite as far fetched. Given their existence, however, it is extremely easy (at least conceptually) to use them to create time machines. The existence of time machines threatens the notion of causality, a fundamental idea in physics.

While it is possible that the universe conspires, perhaps through quantum effects, to prevent the existence of time machines – which is Stephen Hawking's Chronology Protection Conjecture, this has yet to be established.

P68. *Unless the Chronology Protection Conjecture is true, the existence of traversable wormholes is tantamount to the existence of time machines.*

The existence, moreover, of the closed timelike curves implied by the existence of time machines violates another of the assumptions required to prove the last of the Penrose-Hawking Singularity Theorems. This further undermines the theorem.

The Apparent Violation of the Second Law of Thermodynamics

The Second Law of Thermodynamics implies that it is impossible without an expenditure of energy for heat to flow from a cold object to a hot object. A scalar field capable of producing fluxes of negative as well as positive energy could be used to violate this rule. Imagine a solution to the scalar field equations in which a beam of negative energy bombards a cold object, while an equally intense beam of positive energy irradiates a hot object (Figure 7.2). The negative energy would cool the cold object. The positive energy would warm the hot object. The net effect would be a transfer of heat from the cold object to the hot object without an expenditure of energy. Energy is not expended, because the positive energy of the positive beam is precisely canceled by the negative energy of the negative beam.

P69. *Violations of the NEC by classical matter fields appear to imply violations of the Second Law of Thermodynamics.*

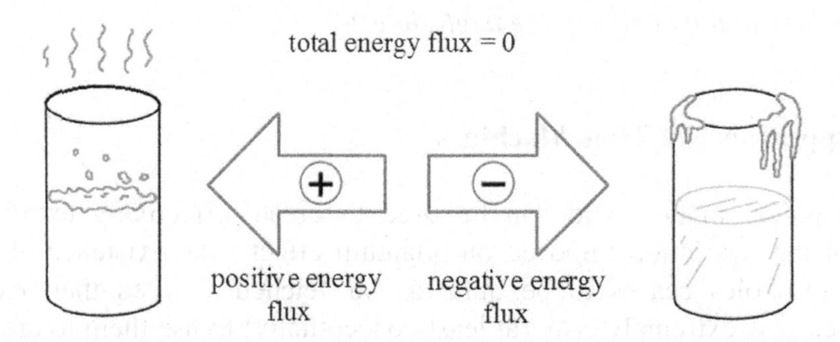

Figure 7.2. Negative energy fluxes violate Second Law. The possibility of a negative energy flux of a scalar field violates the Second Law of Thermodynamics by transferring heat from a cold object to a hot object. Energy is conserved during the transfer, however, because the positive energy beam and the negative energy beam are of equal intensity.

Achieving the Remaining Traversability Conditions

Ensuring that wormhole matter violates the Null Energy Condition is a necessary condition for traversability, but it is not sufficient. We must still satisfy the six criteria that define traversability:

1. No event horizon.
2. No harmful radiation.
3. No harmful tidal forces or accelerations.
4. No harmful interactions with wormhole matter.
5. Stability.
6. Short traversal time.
7. Endurance.

Criteria 1 and 2 are satisfied by ensuring that our wormhole is not also a black hole. This is guaranteed by the presence of sufficient exotic matter, which eliminates event horizons and the radiation-emitting singularities within them.

Criterion 5 is discussed in the next chapter.

Criterion 7 might at first be confused with criterion 5. They are different, however. Consider for example a large wave in the middle of the Pacific Ocean. The wave endures for, say, a half hour. During that time it is completely stable against perturbations caused by its interaction with ships. It nevertheless dissipates after only thirty minutes. Such a wave would demonstrate the independence of stability and endurance. For this wave stability would be high and endurance low. By contrast, we could imagine a wave that would in the absence of perturbations endure for twelve hours, but in their presence immediately collapse. This would be an example of low stability with high endurance.

By only considering wormholes that are static, we can ensure that criterion 7 is met. We should nonetheless retain an explicit cognizance of this criterion and our strategy for its satisfaction. It might turn out to be far easier, for example, to create a wormhole that endures for a mere century, than it would be to create one that endures forever. The economics of such a case would compel us to relax our tentative requirement for unlimited endurance.

Criteria 3, 4, and 6 remain. In discussing these and other traversability constraints we should be aware of the following:

P70*. As a consequence of the Einstein field equations, constraints on wormhole geometry also constrain wormhole matter.*

Limiting Tidal Forces

Tidal forces are what you would experience as you fall in a non-uniform gravitational field. The greater the non-uniformity (i.e. the spatial gradient) of the field, the stronger the tidal forces. Imagine that you are falling toward a wormhole. Its gravitational field, like that of any other object, decreases with the square of the distance from it. Suppose that you are falling toward the wormhole feet first. The gravitational field at your feet is stronger than that at your head. This means that your feet are attempting to accelerate toward the wormhole faster than is your head. The difference in the accelerations of your feet and your head is directly proportional to the tidal force to which you are exposed. If this tidal force much exceeds your weight, you will die. [Recall that execution by hanging operates on a similar principle.]

In addition to the radial tidal force described above, there is a transverse tidal force that acts in directions perpendicular to radial lines. Imagine two baseballs falling toward the wormhole. Suppose that they are the same distance from the wormhole, and that they are a few meters from each other. As the baseballs fall in toward the wormhole, they follow radial lines that converge. Imagine a light spring connecting the baseballs. The spring compresses as the baseballs fall inward and thus converge on each other. The force compressing the spring is the transverse tidal force. It acts on any inward falling object including the body of a traveler. Should a traveler stop moving in a radial direction, however, he will cease to experience a transverse tidal force. When he resumes his radial motion, the faster he moves, the stronger will be the transverse tidal force acting on him.

In order for a wormhole to be traversable by a human traveler, the tidal forces that he will experience as he passes through it must be too weak to harm him. When this constraint is applied to the radial tidal force, it leads – in the case of a Morris Thorn wormhole -- to a somewhat complicated inequality involving the wormhole's shape function and its red shift function. In other words for a given choice of shape function, the red shift function cannot be chosen arbitrarily but must satisfy this inequality.

When the constraint is applied to the transverse tidal force, it imposes a limit on the maximum speed with which a traveler may safely pass through the wormhole.

P71. *Tidal forces can be restricted to levels tolerable by humans, when a wormhole's geometry is chosen so that its red shift and shape functions are related in a particular way.*

Limiting Traversal Time

The speed limit imposed by the need to restrict transverse tidal forces (in order to protect the traveler) is used to determine the minimum time required to traverse the wormhole from Universe *A* to Universe *B*. For the wormhole to be of use to humans in uniting universes (or distant regions), the traversal time must be a small fraction of the human lifespan. The traversal begins in Universe *A* at a position where the gravitation due to the wormhole is no stronger than it is at the surface of the earth. The traversal continues until a similar position is reached in Universe *B*. The traveler's motion is further constrained by requiring that he not be crushed by excessive acceleration of his spaceship. This latter constraint turns out – at least in the case of a Morris Thorne wormhole – to impose a relationship between the wormhole's red shift function and the traveler's radial speed. Despite these constraints Morris and Thorne were able to find examples of wormhole with traversal times on the order of seconds, hours, and months. Hence, despite the imposition of four conditions, the problem is not over constrained:

P72. *The constraints on radial and transverse tidal forces and that on the traveler's acceleration do not prevent the constraint on traversal time from being met.*

Limiting Interaction with Wormhole Matter

A spherically symmetrical wormhole solution requires a traveler to pass through a shell consisting of both normal and exotic matter. Even the presence of normal matter at sufficiently high densities can pose a problem for the traveler. For exotic matter, the nature of its coupling to the normal matter of the traveler's own body or spaceship is unknown. It might well be exceedingly painful even at low densities. For this reason human-traversable wormholes should allow travelers to pass through them without coming into physical contact with any of the wormhole's supporting matter.

Morris and Thorne proposed that this could be achieved by perturbing a spherically symmetric wormhole. They would puncture their wormhole (and any concentric, negative-energy-generating Casimir conducting shells that it might contain) with a matter-free tube just wide enough to support the clear passage of a human traveler. Unfortunately, this modified wormhole solution no longer solves the Einstein field equations exactly.

Matt Visser proposed instead that spherical symmetry be completely abandoned. This allowed him to devise thin-shell wormholes – guaranteed to be solutions to the Einstein equations – that permitted traveler's to pass through without directly interacting with wormhole matter. The traveler would only interact with such matter through its effect on the curvature of spacetime. As with all thin-shell wormholes, the nature of the matter required to generate the required densities and pressures remains unspecified. Nevertheless, we see that

P73. *It is possible to find wormhole solutions to Einstein's equations in which travelers do not directly encounter wormhole matter – exotic or otherwise.*

1998 – The Year the Universe Changed

A Big Surprise

Traversable wormholes cannot exist without exotic matter to hold them open. The prevalence, then, of exotic matter determines the degree to which macroscopic traversable wormholes are more than hypothetical. In 1998 measurements of supernova luminosities forced two independent groups of experimenters to reach a startling conclusion. They found that, contrary to the conventional belief, the expansion of the universe is not slowing down. It appears instead to be *accelerating*. This, as we shall see, is the first indication that the universe might actually be awash in exotic matter.

It might turn out, of course, that such cosmic exotic matter – called phantom energy or superquintessence – has nothing to do with the accelerating expansion. Observations, however, have yet to rule this out. It remains one of several possible explanations that physicists continue to consider. One class of these explanations, of which phantom energy is an example, assumes that the invisible *dark energy* – of which 70% of the universe is believed to be composed – exerts a *negative* pressure. In general relativity, negative pressure has an antigravitating effect. It acts to promote the expansion of the universe. If the pressure is weakly negative, its antigravitating effect will be exceeded by the gravitating effects of the corresponding matter's positive energy density. The expansion of the universe will slow. If the pressure is strongly negative, its antigravitating effects will win out over the effect of the concomitant energy density. The expansion of the universe will accelerate. Whether the universe will accelerate or not depends, then, on the negativity of the pressure for a given density of exotic matter. That negativity is quantified by the ratio w of pressure to density. Because the cosmic exotic matter is assumed to be extremely dilute, we can use this ratio to express the relevant equation of state,

$$\text{Pressure} = \boldsymbol{w} \times \text{density} \quad .$$

According to the rules of relativistic cosmology, w must be less than -1/3 for the expansion of the universe to accelerate. We may classify the leading negative-pressure dark energy models according to their associated values of w as follows.

Model	*w* range
quintessence	$-1 < w < -1/3$
cosmological constant	$w = -1$
phantom energy	$w < -1$

The quintessence models assume the existence of a scalar field that generates negative pressure. This pressure is a consequence of the field not having yet reached its minimum vacuum energy, which it slowly approaches but will not reach for perhaps billions of years. The cosmological constant approach exploits the slight modification to general relativity that Einstein used in his misguided attempt to stabilize the universe against expansion. Phantom energy models, of interest to us because phantom energy is actually exotic matter, are still consistent with observational bounds on the lower limit of w. One class of phantom energy models even predicts that the exoticity of phantom energy increases with time (i.e. w becomes progressively more negative). Were this idea to pan out, the task of collecting exotic matter for wormhole construction would ease with each passing epoch. However, observational data seems to indicate that the accelerated expansion of the universes is most likely due to a cosmological constant. If dark energy actually is phantom energy, it is just barley so, with w being only slightly less than -1.

From Big Bang to Big Rip

An ever accelerating expansion of the universe poses a problem. To understand this, we must consider an observed feature of the universe called *Hubble's Law*. In the mid 1920s it was known that the frequency of the light that we receive from distant galaxies is shifted downward in the direction of the red end of the spectrum. In 1929 Edwin Hubble discovered that this *redshift* of a galaxy's light is proportional to the galaxy's distance from us. Hubble then supposed that a galaxy's redshift was due to the component of its velocity in the direction away from earth. In other words, he guessed that the redshift is just a motion-induced Doppler shift. As such the redshift is proportional to the aforementioned component of velocity. Hubble's Law, then, is that the speed at which a galaxy recedes from Earth is proportional to its distance from Earth. More simply, it is that the rate of separation is proportional to the separation, i.e.

$$\text{separation rate} = H \times \text{separation}$$

where H is known as the *Hubble constant.*

The thing about Hubble's Law is that it does not just apply to the separation between Earth and distant galaxies. It applies to the separation between any two bodies. The reason that we never notice this effect in our quotidian affairs is that the Hubble constant is exceedingly small: 70 kilometer/second per *megaparsec*. A megaparsec is over 10^{19} kilometers.

The problem that results from an accelerating expansion is that it causes the Hubble "constant" to increase. When the acceleration is due to quintessence or a cosmological constant, this increase is such that the Hubble constant never exceeds a maximum value. But when the acceleration is due to phantom energy, the Hubble constant increases without end. This makes Hubble's Law a death sentence.

The Hubble constant will, for example, become so large that the rate at which it causes the earth and the moon to separate will exceed the rate at which gravity causes them to fall toward each other. The Earth-Moon system will be ripped apart. By the time that this happens, however, systems involving greater separations, and therefore greater susceptibility to the effects of Hubble's Law, will have long since been ripped to shreds. Our local group of galaxies, our galaxy itself, and our solar system will no longer be gravitationally bound. As the Hubble constant continues to increase, the ripping effects of Hubble's Law will be noticed at progressively smaller separations. The earth will be ripped into little chunks. The passengers of the space ark that left Earth in a futile attempt to escape will only have bought themselves a few minutes, before their ship suffers the same fate. Those that donned space suits will survive for another few seconds, before being rent asunder. Shortly thereafter, the ripping will destroy every molecule of their corpses, then every atom, then every nucleus, and every subnuclear particle.

Robert Caldwell of Dartmouth together with Marc Kamionkowski and Nevin Weinberg, both of Caltech, published this scenario for cosmic doom in 2003. They call it, appropriately, *The Big Rip*. It is the natural fate of a universe dominated by exotic matter. While it is possible to devise phantom-energy-featuring models that that avoid this fate, they tend to be more complicated and might to some seem a bit contrived.

===/===

For wormhole physics, then, the greatest implication of the events of 1998 is that exotic matter might not in fact be exotic. It might be the most common constituent of the universe.

8. Wormhole Stability

Before we explore the subject of wormhole dynamics in general, let us first consider this subject's most important question. Are traversable wormholes unstable? A wormhole solution to Einstein's gravitational field equations is said to be unstable if a slight perturbation of its spacetime metric or its matter field ultimately leads to the wormhole's collapse or to its runaway expansion. Investigations of wormhole stability fall into two categories that correspond to the two types of wormholes considered. The first considers the usual *thick-shell* wormholes in which the matter is distributed within an extended region centered about the wormhole's throat. The second probes the stability of *thin-shell* wormholes in which matter is concentrated on infinitesimally thin surfaces or line elements.

Thick-Shell Wormholes

A consensus has recently emerged concerning the stability of the potentially traversable wormhole solutions of the thick-shelled (i.e. smooth) sort that have been the focus of most recent research. It appears to be that these solutions are unstable. The following table summarizes the conclusions extant in the early literature.

Authors	Year	Scalar Field	Perturbation	Stable?
Armendáriz-Picón	(2002)	massless ghost massive ghost	metric & field metric & field	yes in some cases
Shinkai & Hayward	(2002)	massless normal massless ghost	field field	no no
Bronnikov & Grinyok	(2002)	massless normal*	metric & field	no

*Non-minimally coupled

The scalar fields in these researches had either intrinsically positive energy, in which case I call it "normal", or intrinsically negative energy, in which case it's called a "ghost". The scalar field considered by Bronnikov and Grinyok differed from the others in that it was non-minimally coupled, which is to say that it was more strongly connected to the gravitational field by means of a direct coupling to the field's curvature scalar. The discrepancy between the results of Armendáriz-Picón and those of the others might be due to his restriction to linear perturbation and to the absence of any normal matter in his wormhole. In the latter case the tendency for his wormhole to contract would be weakened. The absence of positive energy matter would mean that his wormhole's contraction could only be driven by the geometry's gravitational self interaction. This could have improved stability to the point where instabilities could only have been detected had the analysis retained terms nonlinear in the perturbation, which it did not.

In appears, then, that a smooth seemingly traversable wormhole is a delicate balance between the attraction due to its positive energy matter (and geometrical self interaction) that attempts to collapse it inward, and the repulsion due to its exotic matter that attempts to expand it outward. If a slight amount of negative-energy matter is sent into the wormhole, it could expand outward indefinitely. If a similar quantity of positive energy matter is sent in, the wormhole could collapse to a black hole. The wormhole's geometry can be perturbed in such a way as to induce either result.

Nevertheless, there are conditions under which thick-shell wormholes can be stable. These require the existence of exotic matter whose gravitational repulsion is sufficiently strong at high densities to counteract the wormhole's tendency to inflate, yet sufficiently weak at low densities to prevent the wormhole from inflating. An example of such exotic matter might be the cosmic fluid conjectured to be responsible for the accelerating expansion of the universe. Even in this case, the wormhole is in general unstable. It is only stable for throat radii within a certain range and for similarly bounded values for the exotic matter's speed of sound.

P74. *Thick-shell wormholes are in general unstable. They can be stable, however, if they contain sufficient exotic matter that obeys a peculiar type of equation of state.*

Thin-Shell Wormholes

Studies of thin-shell wormholes seem to confirm these findings. It is difficult, however, to imagine how thin-shell wormholes could come into existence either naturally or artificially. They are *bona fide* solutions to the gravitational field equations, but they are characterized by discontinuities in the gradient of the metric and corresponding infinite curvatures and matter densities. These infinities, while seemingly problematic, are actually innocuous in that spatial integration over them is finite. In other words, the regions of infinite density and curvature are actually *infinitesimally* thin. Such an idealization is as reasonable as the familiar "surface charge" and "point particle" of elementary physics.

In order to "create" a thin shell wormhole, proceed as follows (Figure 8.1). Consider two identical spacelike hypersurfaces – one for Universe A, the other for Universe B. Cut identical holes – of any shape in both hypersurfaces. Glue the hypersurfaces together at the boundaries of the holes. A traveler may now pass from Universe A, across this boundary, and into Universe B.

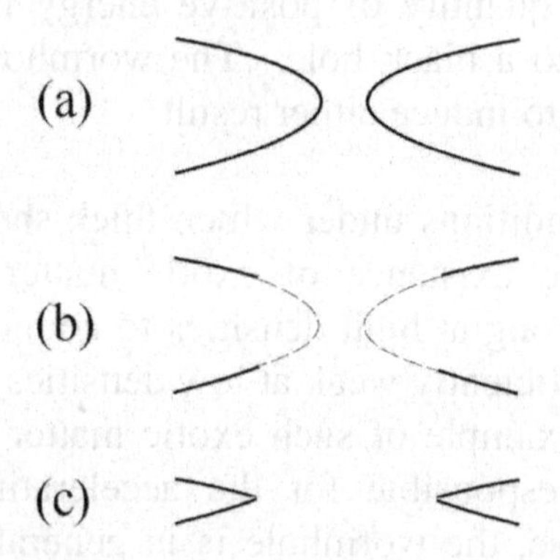

Figure 8.1. Creating a thin-shell wormhole from Schwarzschild solution. (a) Start with the Schwarzschild extended black hole solution. (b) Cut out everything within some radius greater than the event horizon. (c) Paste the remaining pieces together and enforce junction conditions to obtain a thin-shell wormhole with a characteristic kink at the throat.

Inserting the resulting geometry into Einstein's equations generates the distribution of energy and pressure due to matter required to create this geometry's curvature. Studies by Visser and others have shown that thin-shell wormholes can be stable – at least in principle. Their stability depends on ensuring that the wormhole possesses sufficient exotic matter to prevent it from collapsing.

It also depends on the particular way in which the density of exotic matter depends on its pressure. This relationship is known as an equation of state. The equation of state of exotic matter is unknown. It can be guessed, however. For certain seemingly unlikely guesses thin-shell wormholes are stable. The importance of this finding depends entirely on the reasonableness of these guesses.

While thin-shell wormholes can have any shape, the spherically symmetric ones are of greatest interest. These can be used to model the physics in the sort of wormholes that are presumed to occur naturally. These models are normally constrained to have a single degree of freedom – the radius of the wormhole's throat. This makes analyzing the stability of a thin-shell wormhole exactly like analyzing the stability of a particle subject to the influence of a potential field in one dimension. Determining the stability of such a particle is easy. Just graph the potential function under whose influence it moves. Draw a horizontal line to represent the total energy of the particle (Figure 8.2). Realize that the particle can only travel in regions where its total energy exceeds the value of the potential. Observe that the force on the particle at any point is the negative of the slope of the potential function. Obtain the kinetic energy of the particle at any point by taking the difference between the line of total energy and the potential function. Conclude that the particle behaves in this force field as if it were rolling on hills and valleys formed by the graph of the potential function.

The shape of effective potential function V(r) that governs the behavior of the wormhole's throat depends, as stated above, on the equation of state of exotic matter. It also depends on the geometry from which the thin-shell model was constructed. For example, Visser constructed one of his thin-shell models by cutting identical holes in the Schwarzschild geometries in two universes and pasting the holes together at their rims. This resulted in the thin-shell model retaining the tendency to collapse irreversibly when the wormhole's throat contracts within the radius of the Schwarzschild event horizon. This collapse to a radius of zero results as a consequence of the negative divergence of the potential functions at $r = 0$ shown in Figure 8.2c. At large radii, however, it is the equation of state of the wormhole's exotic matter that entirely determines the throat's behavior. For an equation of state corresponding to a sufficiently positive surface pressure due to this matter, the wormhole will inflate. The potential's character – its negative divergence with increasing throat size – forces this runaway inflation to continue and accelerate. The only thing that might prevent the inflating wormhole from devouring the entire universe is the universe's own expansion at a sufficiently rapid rate.

P75. *Thin-shell wormholes are in general unstable. They can be stable, however, if they contain sufficient exotic matter that obeys a peculiar type of equation of state.*

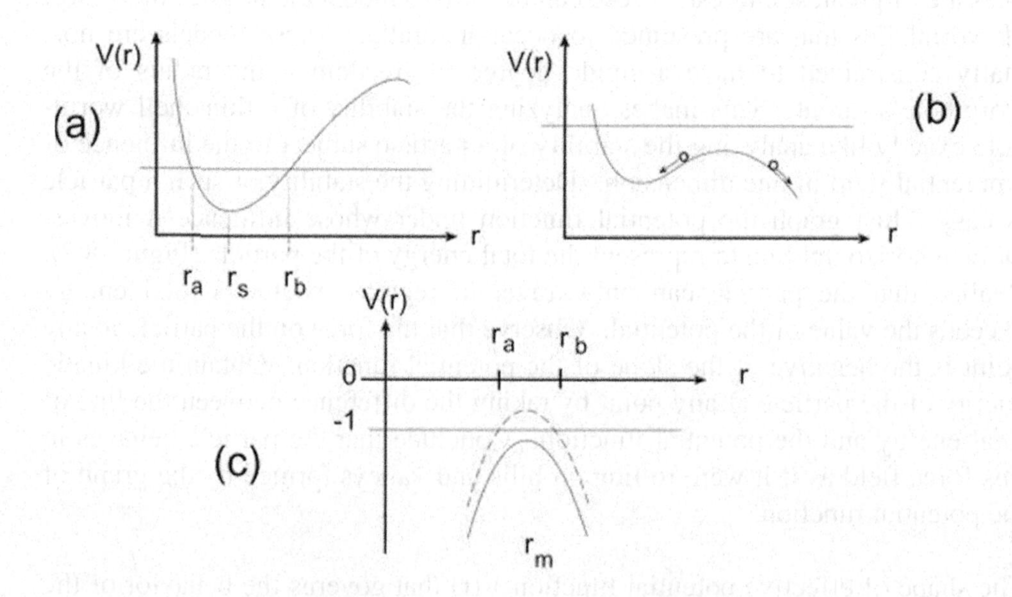

Figure 8.2. Stability and Potential. (a) A particle with total energy E in potential $V(r)$ may only be found between r_a and r_b. Because the potential at r_s is a minimum, the system is stable in that the particle oscillates about this point. (b) The motion of the system (particle or wormhole throat) can be determined by imagining the ball rolling on the potential graph under the influence of a uniform gravitational field. (c) $V(r)$ is the effective potential that arises in the Einstein field equation for a thin-shell wormhole, where r is the throat radius. Solid line shows a potential with a maximum at r_m where the wormhole is unstable against collapse or inflation. Dashed line represents another possible effective potential in which a throat radius of the thin-shell wormhole would have to either: be less than r_a and collapse to zero, or be greater than r_b and inflate to infinity. Radii between r_a and r_b would in this case be forbidden.

9. Wormhole Dynamics

What is wormhole dynamics?

Wormhole dynamics is the study of the ways in which wormholes change with time. What it means for a wormhole solution, or any other spacetime, to change with time requires a bit of explanation. Recall that the definition of time in general relativity is tantamount to the slicing of spacetime into a sequence of spacelike hypersurfaces. Recall that a particular choice of such hypersurfaces is called a foliation. Each spacelike hypersurface of the foliation may be assigned a number, which partially defines a coordinate system in which events located in a particular hypersurface may be said to occur at the same time.

We are free to foliate spacetime however we like. However, upon doing so, we will encounter a slight problem. For any foliation in which a given spacetime appears to be static it is always possible to choose another foliation in which this spacetime appears to change with time. In other words, for any foliation in which the spacetime metric is time-independent, there are other foliations for which the metric depends explicitly on the new foliation's time coordinate. This complicates our study of wormhole dynamics in that it leaves us unable to immediately distinguish static from dynamic spacetimes.

Fortunately, there is a solution to this problem. It relies on the existence of a special vector field called a *Killing vector field,* named after the German mathematician Wilhelm Killing (1847-1923). A Killing vector field is one along which the metric does not change. If this vector field is timelike, it is possible to define foliations of spacetime that are everywhere perpendicular (orthogonal) to the Killing field. The metric will not depend on the time-coordinates defined by this foliation. It will be manifestly static. Whether a spacetime is static, then, depends upon whether it possesses a globally timelike Killing vector field, i.e.

> A **static spacetime** is one for which there exists a globally timelike Killing vector field, whose existence permits a foliation in which the metric is manifestly time-independent.
>
> A **dynamic spacetime** is one that is not static.

The existence, then, of a foliation that makes the metric time-dependent does not necessarily imply a dynamic spacetime. A spacetime is dynamic only if it does not admit a globally timelike Killing vector field.

Defining a Wormhole's Throat

Now that we understand what it means for a spacetime to be dynamic, let us consider the simplest example of a dynamic wormhole. This is the case of a wormhole that is of the Morris-Thorne style except that the size of its apparent throat changes with time. I shall usually refer to such a wormhole as a "dynamic MT wormhole". For any foliation of this wormhole's spacetime, the apparent throat at a particular time is the surface of minimum area in the spacelike hypersurface that corresponds to that time. Were the wormhole static as defined above, this surface would be the familiar throat defined by Morris and Thorne.

I have used the term "*apparent* throat" for the aforementioned surface of minimal area, because in the dynamic case something unexpected happens. The throat bifurcates.

To understand this we must reconsider our definition of the throat. There are two rival definitions between which we must choose – our original geometric definition, or a functional definition. To define the throat geometrically, we identify it with the surface of minimal area for a particular spacelike hypersurface -- i.e. for a particular time. This is the apparent throat that, by assumption, now changes with time. To define the throat functionally, we identify it as the surface at which light beams are defocused – where they cease to converge and begin to diverge. In the static case the geometric and functional definitions identify the same throat. If the wormhole is allowed to change with time, however, this is no longer true.

To see how the geometric and functional definitions position the throat differently in the dynamic case, let us further examine the functional definition. To do so, we must consider how the expansion of null geodesics – the divergence of light beams -- depends on the geometry of spacetime. Imagine a sphere surrounding an empty region in flat space. Position two lasers outside the sphere at a distance several times its radius. Separate the lasers by a few meters, and aim them at the center of the sphere (Figure 9.1b). The beams emitted by these lasers converge at a constant rate as they approach the surface of the sphere, cross it, and continue on toward its center. Beams emitted outward by lasers positioned at the center of the sphere similarly *diverge* at a constant rate. This divergence is called an *expansion.* When the expansion is negative, it describes a convergence, which is sometimes called a *contraction.* The value of an expansion, denoted by θ, is determined in particular circumstances by what is known as the *Raychaudhuri equation.* We will use the following notation to further qualify solutions to this equation, where each item is defined at an unspecified surface.

$\boldsymbol{\theta_+}$ -- the expansion of outgoing beams perpendicular to the surface

$\boldsymbol{\theta_-}$ -- the expansion of ingoing beams perpendicular to the surface

$\mathbf{D}\boldsymbol{\theta_+}$ -- the change in θ_+ per unit distance that occurs in leaving a surface perpendicularly in an outward direction (1)

$\mathbf{D}\boldsymbol{\theta_-}$ -- the change in θ_- per unit distance that occurs in approaching a surface perpendicularly in an inward direction

The flat-spacetime situation described above – in which the expansion remains constant as beams perpendicularly intersect any sphere – differs from the situation that occurs in the presence of wormholes. As shown in Figure 9.1a, ingoing beams converge as they approach a static wormhole's throat ($\theta_- < 0$), travel in parallel as they cross it perpendicularly ($\theta_- = \theta_+ = 0$), and diverge as they leave it ($\theta_+ > 0$).

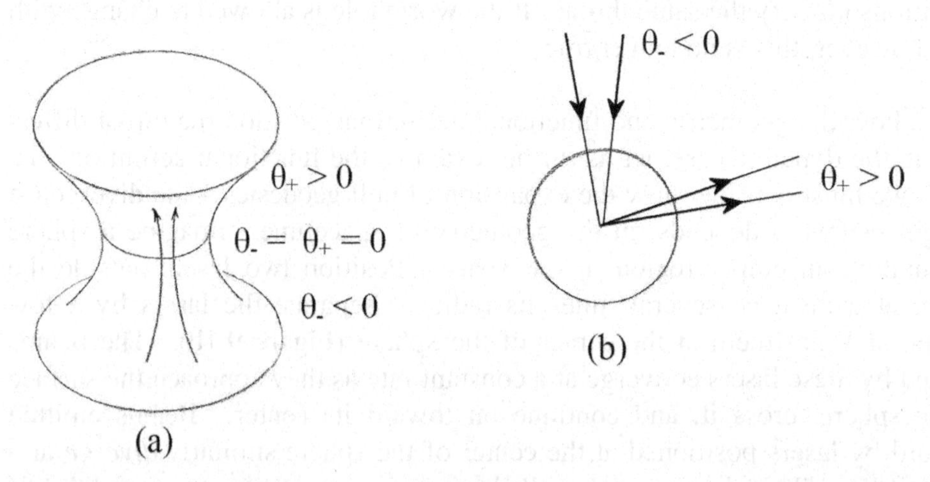

Figure 9.1. Divergence angle θ. (a) Light beams converge as they approach the throat of a static wormhole ($\theta_- < 0$), travel in parallel as they cross the throat ($\theta_- = \theta_+ = 0$), and diverge as they leave the throat ($\theta_+ > 0$). (b) Situation in flat space. Ingoing beams do not cease to converge as they cross any closed surface ($\theta_- < 0$). Nor do outgoing beams cease to diverge ($\theta_+ > 0$).

It will be useful to think about the expansion of null geodesics in another way. Consider a spherical surface in flat space. Suppose that omnidirectional light pulses are simultaneously emitted from every point on this surface, as shown in Figure 9.2. As the wave fronts of the light pulses -- emitted from the surface S_O from points such as *a*, *b*, *c*, and d -- travel outward, they create the surface S_+. As they travel inward, they create S_-. Immediately after the pulses are emitted, the relative areas of these surfaces depend on the value of the expansion θ on S_O as follows.

$$\begin{aligned}
\theta_+ > 0 &\Leftrightarrow \text{Area of } S_+ > \text{Area of } S_O \\
\theta_+ = 0 &\Leftrightarrow \text{Area of } S_+ = \text{Area of } S_O \\
\theta_+ < 0 &\Leftrightarrow \text{Area of } S_+ < \text{Area of } S_O \\
\theta_- > 0 &\Leftrightarrow \text{Area of } S_- > \text{Area of } S_O \\
\theta_- = 0 &\Leftrightarrow \text{Area of } S_- = \text{Area of } S_O \\
\theta_- < 0 &\Leftrightarrow \text{Area of } S_- < \text{Area of } S_O
\end{aligned} \qquad \textbf{(2)}$$

As Figure 9.2 makes clear, these relations imply that on any sphere in flat spacetime, $\theta_+ > 0$ and $\theta_- < 0$ (as was shown in Figure 9.1b). As mentioned above, matters differ in the vicinity of a static wormhole's throat. These relations allow us to interpret the aforementioned vanishing of θ_- and θ_+ at the throat itself in a slightly different way. Not only do beams emitted outward from the throat travel (for an infinitesimal distance) in parallel, the surface area of the associated outgoing wave front remains (for an infinitesimal distance) the same as that of the throat (i.e. $\theta_+ = 0$). We can similarly conclude that the area of the wave front produced by outgoing pulses emitted from a surface beyond the throat exceeds the area of this surface (i.e. $\theta_+ > 0$ just beyond the throat). And the area of an identically produced ingoing wave front from the same surface is less than the area of this surface (i.e. $\theta_- < 0$ just beyond the throat).

Irrespective of this interpretation of the meaning of expansion θ in terms of the changes in surface area of wave fronts, we may use the definitions (1) to more clearly express the functional definition of a static throat. A static wormhole throat is a surface at which perpendicularly incident light rays defocus, in other words a surface at which

$$\theta_+ = 0 \quad \text{and} \quad D\theta_+ > 0 \tag{3a}$$

$$\theta_- = 0 \quad \text{and} \quad D\theta_- > 0. \tag{3b}$$

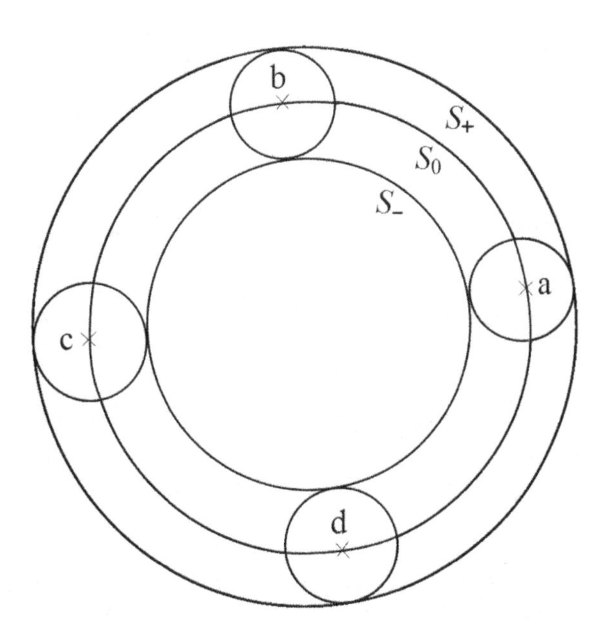

Figure 9.2. Ingoing and outgoing waves. Pulses emitted from the surface S_0 (at points such as a, b, c, d) create ingoing wave front S_ and outgoing wave front S_+.

That θ_- vanishes at the throat and increases as the throat is approached implies that θ_- is negative on approaching the throat. Hence, this definition is equivalent to the aforementioned sequence encountered by ingoing light as displayed in Figure 9.1a: contraction ($\theta_- < 0$), followed by a cessation of contraction in the absence of expansion ($\theta_- = \theta_+ = 0$), followed by expansion ($\theta_+ > 0$).

The Two Throats of a Dynamic Wormhole

We are now prepared to see how this functional definition bifurcates the throat of a dynamic wormhole. First consider the static case again, this time using Figure 9.3. The velocity vector of the light beams shown in this figure must be tangent to the surface of the embedding diagram. For an ingoing beam the radial component of this vector is negative – meaning inward (Figure 9.3a). This inward motion corresponds to a contracting wave front comprised of identical ingoing beams emitted from every point on a surface larger than the

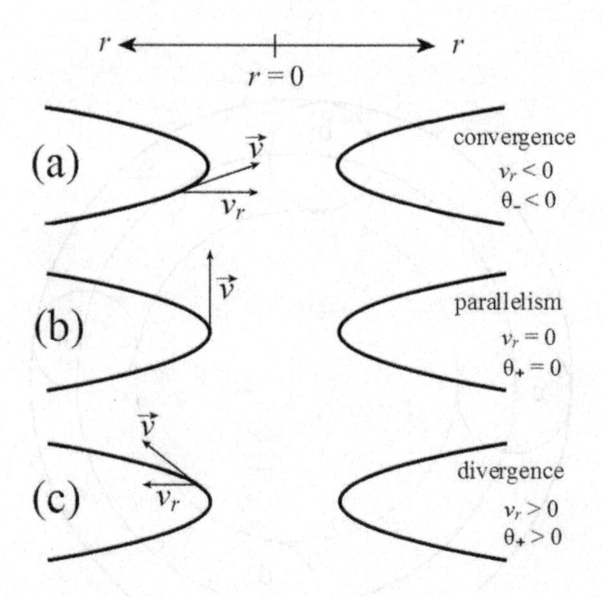

Figure 9.3. Throat of static wormhole. (a) Light rays that approach throat from "lower" universe converge. (b) Rays travel in parallel at the throat. (c) Rays diverge as they leave the throat. Radial component of v_r velocity vector in embedding space goes from negative (inward), through zero (at throat), to positive (outward).

throat, i.e. $\theta_- < 0$. As the beam reaches the geometrically defined throat, its radial velocity becomes zero (Figure 9.3b). This corresponds to a wave front that neither contracts nor expands, i.e. $\theta_- = \theta_+ = 0$. When the beam proceeds beyond the throat, its radial velocity becomes positive – meaning outward (Figure 9.3c). This corresponds to an expanding wave front, i.e. $\theta_+ > 0$.

Now consider the dynamic case, specifically the case in which the throat contracts with a fixed effective radial velocity (Figure 9.4). In this case a light beam's radial velocity is the sum* of two contributions. The first is that that results, as in the static case, from the radial motion of the beam within space. The second is the radial velocity due to the contraction of space itself. In the static case the throat is located where the light beam's radial velocity is zero. To apply the same definition to the dynamic case, we seek, then, the surface at which an outgoing light beam's *total* radial velocity vanishes. This occurs on the surface at which a light beam's outward radial velocity precisely cancels the inward radial velocity due to the wormhole's constriction as in Figure 9.4b. Notice that this occurs at *two* surfaces, one for each outward direction. This is the sense in which the throat, formerly located at the minimal surface, bifurcates in the dynamic case.

Notice as well that these throats meet the first functional definition of relations (3a). The vanishing of the radial component of the light beams total velocity indicates that an associated outgoing wave front at the throat has yet to begin its expansion. In other words, $\theta_+ = 0$ at the throats. Moreover, it is clear that the outgoing beams, upon crossing the throat, immediately attain a positive (outward) value for the radial component of their total velocity. This is tantamount to outward going, expanding wave fronts: $D\theta_+ > 0$. Unlike the static case, however, it is no longer true that $\theta_- = 0$ at the throats. A glance at Figure 9.4b reveals that the radial velocity of an ingoing light beam would reinforce that of the contracting throat. The resulting inward direction of the beam's total radial velocity would correspond to a shrinking area of the corresponding wave front. Or, by the last of the relations (2), $\theta_- < 0$. Hence, relations (3b) are *not* met by an expanding MT wormhole. However, as we will soon see, throats of a contracting MT wormhole do satisfy (3b).

These throats are the new locations at which violations of the energy condition occur in the wormhole. While this follows immediately from the Raychaudhuri equation, we can also see it intuitively. These throats are the only surfaces at which light beams defocus, the only places where the contracting wave fronts of outgoing pulses cease to contract and begin to

---

* Velocity addition rule of special relativity does not apply, because one of the velocities is due to the contraction of space itself.

expand. This suggests that the wormhole's exotic matter might no longer be concentrated at the surface of minimal area, which was the geometrically defined location of the throat in the static case. Rather, this matter might extend to shells centered at the functionally defined dynamic throats.

Similar comments apply to an expanding throat (Figure 9.5).

The Region between the Throats

The bifurcation of the throat in the dynamic version of the Morris-Thorne wormhole endows it with a feature unseen in its static sibling – an interior region that lies between the two throats. This region is characterized by surfaces from which the wave fronts of both outgoing and ingoing pulses contract, i.e. θ_+ and θ_- are both negative. We can further understand this region by considering the Penrose diagram of a dynamic wormhole. Figure 9.6a (cf. Figure 9.7a) shows such a diagram for a dynamic traversable wormhole. In this example the geometric throat changes it size, but the functional (outer) throat remains unchanged. This wormhole differs from the static Morris-Thorne class in that it 1) comes into existence at some finite time before $t = 0$, 2) expands its surface of minimal area at a finite rate, 3) ceases this expansion and begins a reversing contraction at $t = 0$, 4) continues to contract its surface of minimal area so as to reverse its original expansion, and 5) goes out of existence at a finite time after $t = 0$. This sequence might seem familiar. It is precisely the life cycle of the Schwarzschild wormhole, whose Penrose diagram is reproduced in Figure 9.6b (cf. Figure 9.7b). This similarity between a dynamic traversable wormhole and the maximally extended Schwarzschild solution suggested a connection. In the late 1990s this led Sean Hayward, then of Ewha Womens University in Seoul, to suppose that black holes and wormholes are different phases of the same object.

Unlike the travelers who enter the black hole interior of a Schwarzschild wormhole, those who enter the corresponding interior region of a dynamic MT wormhole are not trapped there. This is interesting because surfaces in this interior region are characterized, as mentioned above, by

$$\theta_+ < 0 \text{ and } \theta_- < 0. \qquad (4)$$

This condition happens also to pertain to the surfaces within a black hole's interior. It is the definition of what is called a *trapped surface*. This concept,

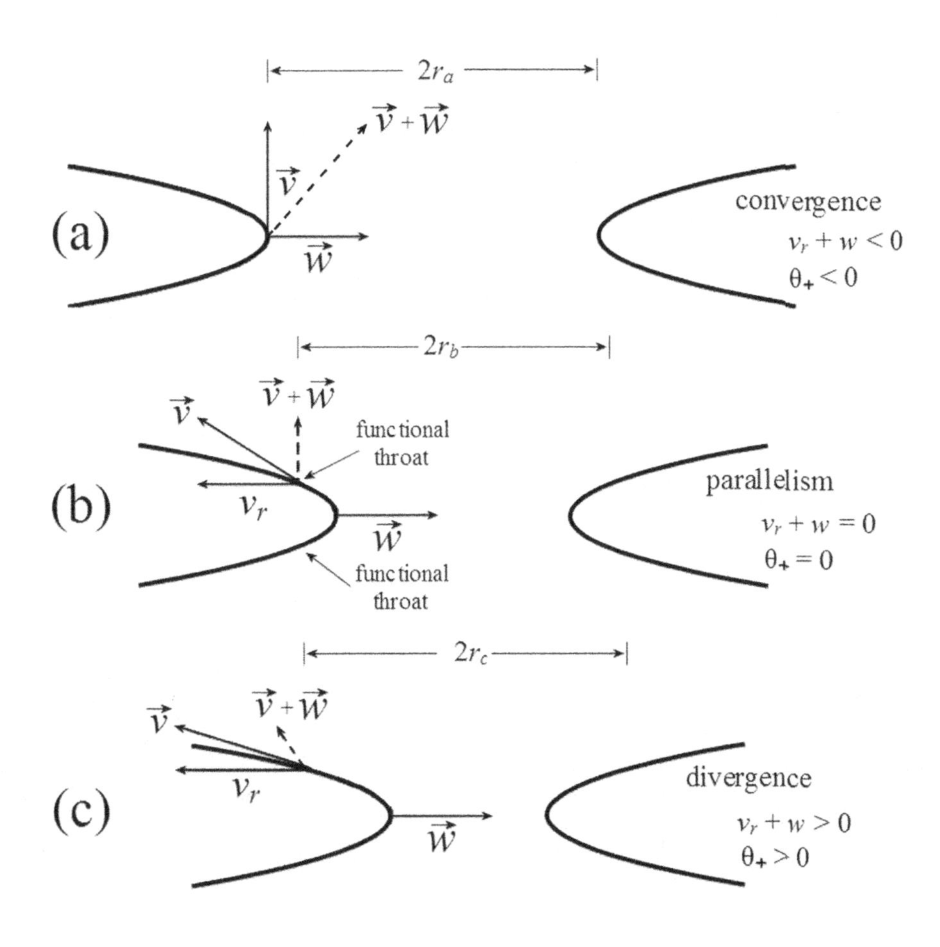

Figure 9.4. Throats of a contracting wormhole. Light ray is emitted from geometric throat of a wormhole contracting with effective radial velocity $\vec{w}$. Light has velocity $\vec{v}$ (with v = c). (a) Radial component of light ray's resultant velocity is $\vec{v} + \vec{w}$ is inward (negative) indicating convergence. (b) Radial component is zero indicating parallelism. (c) Radial component is outward (positive) indicating divergence. Surface area of outgoing composite wave front, decreases (from $4\pi r_a^2$ to $4\pi r_b^2$) until wave front reaches functional throat, then increases (from $4\pi r_b^2$ to $4\pi r_c^2$). [Note that the velocity addition rules of special relativity do not apply, because $\vec{w}$ represents the contraction of space itself.]

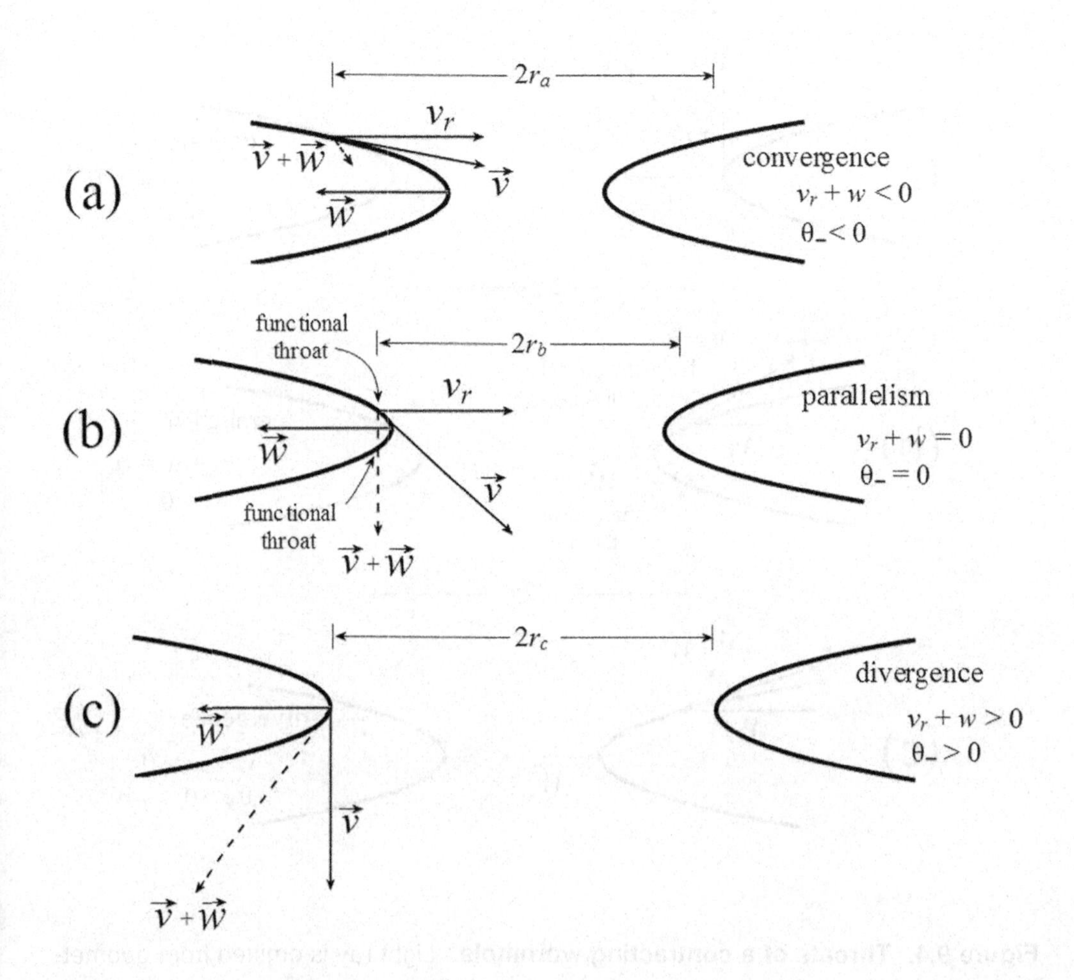

Figure 9.5. Throats of an expanding wormhole. Light ray approaches a wormhole that expands with effective radial velocity $\vec{w}$. Light has velocity $\vec{v}$ (with v = c). (a) Radial component of light ray's resultant velocity $\vec{v} + \vec{w}$ is negative (inward) indicating convergence. (b) Radial component is zero indicating parallelism. (c) Radial component is positive (outward) indicating divergence. Surface area of ingoing composite wave front decreases (from $4\pi r_a^2$ to $4\pi r_b^2$) until wave front reaches functional throat, then increases (from $4\pi r_b^2$ to $4\pi r_c^2$). [Note that special relativistic velocity addition rules do not apply, because $\vec{w}$ describes the expansion of space itself.]

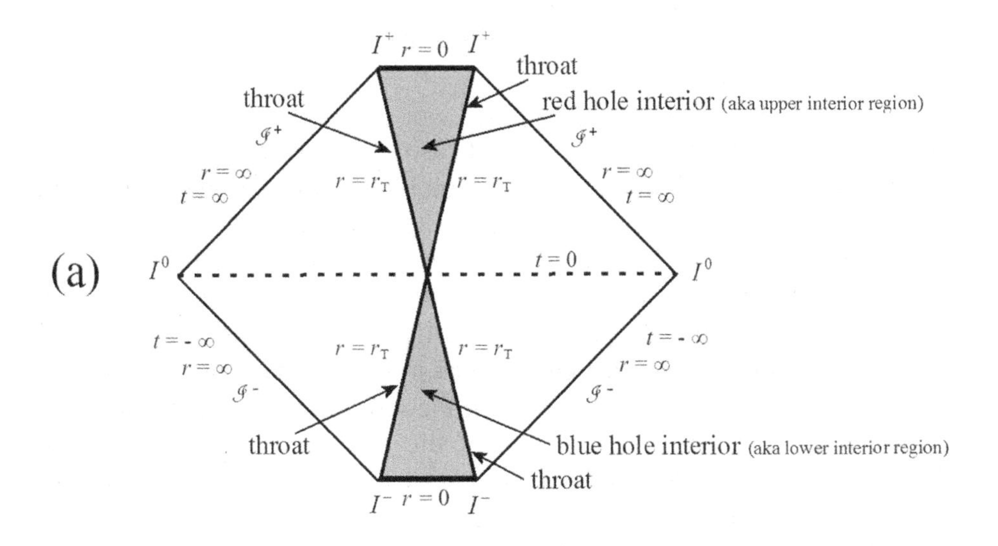

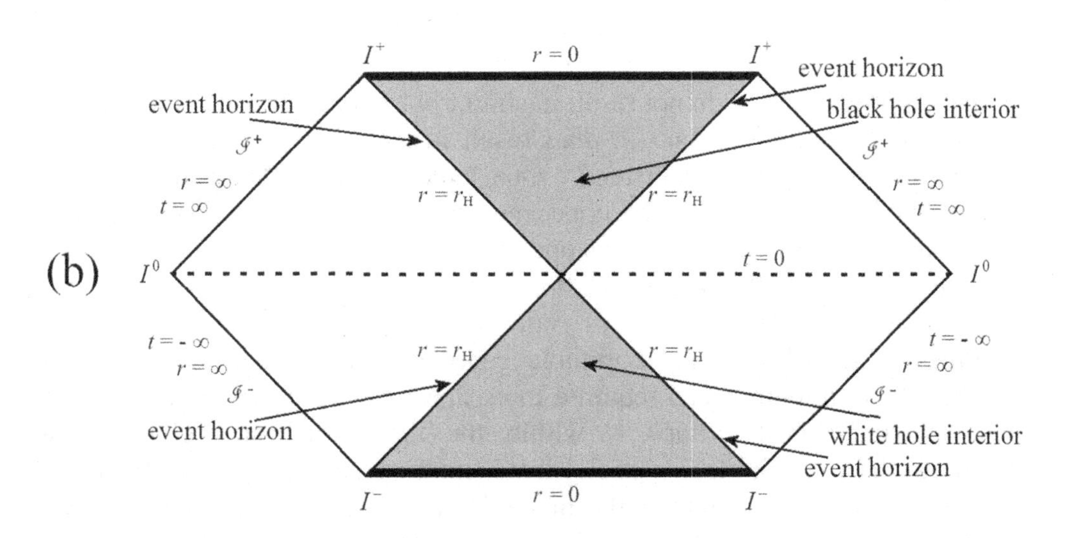

Figure 9.6. Traversable and untraversable dynamic wormholes. (a) Dynamic traversable wormhole expands from nothing, reaches maximum size at t = 0, contracts to nothing. Double throats border interior regions containing hindered surfaces. (b) Dynamic untraversable wormhole (a.k.a. maximally extended Schwarzschild solution) expands from nothing, reaches maximum size at t = 0, contracts to nothing. Double horizons border interior regions containing trapped surfaces.

due to Roger Penrose, defines a surface as "trapped" if the composite wave fronts of *both* outgoing and ingoing pulses emitted from the surface have less area than the surface itself. In other words, a surface is trapped if it is impossible to send from it a signal that will escape to infinity. How is it that the same condition -- relations (4) -- does not result in similarly trapped surfaces for the dynamic MT wormhole?

The answer is that, unlike the Schwarzschild case, it is perfectly possible in the case of a dynamic wormhole for light to escape its interior region even though the composite wave front (Figure 9.2) of outgoing light decreases. Take another look at Figure 9.4. Imagine that the wave front is launched from the surface of minimal area, as shown in Figure 9.4a. Because of the wormhole's contraction, the total radial component of the light rays comprising the wave front is *negative*. This will result in an inward-moving wave front whose area must decrease. This decrease continues until the radial component of the wave front's total velocity reaches zero at one of the throats, as shown in Figure 9.4b. Beyond this throat the radial component of the wave front's total velocity is positive. The wave front's area increases, then, as it speeds away from the wormhole, as shown in Figure 9.4c.

A further answer to the question of why relations (4) trap surfaces in the Schwarzschild case but do not result in similarly trapped surfaces in the case of a dynamic wormhole is this: It *does* result in a similarly trapped surface, if "similarly" is properly understood. One look at Figure 9.6a tells us that a particle emitted from within the upper interior region cannot escape to infinity unless it moves fast enough. The slope of its world line must be more horizontal than those of the boundaries of this region – the wormhole's throats. The slope of the throats' world lines reflects the rate at which the wormhole contracts – the slower the wormhole contracts, the steeper the slope, the smaller the minimum speed required to escape from its upper interior region. Just as condition (4) traps -- within the upper interior region of the Schwarzschild wormhole -- particles that cannot exceed the speed of light, this condition also traps -- within the upper interior region of the dynamic MT wormhole -- particles that cannot exceed the speed at which this wormhole contracts.

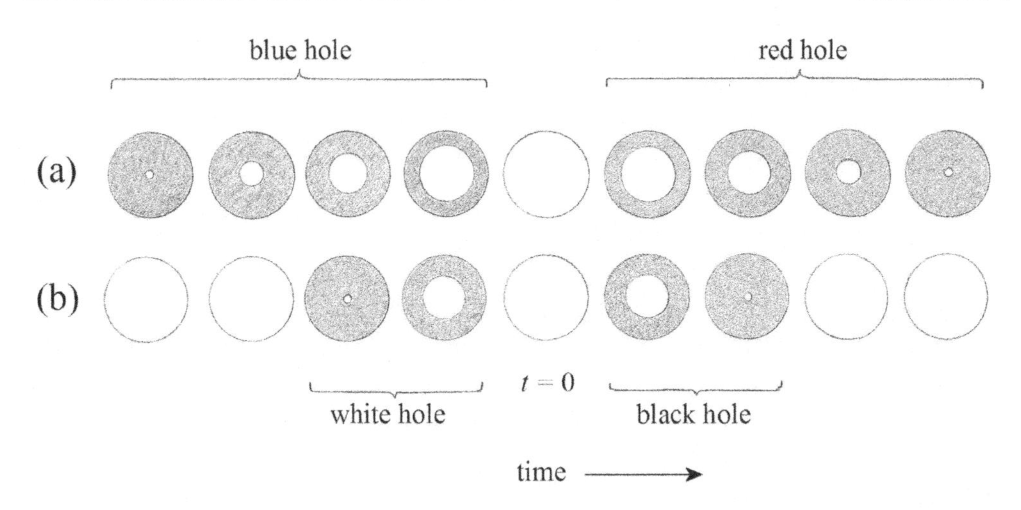

Figure 9.7. Cross sections of traversable and untraversable wormholes. In both cases shown, inner ring is surface of minimal area – the geometric "throat". (a) Outer ring is functional throat. Geometric throat of traversable wormhole slowly expands, momentarily halts its expansion at t = 0 (when functional and geometric throats coincide) and slowly contracts (cf. Figure 9.6a). (b) Outer ring is event horizon. Geometric throat of untraversable wormhole rapidly expands, momentarily halts its expansion at t = 0 (when event horizon and geometric throat coincide), and rapidly contracts (cf. Figure 9.6b). Contraction phase occurs so rapidly that light moving outward within the interior region (shaded) is nevertheless dragged inward.

Interior Surfaces Not in General Trapped

If a dynamic wormhole is traversable, this speed of contraction does not exceed that of light. Hence, pulses of light emitted from surfaces within the interior regions of a dynamic MT wormhole normally escape to infinity. Such surfaces are not so much trapped, then, as they are *hindered*, which I define as follows

> A **hindered surface** is one that 1) is contained within the region bounded by the throats of a contracting wormhole, 2) is a place from which a particle must travel above a certain nonzero speed in order to reach the region exterior to the throats, 3) satisfies $\theta_+ < 0$ and $\theta_- < 0$.

So the upper interior regions of Figure 9.6a contain hindered surfaces, because particles can leave the region only if they exceed a certain speed. What if the speed required exceeds the speed of light? Then the hindered surfaces become trapped. If we suppose that the event horizons of the Schwarzschild wormhole are actually its throats, we may define -- in reasonable accord with the usual definition -- the following.

> A **trapped surface** is a hindered surface from which a particle must exceed the speed of light in order to escape from the region interior to the throats of a wormhole.

Trapped surfaces are not restricted to the black hole interiors of the Schwarzschild wormhole. Some of the hindered surfaces within the upper interior region of the dynamic MT wormhole are also trapped (Figure 9.8). These are the surfaces from which an occupant of the interior region would have waited too long to make his escape. The punishment for his procrastination is a crushing death in the singularity that forms as the wormhole pinches off.

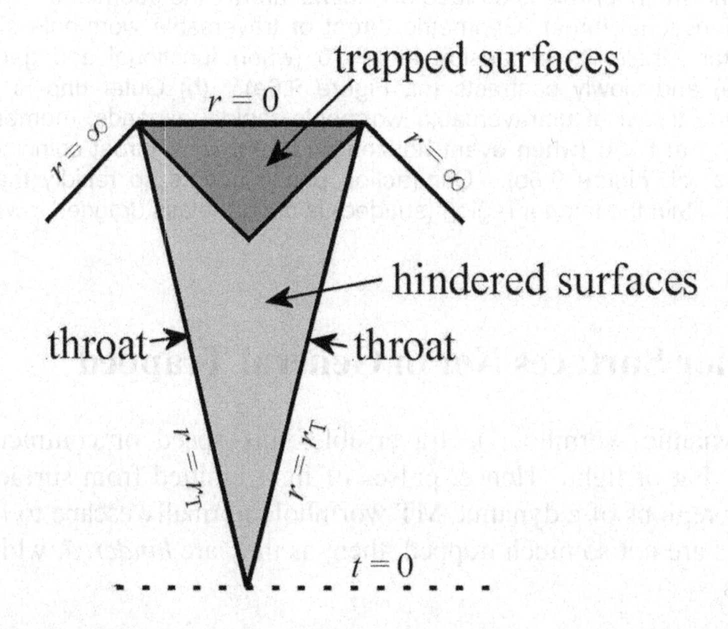

Figure 9.8. Trapped and hindered surfaces. Most of the surfaces within the red hole interior of a dynamic wormhole are merely hindered, but some of the later surfaces are also trapped.

Red Holes and Blue Holes

Outgoing light from a hindered surface must by definition prevail against the wormhole's inward contraction. Consequently, it is redshifted in the same way that the light from distant galaxies is redshifted by the expansion of the universe. This redshift occurs in addition to the usual gravitational redshift that applies even to static wormholes. The magnitude of the contraction-induced redshift depends on the duration of the wormhole. The shorter the wormhole persists, the greater will light from its hindered surfaces be red-shifted. If it persists for a sufficiently short time, its effective speed of contraction will reach the speed of light. Light emitted from its hindered surfaces will be infinitely redshifted. Such surfaces will have become trapped, and the upper interior region of the dynamic wormhole will have become akin to the black hole interior of the Schwarzschild solution. This is why I will call the upper interior region of the dynamic MT wormhole a "red hole" interior.

> A **red hole** interior is a region of a dynamic wormhole solution whose surfaces are all hindered and from which outgoing light acquires a contraction-induced redshift.

In short, the red hole sector of the dynamic MT wormhole is a kinder and gentler version of the black hole sector of the maximally extended Schwarzschild solution.

A similar relation pertains to this solution's white hole sector and what I shall call the "blue hole" sector of the dynamic wormhole defined as follows.

> A **blue hole** interior is a region of a dynamic wormhole solution whose surfaces are all anti-hindered and from which outgoing light acquires an expansion-induced blueshift.

Just as the surfaces within a white hole are said to be anti-trapped, so are those within a blue hole anti-hindered. Both surfaces satisfy the relations

$$\theta_+ > 0 \text{ and } \theta_- > 0 \qquad (5)$$

Just as light from a trapped surface within a black hole cannot reach the outside, light from the outside cannot reach an anti-trapped surface within a white hole. Just as light from a hindered surface within a red hole *can* reach the outside, light from the outside *can* reach an anti-hindered surface within a blue hole. Just as the light that leaves a red hole will be redshifted by the wormhole's contraction, the light that *enters* a blue hole will be redshifted by the wormhole's expansion. Just as the light that enters a red hole will be blueshifted by the wormhole's contraction, the light that *leaves* a blue hole will be blueshifted by the wormhole's expansion. Just as a red hole may be considered a less severe version of a black hole, a blue hole may be considered a softened version of a white hole. These considerations and the definition of a hindered surface suggest the following definitions,

> An **anti-hindered surface** is one that 1) is contained within a region bounded by the throats of an expanding wormhole, 2) can only be reached from the region exterior to the throats by particles that travel above a certain nonzero speed, 3) satisfies $\theta_+ > 0$ and $\theta_- > 0$.
>
> An **anti-trapped surface** is an anti-hindered surface that can only be reached from the region exterior to the throats of a wormhole by particles that exceed the speed of light.

So it appears that a Schwarzschild wormhole is a dynamic MT wormhole on steroids. Its black-hole and white-hole interiors have dynamic-MT counterparts as do the surfaces within these regions. These regions are bounded by marginal surfaces: event horizons in the Schwarzschild case, throats in the case of a dynamic-MT wormhole. When these surfaces are horizons, they prevent the passage of particles in one direction. When they are throats, they merely hinder such passage – exacting a redshift, or requiring a minimal speed. These surfaces are examples of the various sorts of hypersurfaces encountered in wormhole physics. These are classified below.

Classifying Surfaces

By Location

The surfaces of interest in wormhole physics are of three categories of closed surfaces centered on the wormhole's geometric throat: exterior, marginal, or interior. Exterior surfaces are those timelike hypersurfaces exterior to a wormhole's functional throat or event horizon. These pass through the overwhelming bulk of space including our home in the solar system.

Marginal surfaces lie at the boundaries between interior and exterior regions, between exterior regions of differing character, or between interior regions of differing character. The "character" of a region is determined by the behavior of bundles of light rays. As we have seen, such ingoing or outgoing bundles may converge or diverge. Marginal surfaces are places where character changes: converging bundles begin to diverge or diverging bundles begin to converge. In other words, the expansion θ changes sign at these surfaces, i.e. θ is zero there. An event horizon is an example of a marginal surface.

Interior surfaces are those separated from exterior surfaces by a marginal surface. The propagation of light in at least one direction is hampered. In the case of black or red holes outgoing light is hindered. For white or blue holes, the hindered light is ingoing. While particles attempting to leave a black or red hole are hindered, those attempting to leave a white or blue hole are assisted. For this reason, I have called the interior surfaces of white and blue holes "anti-hindered".

By Penetration Speed

We can assign to members of each of these three categories – exterior, marginal, and interior – a "penetration speed". This is the minimum speed at which a particle must travel in order to pass through a surface. One obtains this speed from the wormhole's Penrose diagram. Hence, it is relative to a reference frame in which the wormhole's center of mass is stationary. It varies with the direction of passage. There is no minimum speed for inward passage through the event horizon of a Schwarzschild black hole, for example. Technically, however, there is an *infimum* speed of zero, because a particle must be moving to penetrate the surface in the inward direction. In the outward direction, this horizon can only be penetrated by particles that somehow exceed the speed of light. That is to say, not at all. For marginal surfaces that are throats instead of horizons, less stringent penetration criteria apply. The penetration speed does not in this case exceed the speed of light. For example, a particle within a contracting wormhole (a red hole) cannot penetrate its functional throat – the softened version of an event horizon – unless the speed at which the particle travels outward exceeds the wormhole's effective speed of contraction.

By Transmission Direction

An ambiguity remains in the classification of surfaces – the distinction between ingoing and outgoing rays. This pertains to surface classification, because surfaces are classified on the basis of the values of expansions θ of ingoing and outgoing rays. In the static, spherically symmetrical case, a distinction between these presents itself naturally. An ingoing ray became an outgoing one the moment it crossed the wormhole's throat. In the dynamic case, however, there are two throats (with the geometric "throat" no longer performing this function). An ingoing ray becomes an outgoing one somewhere in the region bounded by the two throats. But where? For our purposes, it suffices to assume that such a surface exists within the interior region.

We need not specify it, although the geometric (nonfunctional) throat seems to be a reasonable guess.

By Temporal Orientation

There is a bit more terminology in common use. In black hole physics a surface is said to be "future trapped" if particles on it are destined to encounter a spacelike singularity or an interior event horizon. Such surfaces are found within (the Type B regions of) black holes. A surface is called "past trapped" (or "anti-trapped") if particles in it have such an encounter in their past. These surfaces exist within (the Type B′ regions of) white holes. A marginal surface forming a boundary between an interior region of anti-trapped surfaces and an exterior region containing surfaces that are not trapped would be a "marginally anti-trapped outer horizon". Such a surface could also be called a "past outer marginal surface".

We will use similar terminology with regard to dynamic wormholes. Hence, a "future hindered" surface is the red-hole analog of a black hole's future trapped surface. A future hindered surface is one on which particles are destined to encounter a singularity, *if they remain between the throats.* Much as every point on one of a black hole's future-trapped surfaces *does* have an encounter with a singularity in its future, every point on one of a red hole's future-hindered surfaces *might* have an encounter with a singularity in its future. Similarly, a "past hindered" (or "anti-hindered") surface is the blue-hole analog of a past-trapped (or anti-trapped) surface of white hole. It is one that can be reached by particles that emerge from a past singularity *and remain between the throats.* Much as every point on one of a white hole's past-trapped surfaces *does* have an encounter with a singularity in its past, every point on one of a blue hole's past-hindered surface *might* have an encounter with a singularity in its past. In other words, future-trapped and past-trapped surfaces have the strongest possible causal connection to singularities in the future and past respectively. Future-hindered and past-hindered surfaces are in weaker causal contact with their respective future and past singularities.

While we are discussing terminology I should mention that I have been a bit lax in my use of the term "event horizon". A true event horizon cannot be determined on the basis of local measures such as the value and rate of change of the expansion θ. One cannot know the event horizons of a spacetime unless the entire spacetime – from the indefinitely far past to the indefinitely far future – is known. Then one creates a Penrose diagram for spacetime, and one determines the boundaries between regions that are out of causal contact.

These globally determined boundaries are the event horizons. The locally determined boundary surfaces that we have called "throats" in the case of dynamic wormholes and, loosely, "event horizons" in the case of black or white holes have been dubbed "trapping horizons" by Sean Hayward.

In the case of black holes, trapping horizons that we have specified are surfaces that will become event horizons, once things have settled down. For example, the trapping horizon of a star collapsing to become a black hole gradually approaches the event horizon of the black hole that ultimately forms. In other words, we can safely refer to the trapping horizon of black hole (or white hole) as the "event horizon", after the space exterior to the black hole has become static. Were this space to change again as a consequence of, for example, the black hole swallowing a star, the trapping horizon would no longer equal the event horizon of the newly enlarged black hole. As the dynamic phase passes, the trapping horizon would gradually approach the event horizon until they again coincide.

A Classification Table

The above considerations are summarized in the following table. The plus and minus signs appear wherever the indicated quantity is positive or negative respectively. Empty cells imply that the unspecified value is inapplicable or superfluous to the classification.

θ_+	θ_-	$D\theta_+$	$D\theta_-$	Penetration Speed*	Surface Type	Surface Location
+	-	0	0		unhindered	wormhole exterior
0	0	+	+		marginal	static wormhole throat
0	0	-	-		marginal	static wormhole belly
0	-	+		$> c$	future outer marginal	black hole outer horizon
0	-	+		$\leq c$	future outer marginal	red hole throat
+	0		+	$> c$	past outer marginal	white hole outer horizon
+	0		+	$\leq c$	past outer marginal	blue hole throat
0	-	-		$> c$	future inner marginal	black hole inner horizon

0	+		-	> c	past inner marginal	white hole inner horizon
-	-			> c	future trapped	black hole interior
-	-			≤ c	future hindered	red hole interior
+	+			> c	past trapped (anti-trapped)	white hole interior
+	+			≤ c	past hindered (anti-hindered)	blue hole interior

*in direction in which motion is limited

The Interconversion of Black Holes and Wormholes

Converting a Black Hole into a Wormhole

It would seem that Einstein's field equations – the equations of motion for the geometry of space – must allow the conversion of a black hole into a wormhole. As long as we are free to leave the stress-energy tensor unconstrained, we can surely plug a time-dependent geometry describing such a conversion into the left side of the Einstein equations. Doing so will result in *some* (time-dependent) value for the stress-energy tensor on the right side. This, after all, is precisely what Morris and Thorne did for a special case of a particular type of time dependence, namely none -- the static traversable wormhole. The problem with this approach, as perfectly understood by Morris and Thorne, is that in a dynamical system the stress-energy tensor cannot be assumed to be unconstrained. The stress-energy participates in the physics. This participation results from specifying an equation of state for matter. Doing so causes the stress energy to depend on geometry. The Einstein equations thus cease to be formulas for determining the stress energy. Rather, they become a riddle, the answer to which is a particular time-dependent geometry. Because geometry now appears in both sides of the equations, the equations must now be *solved.* A time-dependent geometry must be found for which both sides have the same value.

Such a situation is not of course unique to general relativity. Consider the following analogy. A ball is attached to an anchored spring as shown in Figure 9.9a. The position of the ball is the analog of a wormhole's geometry, while the restoring force of the spring represents gravitational contraction. Suppose that we wish to consider a particular motion of the ball, the one in which it remains stationary at the nonequilibrium position a_1, as shown in

Figure 9.9b. When we plug this desired (non) motion into Newton's equation, we obtain the force F required to achieve it. This is analogous to plugging the geometry of a static traversable wormhole into Einstein's equation and thus obtaining value of the stress-energy required to sustain it. In a dynamical system the geometry determines the stress energy, as before, but the stress energy also determines the geometry. As a result the geometry can no longer be specified arbitrarily. It is constrained to be a solution of the system's dynamical equations. This is analogous to the situation depicted in 9.9c. There the force on the ball becomes the tangential component F_g of the ball's weight on a sloping surface. The ball will no longer remain stationary at a distance a_1 from the anchor block as before. Its new position is a_2. In other words, the ball is unstable at position a_1. The force F required to balance the ball at a_1 differs from the position-dependent force F_g that now acts on the ball at a_1. This is analogous to the instability of a traversable wormhole. Just as the ball no longer retains its initial location when the force acting on it depends on position, a Morris-Thorne wormhole will not retain its diameter when its stress energy tensor depends on geometry. Finally, the motion of the ball can be influenced by other forces. For example, when the ball is blown by air currents (as shown in Figure 9.9d), its equilibrium position changes. This is analogous to the way in which in-falling matter can induce a wormhole to contract.

With our understanding of the difference between wormhole statics and dynamics improved by these considerations, let us turn to the task at hand. We need to find a solution to Einstein's equations that correspond to the conversion of a black hole into a traversable wormhole. We can proceed as follows.

First, we return to Visser's Schwarzschild-based, thin-shell wormhole – the one we considered briefly when discussing wormhole stability. Recall that it is constructed by starting with two Schwarzschild black holes – one in each of the universes to be connected. A spherical hole centered about the singularity is cut out of each black hole. The surfaces of the holes are then identified (Figure 9.10). Freezing all degrees of freedom except for the radius of this surface – the throat containing all of the wormhole's exotic matter -- simplifies the dynamics. Finally, we choose an equation of state for the exotic matter. This results in a simple, one-dimensional dynamical system whose motion is no more difficult to ascertain than that of the ball in the analogy above. For seemingly reasonable choices for the equation of state for exotic matter, the shape of the effective potential that governs the motion of the

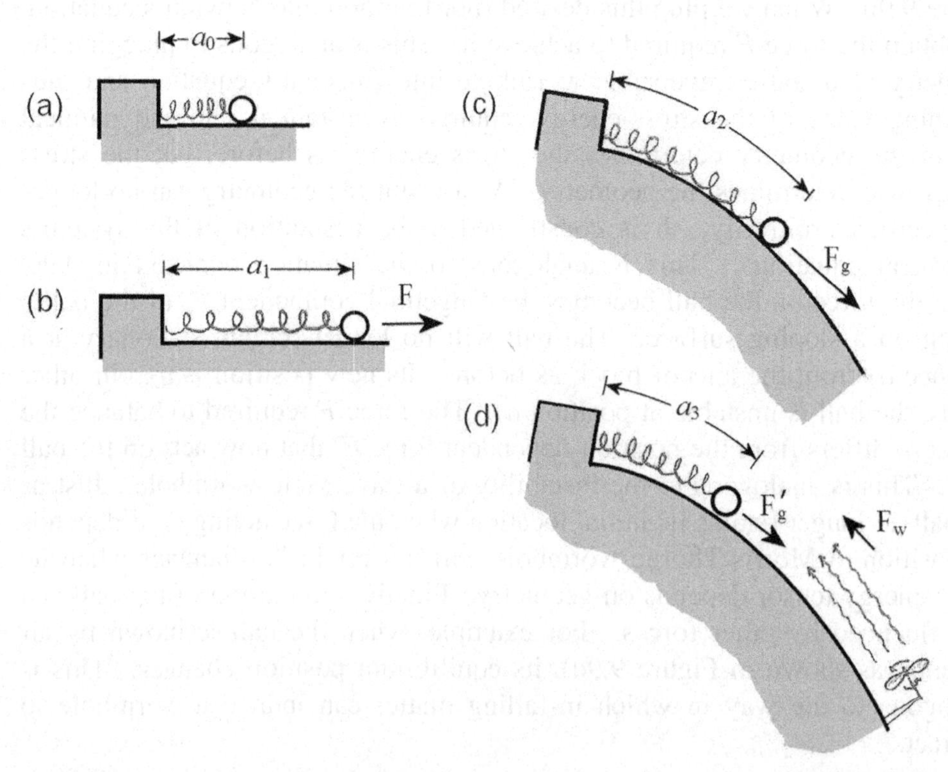

Figure 9.9. Analogy for wormhole dynamics. (a) Equilibrium position of ball is a_0. (b) Keeping ball at position at a_1 requires force F. [Analogous to unstable static wormhole] (c) When expanding force depends on position, a_1 is no longer solution; a_2 is. [Analogous to wormhole with geometry-dependent stress-energy tensor] (d) External force F_w from fan changes solution from a_2 to a_3. [Analogous to wormhole whose stress-energy tensor is augmented by in-falling matter]

system – the expansion or contraction of the surface – is roughly that of an inverted parabola, as was shown in Figure 8.2c. Consequently, the surface will either contract to zero or ceaselessly expand. Moreover, Einstein's equations for this model make it clear that there is a simple way to stabilize the wormhole against such behavior. For a given radius of the wormhole's throat, all one need do is to cause the density of the negative-energy matter there to change sufficiently rapidly. An increasing negative-energy density acts as an outward, expansion-inducing force. A decreasing negative-energy density has the opposite affect. This toy model suggests, then, that the size of a wormhole can be controlled by fluxes of either negative-energy or positive-energy matter. It suggests in particular that a black hole can be converted into a traversable wormhole by causing it to ingest a quantity of negative-energy matter sufficiently great to enlarge its throat beyond the radius of its event horizon. Once enlarged, this model also indicates that the wormhole could be maintained against collapse by a suitably intense constant influx of negative-energy matter.

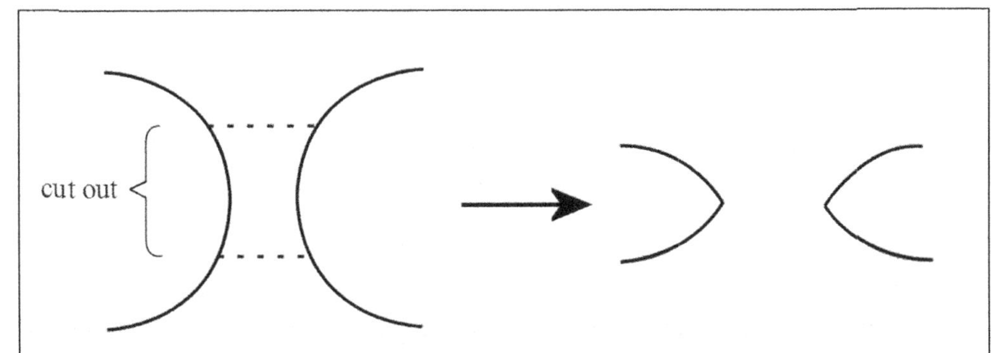

Figure 9.10. Schwarzschild-based thin-shell wormhole. (a) Cross section of embedding diagram of Schwarzschild solution. Dotted line shows locations where spherical holes are cut out. (b) Holes are glued together to create thin-shell wormhole. All exotic matter is concentrated at the kink surface. Geometry remains Schwarzschild everywhere else.

The next step toward finding a solution that demonstrates black-hole-to-wormhole conversion might be to use the intuition gained from the thin-shell model to direct a numerical calculation. Such a calculation would attempt a conversion through simulated bombardment of a black hole with negative-energy matter. While this task would be straight forward, numerical calculations tend to obscure the underlying physics. It is difficult to see how particular aspects of the solution depend upon parameters of the problem – densities, radiation intensities, pressures etc. -- because these parameters are necessarily reduced to numbers before the computation can proceed. Each numerical solution to the equation of motion, then, is specific to a particular choice of parameter values. One can, by contrast, achieve maximal clarity about the physics of the system by finding what is known as an *analytic solution* to the equations of motion. Such a solution expresses the motion – in this case the time-dependent geometry of space-- as a formula in which the various parameters appear as unspecified variables or functions. Finding an analytic solution is in general impossible. It can, however, be achieved in special cases, some of which we have already discussed – the Schwarzschild, Reissner-Nordstrøm, and Kerr cases for example.

Nevertheless, in 2001 Sean Hayward and his collaborators, Sung-Won Kim and Hyunjoo Lee, were able to find such a solution in a particular model of *two-dimensional* gravity coupled to scalar fields. One of these scalar fields corresponded to negative-energy matter – a so-called "ghost field". Because the two-dimensional version of the problem is more tractable than its four-dimensional counterpart, they were able to show explicitly that

- A wormhole will collapse to a black hole unless it is stabilized by negative-energy radiation.
- If a black hole is exposed to sufficient negative-energy radiation, it becomes a static wormhole.

As mentioned in the Chapter 8's discussion of wormhole stability, Shinkai and Hayward demonstrated numerically in 2002 the instability of the Morris Thorne wormhole. By 2004 Hayward and Koyama were able to extend the analytic two-dimensional-gravity solution to four dimensions. They confirmed that the above points hold as well in the four-dimensional case.

As in the solution for two-dimensional gravity, their solution in four dimensions involved a negative-energy scalar field – a ghost field. Their solution requires an ingoing, high-intensity, spherical pulse of this field to converge at the speed of light onto a Schwarzschild black hole. The negative energy thus transferred to the black hole converts it into a Morris-Thorne wormhole. This wormhole is then stabilized against collapse by a subsequent low-intensity rain of negative-energy ghost radiation.

They created their solution by patching together three separate solutions. The maximally extended Schwarzschild solution describes the initial black hole to be transformed by the ghost radiation. The Vaidya solution -- published in 1951 by Indian researcher Prahalad Chunilal Vaidya -- describes a spacetime containing the spherically symmetric ingoing and outgoing massless radiation that effected the transformation. Finally, the Morris-Thorne class of solutions describes the static wormhole that resulted from the transformation. They expended the bulk of their ingenuity in getting these solutions to match up across various boundaries. Figure 9.11 shows the corresponding Penrose diagram. According to their solution, one proceeds as follows in order

To convert a black hole into a traversable wormhole.

1. **Find a target.**
   Find an instance of the maximally extended Schwarzschild wormhole solution.

2. **Arrange to hit it.**
   Arrange for the generation in both universes of a thin, high-intensity, spherical shell of ingoing, negative-energy radiation. Each shell must have an arbitrarily large ("infinite") initial radius and must be centered on the Schwarzschild black hole.

3. **Hit it hard from both sides.**
   Synchronously launch in each of the two universes a negative-energy shell of high intensity so that the two shells converge onto the black hole and meet within it.

4. **Follow up with maintenance radiation.**
   Have a continuous flux of low-intensity, negative-energy radiation fall in from infinity on to the wormhole created by step 3 in order to prevent its collapse.

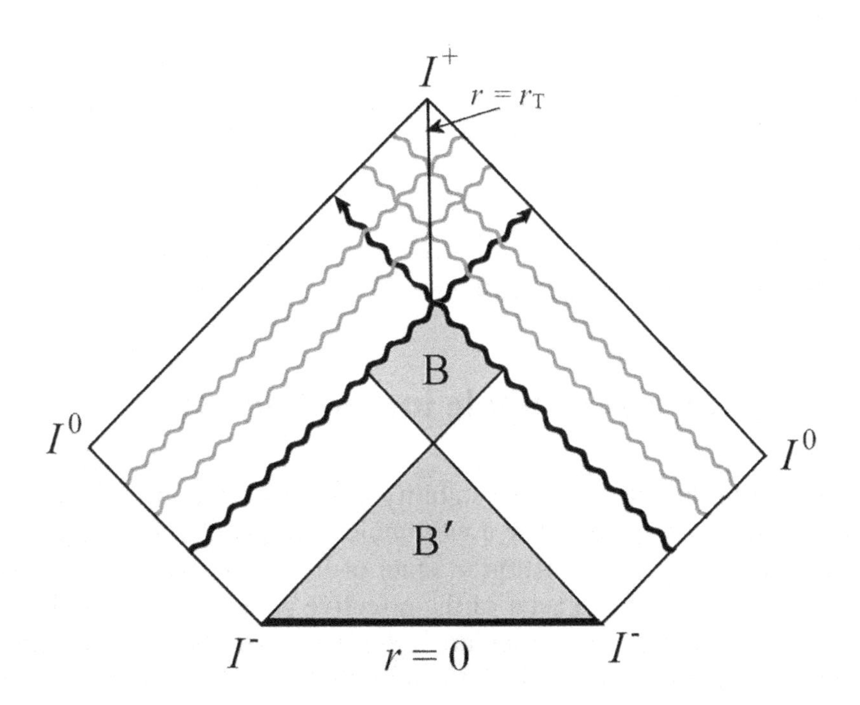

Figure 9.11. Converting a black hole into a static wormhole. Step 1: Hit black hole B with intense pulses of negative-energy radiation (black wavy lines) from both universes. Step 2: Stabilize the resulting traversable wormhole with low-intensity maintenance radiation (gray wavy lines). [From Hayward and Koyama, op. cit.]

This program is, of course, utterly impractical. It nevertheless engenders hope that more practical scenarios are possible. The maximally extended Schwarzschild solution could perhaps be replaced by a more astrophysically realistic black hole. Perhaps asymmetrical negative-energy radiation bombardment from a *single* universe will suffice. It would be up to numerical simulations, informed by this Hayward-Koyama solution, to tell us whether particular wormhole engineering scenarios could successfully convert a black hole.

Another easy way of qualitatively predicting the results of Shinkai and Hayward's numerical calculations would have been to consider Visser's Schwarzschild-based thin-shell wormhole. As mentioned in our discussion of stability, this wormhole either contracts to a throat radius of zero (collapses to black hole) or it inflates (geometric throat radius expands arbitrarily). Moreover, it is clear form this model that a small constriction of the wormhole could be undone by a suitable injection of negative-energy matter. It is also clear that a large injection could alter the model's effective potential function and thus induce inflation (Figure 9.12).

P76. *A black hole that is an untraversable wormhole can be converted into a traversable wormhole by injecting into it a sufficiently large quantity of negative-energy matter.*

P77. *A wormhole can be stabilized against inward collapse by a constant influx of negative energy radiation.*

Converting a Wormhole into a Black Hole

In our discussion of wormhole stability we learned that traversable wormholes are in general unstable. How a wormhole will manifest its instability depends on its size and on the equation of state of its exotic matter. This equation of state determines the character of the effective potential function that drives the system's dynamics. In particular, it determines the location of this potential function's maximal value. If the wormhole is "small" -- in the sense that it is smaller than the size at which this function is maximal -- it will collapse and become a black hole. If it is similarly "large", it will ceaselessly inflate. This means that for a wormhole to remain stationary it must be acted on by a force that either prevents it from collapsing or inflating. Just as a upward current of air can prevent a ping-pong ball from falling, a current of negative-energy matter can prevent a "small" wormhole from becoming a black hole. A

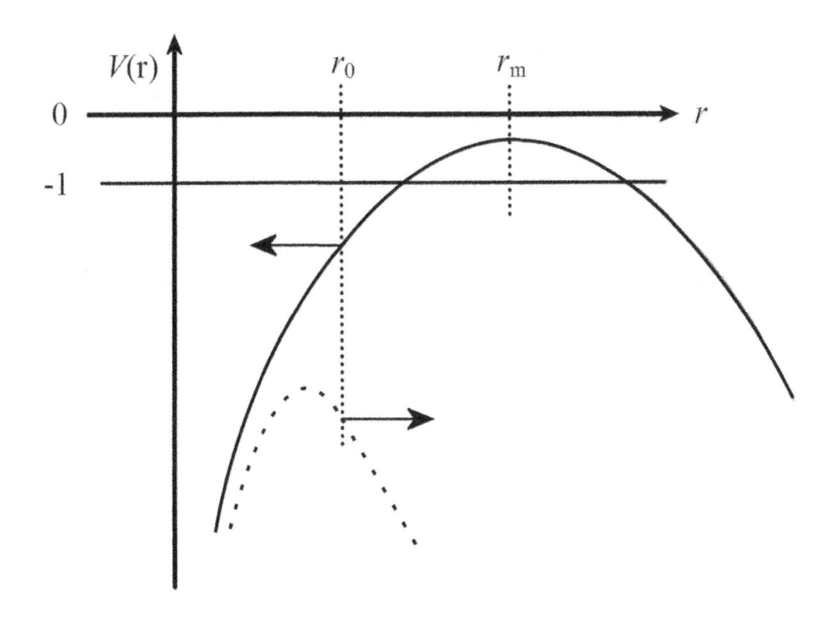

Figure 9.12. Effect of large influx of exotic matter on effective potential of thin-shell wormhole. Solid line represents initial effective potential function $V(r)$. Note that initial potential causes wormhole with throat radius r_0 to contract. Massive injection of exotic matter changes effective potential to that represented by dashed line. New potential causes wormhole with throat radius r_0 to inflate.

current of positive-energy matter would similarly prevent a "large" wormhole from inflating.

The first step in converting a traversable wormhole into a black hole, is to determine whether it is "large" or "small" in the above sense. In other words, we need to know whether the force keeping it stationary is preventing the wormhole from inflating or from collapsing. If it is the former, we must increase the magnitude of the inflation-preventing force until the wormhole shrinks to the point that it becomes "small" in that its propensity to collapse replaces its tendency to inflate. At this point we may turn off the inflation-preventing force, the current of positive-energy matter. The wormhole will collapse inward until it becomes a black hole as shown in Figure 9.13. If the force keeping it stationary had instead been acting to prevent such a collapse, all we would have had to do is to turn this collapse-preventing force off in order to achieve the same result.

The currents of negative matter and positive matter that I have called "forces" are not, strictly speaking, forces in that they do not have the proper units. I shall continue to refer to them in this way, however, in order to preserve a useful analogy. The role they play in the (minisuperspace) equation of motion of general relativity is analogous to that played by the forces due to air currents that appear in the equation of motion for a floating ping-pong ball or in that of the ball of the analogy that was illustrated in Figure 9.9.

In summary,

P78. *A traversable wormhole can be converted into a black hole by injecting into it a sufficiently large quantity of positive-energy matter or by eliminating the influx of negative-energy matter that is preventing its collapse.*

Numerical confirmation of the collapse of wormholes into black holes was provided by the aforementioned work of Shinkai and Hayward. Analytic solutions confirming this appeared in the same papers that demonstrated the reverse transformation.

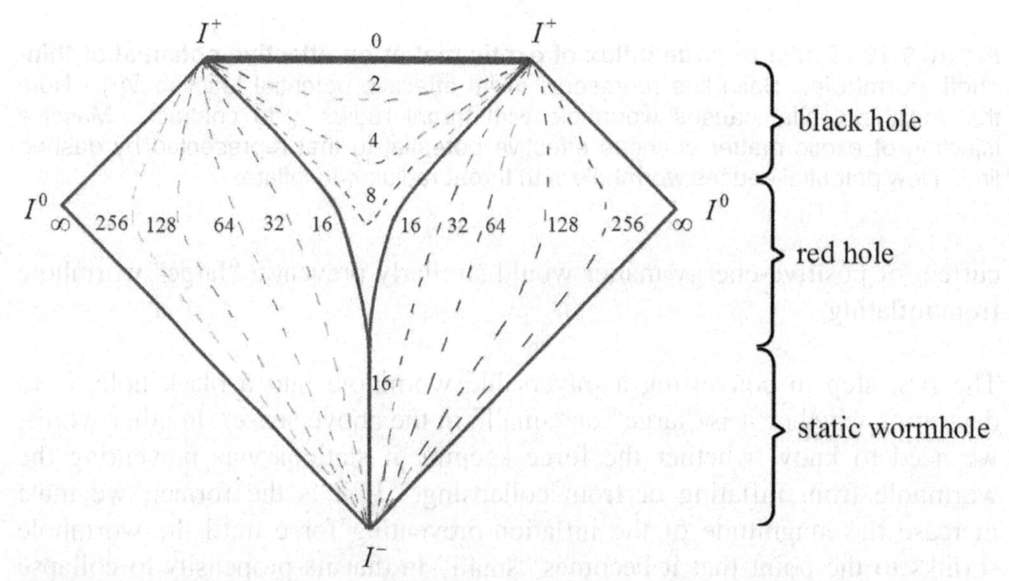

Figure 9.13. Converting a static wormhole into a black hole. Step 1: Turn off maintenance radiation (not shown) that stabilizes wormhole against collapse. No other steps. As usual, contraction of the geometric throat results in dual functional throats. As contraction accelerates light escaping from interior region between the throats becomes progressively more redshifted. Eventually, rate of contraction prevents any light from escaping (redshift becomes infinite). Black hole has formed. In this example the radius of the throat and that of the event horizon it becomes is 16 units.

Wormhole Inflation

The numerical work of Shinkai and Hayward also demonstrated that irradiating a traversable wormhole with negative-energy matter will cause it to "inflate". Inflation as it applies to wormholes is the state in which the size of the geometric throat increases without limit. This increase normally occurs at an ever accelerating rate. As discussed earlier, such an expansion of a wormhole's geometrical throat results in its bifurcation. The surface of minimal area, which we will continue to call the geometric throat, ceases to function as a throat in the sense that it ceases to be a location at which the expansion of light rays becomes zero. Two new functional throats, which coincided with the geometric throat when the wormhole was stationary, now accelerate away from each other, thus creating a rapidly growing interior region. This region is akin to the blue hole that resulted from the wormhole expansion discussed earlier. As this acceleration continues, the throats soon approach the speed of light in the frame in which the wormhole is centered. As this occurs, the character of the interior region changes from that of a blue hole to that of a white hole (Figure 9.14).

The possibility of wormhole inflation was first explored in 1992 by Thomas Roman, then of Central Connecticut State University. He considered the idea that submicroscopic virtual wormholes might be induced to inflate during the inflationary phase of the universe itself. While Roman found that wormhole inflation could occur in such circumstances, we are instead concerned with the inflation of wormholes embedded in space that is not itself inflating. As mentioned above, Visser's Schwarzschild-based thin-shell wormhole (with a suitable equation of state) predicts Shinkai and Hayward's result. To get this thin-shell wormhole to inflate, all we need do is dowse it with a sufficiently large quantity of exotic matter. To perfectly mimic Shinkai and Hayward's result, the parameters of the wormhole need only be selected in order to ensure that it is in a metastable state – a delicate balance between inflation and collapse. The slightest injection of matter would then cause the wormhole to inflate or collapse, depending on whether the matter has negative or positive energy. Alternatively, a non-metastable wormhole could be kept stationary by influxes of either positive (inflation-preventing) or negative (collapse-preventing) matter. In this case the thin-shell model similarly predicts that slight injections of matter in excess of that delivered by these fluxes results as well in inflation or collapse, depending respectively on whether negative or positive matter is injected. In short,

P79. *A wormhole can be induced to inflate by injecting into it a sufficiently large quantity of negative-energy matter.*

P80. *For particular equations of state, a stationary traversable wormhole can be induced to inflate or collapse by the injection respectively of arbitrarily small quantities of negative-energy or positive-energy matter.*

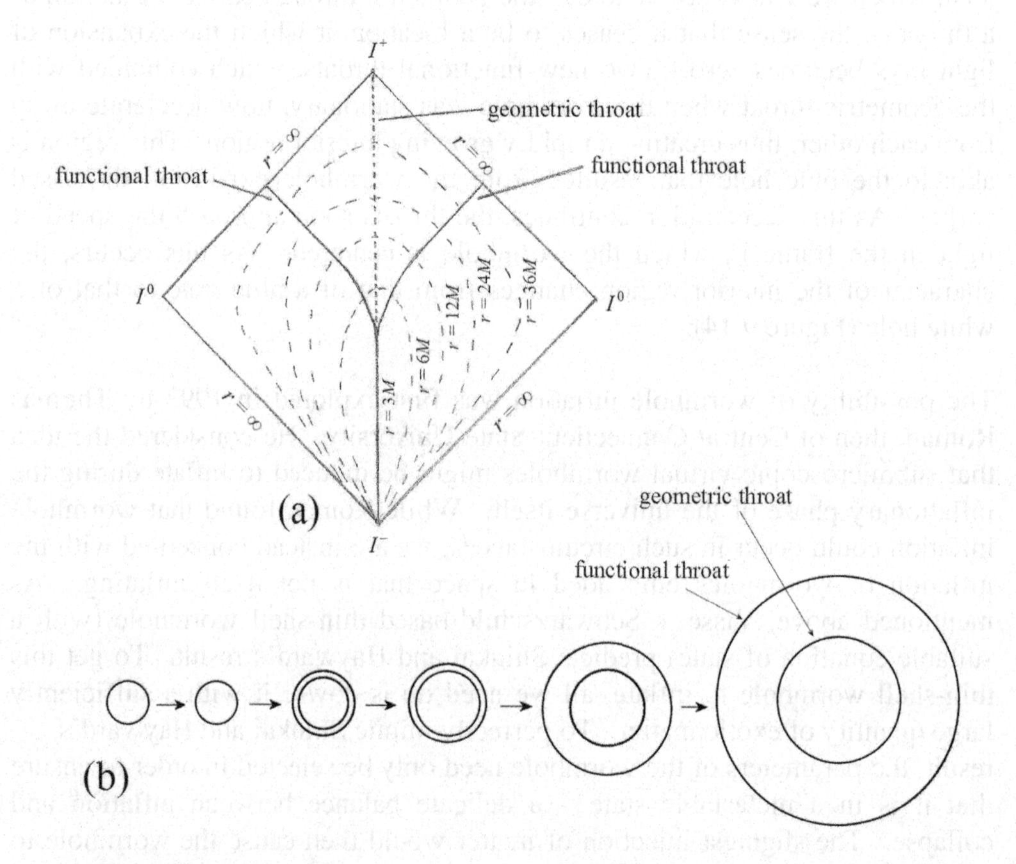

Figure 9.14. Inflating wormhole. (a) Penrose diagram of initial static wormhole that begins to inflate. Throat bifurcates when wormhole becomes dynamic. Functional throats and surface of minimal area (geometric "throat") expand ceaselessly. (b) Cross section of sequential spacelike hypersurfaces containing inflating wormhole as seen in one of the two universes wormhole connects. Region exterior to functional throat is normal space. Region between functional and geometric throats is similar to that of a blue or white hole. Region interior to geometric throat *does not exist.*

Wormhole Maintenance

By now it is clear that wormholes are in general unstable. One could of course cook up an exotic matter model whose equation of state would result in a stable wormhole. These models, according to common intuition, seem unrealistic even for a concept as outlandish as exotic matter. Wormholes can be stationary, then, only through artificial means. A wormhole can be stabilized against contraction by an influx of negative-energy matter, and against expansion by positive-energy matter. If the wormhole is metastable, being as likely to expand or contract, both types of influxes will be required to maintain it. That a wormhole can be stabilized in this way has been confirmed by the aforementioned numerical calculations and analytic solutions. We have, then,

P81. *A traversable wormhole can only be static as a consequence of ongoing influxes of exotic or normal matter that counter its natural tendency to contract or expand.*

Even though a wormhole can be rendered static (Figure 9.15) by an ongoing and *constant* influx of matter, it remains unstable. Stability requires the influx to vary in sign or intensity as needed to counter the effects of perturbations. Consider, for example, a traversable wormhole that is being prevented from contracting by a steady influx of negative-energy matter. If a traveler enters this wormhole, he will increase its mass. This will slightly increase the gravitational force acting to contract the wormhole. The influx of negative-energy matter that had precisely balanced this force no longer does so. The wormhole begins to contract. As this occurs, the contraction-inducing gravitational force increases further. The imbalance between this force and the expansion-inducing negative-energy matter influx worsens. This accelerating imbalance will continue until the wormhole collapses and becomes a black hole. Hence,

P82. *A traversable wormhole made static through the constant influx of exotic or normal matter is still unstable.*

P83. *The geometry of a traversable wormhole can in practice only be maintained through varying influxes of exotic or normal matter calculated to counter changes in this geometry.*

Consequently, from seemingly reasonable exotic matter models,

P84. *Naturally occurring static traversable wormholes are not expected to exist.*

I reemphasize that this conclusion hinges on the validity of current ideas about exotic matter models. Given that neither exotic bulk matter nor exotic radiation has yet to be encountered, these ideas may well change.

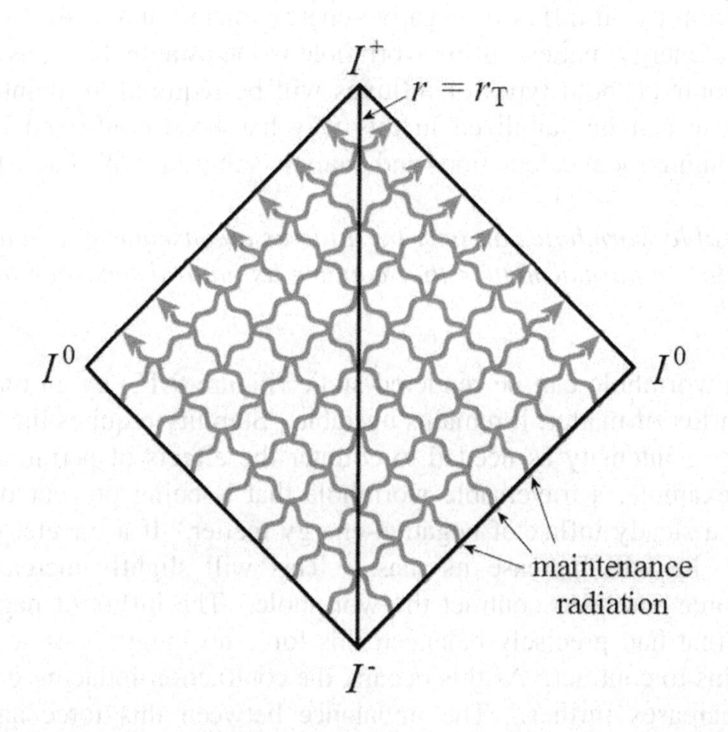

Figure 9.15. Wormhole maintenance. Depending on a wormhole's size and the equation of state of its matter, the wormhole must be prevented from contracting, expanding, or both. This can be accomplished by irradiating the wormhole with negative-energy (contraction-preventing) or positive-energy (expansion-preventing) matter. Stasis can be achieved by a constant flux of radiation. Stability, however, requires the sign or magnitude of the flux to change as needed to counter the effects of perturbations. Figure shows massless negative-energy (ghost) radiation proposed by Hayward et. al. to prevent wormhole collapse. Matter influx need not be massless, however.

Wormhole Enlargement

We have established that wormholes can be induced to inflate if we inject enough negative-energy matter into them. They can be maintained by controlled injection of exotic or normal matter. I should again emphasize, however, that these conclusions rely on certain assumptions about the equation of state of exotic matter. For example, we have assumed that a *positive* tangential pressure results from the presence of negative-energy matter at the wormhole's throat and that the density of matter there decreases as the throat increases -- but not arbitrarily fast. These assumptions about the equation of state are consistent with the model of exotic matter used in recent numerical calculations and analytic solutions. In this model exotic matter is assumed to consist of massless particles whose kinetic energy is intrinsically negative. Because these particles are massless, this model describes negative-energy matter as a kind of ghostly radiation that travels at the speed of light. As I mentioned before, there are other assumptions that we could make. Some of these preclude the possibility of inflation. Others permit gradual expansion instead of runaway inflation. In short, arbitrary assumptions about the equation of state lead to arbitrary wormhole dynamics. In order to retain the capacity to make concrete statements about wormhole behavior, I will continue to assume that the effective potential that controls the expansion of wormhole is shaped roughly as the inverted-parabola potential that arose in Visser's Schwarzschild-based thin-shell wormhole. We can be comforted to some degree by consistency of this toy model with numerical results.

The shape of this potential implies that any attempt to enlarge the wormhole will have to contend with the possibility of runaway inflation. Like the procedure for inducing a wormhole's collapse, the enlargement procedure will depend on the current size of the wormhole relative to the size r_m at which the effective potential reaches its maximum (Figure 9.12). If the wormhole is smaller than r_m , then it is being prevented from contracting by an influx of negative-energy matter. To enlarge the wormhole, we just need to increase this flux. If the desired size of the wormhole is less than r_m, we will allow the wormhole to enlarge until it reaches this size. We will then bathe the wormhole in negative-energy radiation of an intensity calculated to balance the magnitude of the wormhole's collapse-inducing gravitation at this size. If, instead, the desired size is greater than r_m , we will allow the wormhole to enlarge beyond r_m , at which point it will begin to inflate. We can reduce the rate of enlargement by bombarding the wormhole with a flux of positive-energy matter. When the enlargement has proceeded to the desired degree, we will increase the size of this flux in order to prevent further enlargement.

As mentioned above, these intuitive musings inspired by Visser's thin-shell wormhole seem to be born out by the numerical calculations Shinkai and Hayward and the analytic solutions of Koyama and Hayward. The latter induced enlargement by a single large pulse of negative-energy radiation and stopped it with a suitably large pulse of positive energy as shown in Figure 9.16 and further elaborated in Figure 9.17. From this we may conclude that

P85. *A static wormhole that is being maintained against contraction by an influx of negative- energy matter can be enlarged by increasing the influx. The enlargement process can be terminated by sufficiently reducing the influx or by replacing it with an influx of positive-energy matter.*

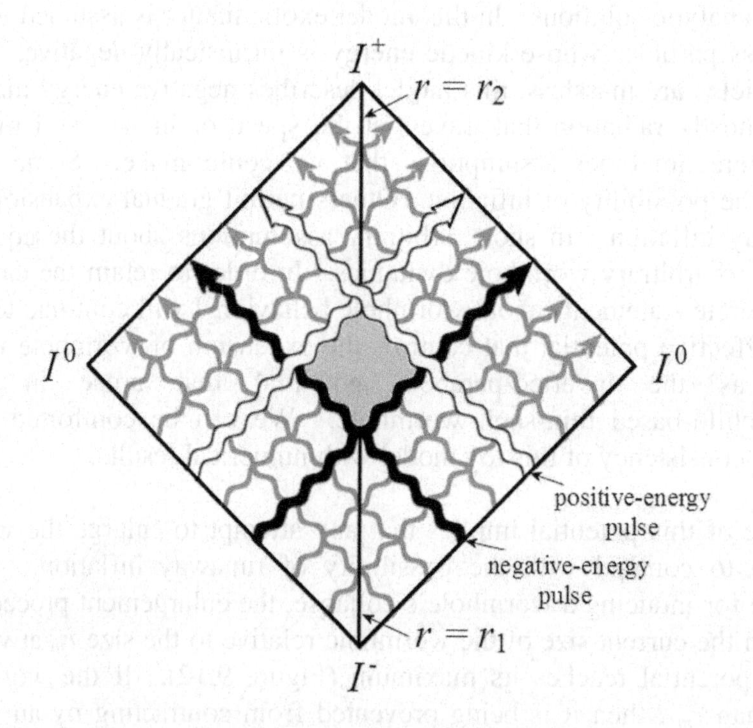

Figure 9.16. Enlarging a static wormhole. Step 1: Begin with a traversable wormhole whose throat radius r_1 is being stabilized against collapse by maintenance radiation (thin wavy lines). Step 2: Hit traversable wormhole with an intense negative-energy pulse (solid thick wavy lines). Step 3: Watch wormhole expand. Step 4: Hit expanding wormhole with a positive-energy pulse calculated to be sufficiently intense to halt expansion (hollow thick wavy lines). Step 5: Stabilize with negative-energy maintenance radiation the enlarged traversable wormhole, whose throat radius is now $r_2 > r_1$.

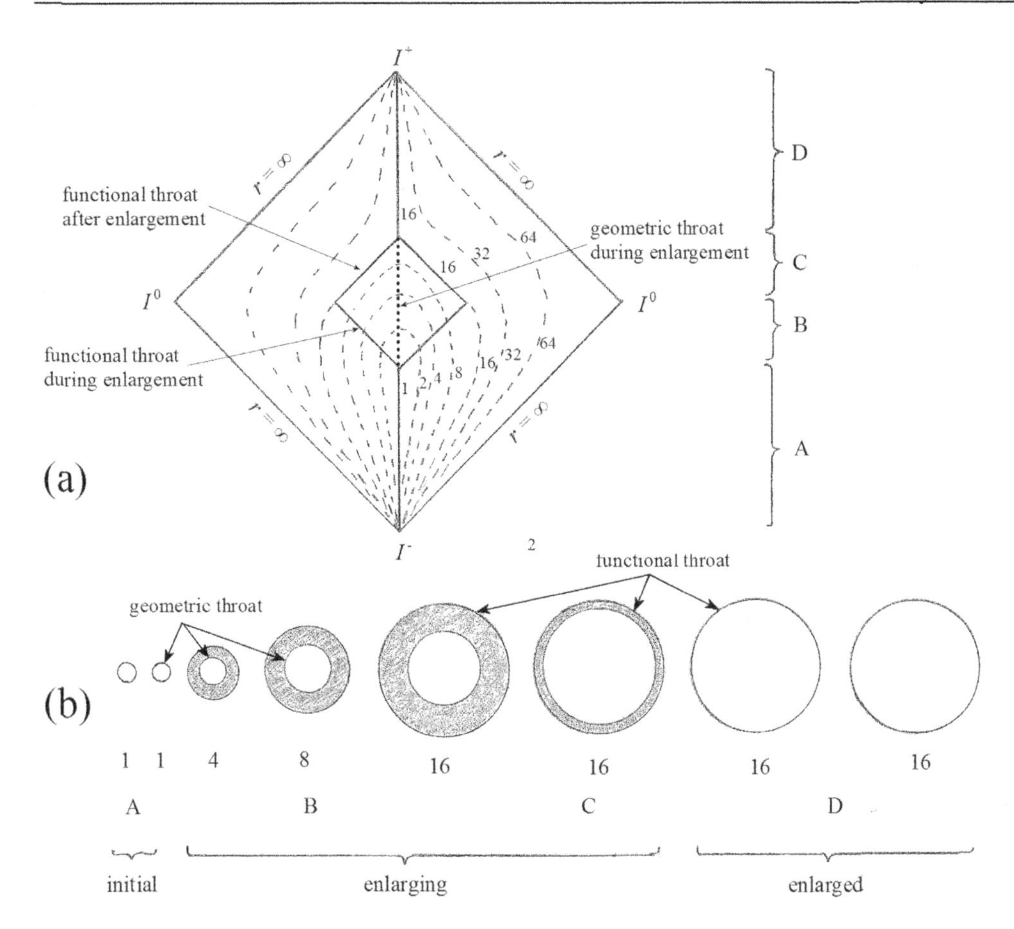

Figure 9.17. Wormhole enlargement. Wormhole throat radius enlarges from 1 unit to 16 units. Four phases of process: A – before enlargement, B – both functional (outer) throat and geometric (inner) "throat" enlarge, C – only geometric throat enlarges to match enlarged functional throat, D – after enlargement. Notice bifurcation of functional throats and distinction between functional and geometric throats only occur during dynamic phases (B & C). (a) Penrose diagram showing dashed lines of constant radial coordinate. Solid central lines show enlarging functional throat. Dotted central line shows enlarging geometric throat. (b) Sequence of cross sections of spacelike hypersurfaces. Region interior to geometric throat does not exist.

P86. *A static wormhole that is being maintained against expansion by an influx of positive- energy matter can be enlarged by reducing the influx. The enlargement process can be terminated by sufficiently increasing the influx.*

==/==

The dynamical situations described in this chapter all involved the absorption of matter by wormholes. This absorption was depicted as occurring symmetrically in both universes connected by the wormhole. While this assumption simplified numerical calculations and efforts to discover analytical solutions, it should not be understood to be essential. Irrespective of details of how matter enters the wormhole, the salient features of wormhole dynamics hold: Positive-energy matter induces contraction, negative-energy matter the opposite.

10. Wormhole Time Machines

How to turn a Wormhole into a Time Machine

If you possess a traversable wormhole, you might not have to lift a finger to turn it into a time machine. Assuming that time travel is in fact possible, there is a good chance that your wormhole is already a functioning time portal. Wormholes, being solutions to the Einstein field equations in spacetime, not only connect separate regions of space, but also join separate regions of time. If your wormhole is not acting as a time machine, you needn't worry. Adjusting it to do so is straight forward – at least in principle.

The following is a retelling of an illustrative tale due to physicist Kip Thorne. Suppose that you have a small wormhole in your living room. Both mouths of the wormhole float a meter above the floor, and they are a few meters apart. Suppose further that your wormhole has an extremely short throat. When you insert your arm into one mouth – which we will call the first mouth -- you immediately see it emerge from the second mouth. That you see it emerge immediately, and not, say, an hour later, is proof that your wormhole is not acting as a time machine (at least not in your reference frame).

After you remove your arm from the first wormhole mouth, enlist the help of your next door neighbor, Jane, who owns a spaceship. Have Jane put the second wormhole mouth aboard her spaceship, which is currently parked in her driveway. Instruct Jane to fly to the Alpha Centauri system four light years away. Ask her to maintain a speed of at least 99.99% of the speed of light for as much of the trip as possible.

While seated in your living room, you look into the wormhole's first mouth. You see Jane, who is now seated at the controls of her spaceship (Figure 10.1).

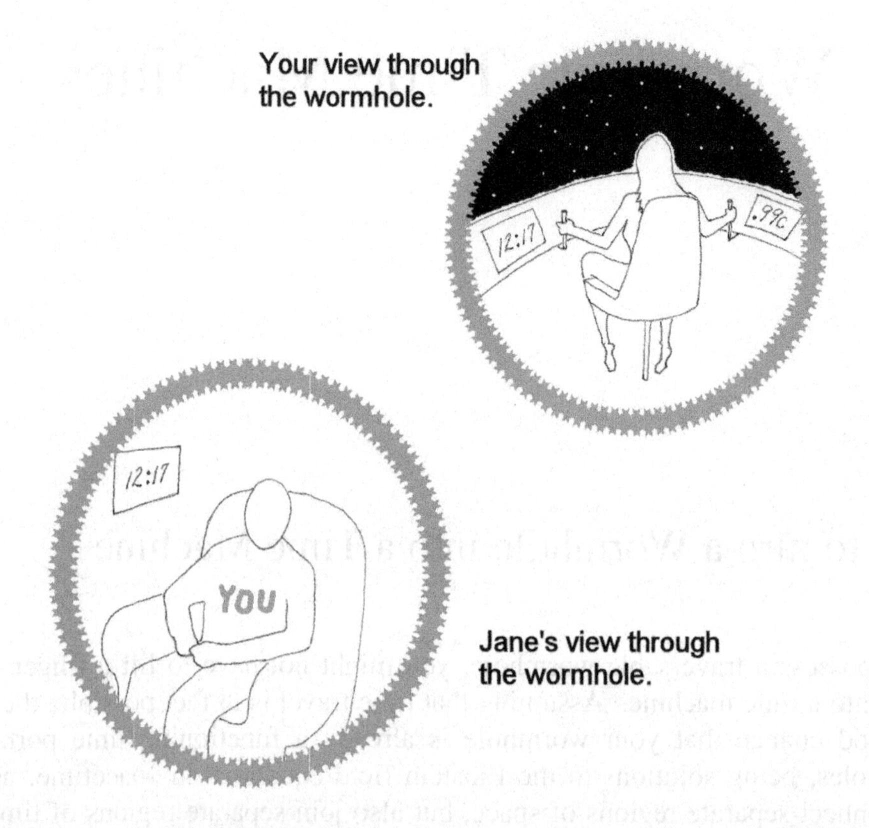

Figure 10.1. Views through a wormhole. One mouth of the wormhole is in your living room, the other in Jane's spaceship. When Jane looks through the wormhole, she sees you. When you look through, you see Jane. Jane's clocks remain synchronized with yours, allowing you to converse with her throughout her trip.

When she looks into the second wormhole mouth, she sees you comfortably seated in your living room (Figure 10.1).

Jane fires her engines, accelerates away from Earth, and within a few seconds you see, by peering through the wormhole at her speedometer, that she has reached 99.99% of the speed of light. Fortunately, Jane let her spaceship dealer talk her into installing a top-of-the-line inertial dampening system, which prevents her from being crushed by the high acceleration.

As Jane speeds towards the Alpha Centauri system and back, you hold a long conversation with her. You notice that the clock in your living room remains synchronized with the dashboard clock in Jane's spaceship. After six hours of conversation, Jane reports that she is approaching Earth. You watch through the wormhole, as she initiates a landing sequence. Through the wormhole you are able to see the scenery that Jane sees, as she looks out through the window of her spaceship. As she lands, you see the second story of her house appear in her spaceship's window, then the top of her garage, and finally the landscaping surrounding her driveway.

You leave your house and walk to Jane's driveway to welcome her home. You are surprised to discover, however, that her driveway is empty. You return to your living room, look through the wormhole, and observe Jane, still seated at her spaceship's controls, doing her post-flight shutdown. Then you notice something. Through the window of Jane's spaceship you see a fully grown Chinese elm tree in precisely the location that Jane planted a sapling last week! Jane's house, also visible in the background, is the wrong color!

In order to investigate you climb into the wormhole and emerge in Jane's spaceship. You turn on her ship's radio, find a commercial all-news station, and soon discover that you both have traveled – by different means -- eight years into the future. You look outside of the window and see a somewhat older version of yourself standing on Jane's driveway, waiting for her to emerge. Rather than meet your older self, you panic and dive back into the wormhole. You emerge from the wormhole into your living room shortly after

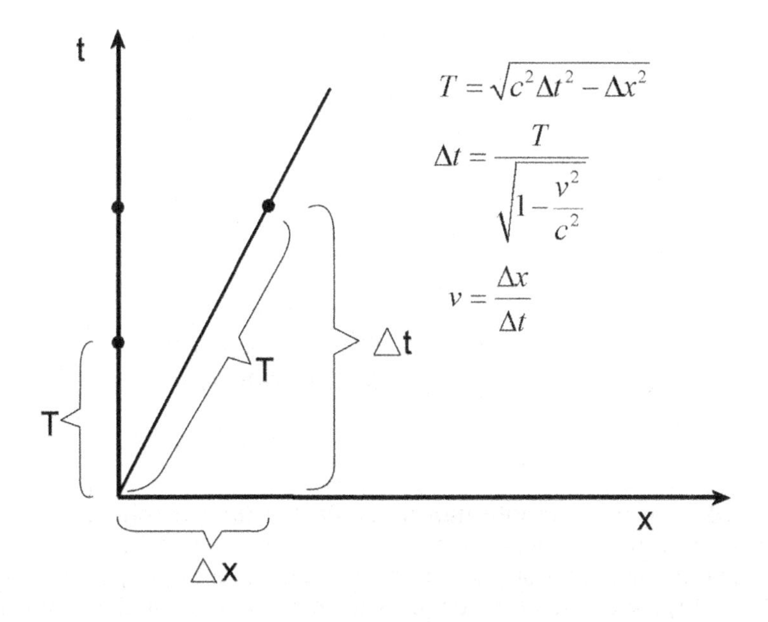

Figure 10.2. Apparent length of an interval depends on motion. The same time interval *T* is shown for a stationary observer and for an observer moving with speed *v*. According to special relativity the moving observer seems to the stationary observer to take longer to experience this time interval. This manifests itself in the diagram as an increase in the apparent length of *T* for the moving observer.

the time you first entered the wormhole – about six hours after the start of Jane's flight. You live your life. Eight years pass. At the expected date and time, you observe Jane's spaceship landing in her driveway. You walk onto her driveway toward her ship and notice a somewhat younger version of yourself standing in the spaceship behind Jane. You watch as your younger self appears to panic and dive into a wormhole mouth.

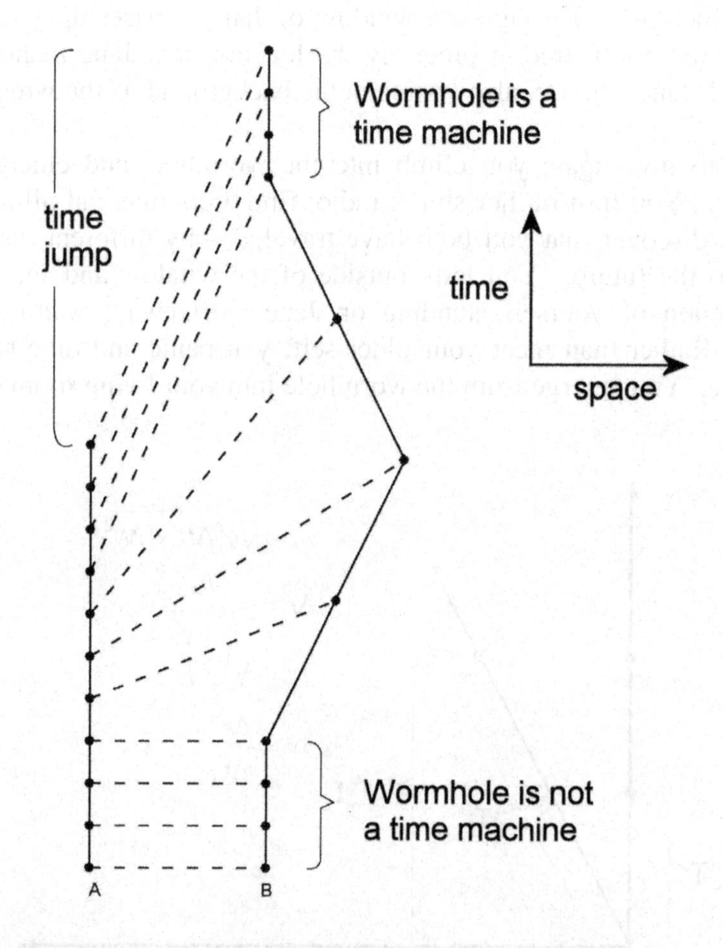

Figure 10.3. Moving a wormhole mouth creates a time machine. A wormhole that is not initially acting as a time machine has mouths *A* and *B*. Mouth *B* moves away from *A* and returns. Dots on each world line represent the elapse of a fixed time interval (e.g. 1 hour) as measured by observers located at each mouth. Dashed lines represent the wormhole connection between the mouths. Net result of the relative motion of the mouths is a non-zero time jump.

The story would have been much the same, if, instead of traveling near the speed of light before returning to Earth, Jane had brought her spaceship into the proximity of a black hole or other massive object before returning. Just as Einstein's theory of special relativity slows down the passage of time for an object that appears to be moving, it also slows it down for an object that enters a deeper gravity well. The faster the relative motion, the stronger the relative gravity, the greater the dilation of time, and the greater the time jump of the corresponding wormhole time machine.

From this story we glean the following principles.

P87. *A wormhole can be turned into a time machine either by 1) rapidly moving one of its mouths relative to the other or by 2) putting one of its mouths into a much stronger gravitational field than that experienced by the other, then bringing the mouths back into close proximity.*

P88. *The time jump supported by a wormhole time machine is given by the difference in the proper time experienced by its mouths in the process of creating the time machine.*

Recall that the proper time experienced by an observer is defined as the time that has elapsed according to his wrist watch.

We can better understand how moving one of the wormhole's mouths relative to the other will introduce a time shift between the mouths, if we consider the spacetime diagram shown in Figure 10.3. The key to this diagram is the realization that the closer an observer's speed approaches that of light (a 45 degree angle), the greater the apparent separation in the diagram between the ticks of the observer's clock. This rule allows the spacetime diagram to reflect the symmetry of the physical world under Lorentz transformations. The clock ticks of a stationary observer are depicted by divisions in the vertical axis. The clock ticks – which are of identical duration as experienced by the moving observer – appear in, accordance with the aforementioned rule, to have a greater separation on the world line of the moving observer (Figure 10.2). Recall that in the story above, Jane's clock continued to be synchronized with the clock in your living room. The wormhole, then, links successive clock ticks as shown in Figure 10.3. This ultimately results in a time shift between the wormhole mouths that defines the time jump permitted by the wormhole time machine.

Any means through which the wormhole mouths can experience different proper times can be used to create a time machine. Hence, we could move the second wormhole mouth (as slowly as we like) into a strong gravitational field. An observer at the first mouth may view through normal space (using, for example, a powerful telescope) the clock of an observer stationed at the second mouth. This clock will appear to be running slowly to the observer at the first mouth. This implies that these observers at the mouths will experience a different elapse of proper time before the mouths are reunited. Just as in the story, the observer at the first mouth will notice no such slow down when he views the second observer's clock through the wormhole. When the second mouth is returned (as slowly as we like) to the vicinity of the first mouth, they will have a relative time shift that will allow them to function as a time machine.

Non-Wormhole Time Machines

It's important to point out that wormholes are not the only means through which theoretical time machines can be constructed. The other methods usually involve huge angular momentum densities or surgery on spacetime.

Angular-Momentum-Dependent Time Machines

The Gödel Universe

In 1949 logician Kurt Gödel and Albert Einstein were good friends and both members of the Institute for Advanced Study at Princeton University. On the occasion of Einstein's seventieth birthday, which occurred that year, Gödel presented him with a novel cosmological solution to Einstein's own equations of general relativity -- the Gödel universe. It supposes that spacetime is filled with a rotating dust whose density happens to be equal to the cosmological constant multiplied by a very particular value. It turns out that if the dust swirls about fast enough, this solution will contain closed timelike curves (CTCs). Any traveler who takes a long circuitous spaceflight will return to her starting point *before* she left. As it happens, the matter in our universe is not observed to be swirling about in the manner required by Gödel's solution. Nor is the ratio of density of this matter to the universe's cosmological constant near the precise value required by this solution. In short, Gödel's universe is not the one in which we live.

The Kerr Black Hole

As we saw in Chapter 5, a rotating black hole described by the Roy Kerr's 1963 solution to Einstein's equation contains closed timelike curves beneath its event horizon – a time machine. One might argue, however, that real black holes, which are the result of stellar collapse, are not described by the entire Kerr solution, particularly not by the part describing closed timelike curves. One might further insist that even if these closed timelike curves exist, they are irrelevant. They cannot possibly effect the lives of anyone who refrains from diving into a rotating black hole.

The Van Stockum Cylinder

In 1937 Dutch physicist Willem Jacob van Stockum published an exact solution to Einstein's equations in which the only matter was assumed to form an infinitely long rigidly rotating cylinder of dust. He had rediscovered a solution published in 1924 by the Hungarian mathematical physicist Cornelius Lanczos. In 1974 American physicist Frank Tipler studied van Stockum's solution and was the first to realize that it contained closed timelike curves. It was another time machine. A traveler, irrespective of his distance from the cylinder, who pilots his spaceship in a loop around the cylinder could return to his starting point before he left. This caused no great stir in the physics community, because physicists also regarded the van Stockum's cylinder as irrelevant due its being unphysical. Infinitely long cylinders cannot exist. No known form of conventional matter could survive rotation at the speeds required to generate the CTCs. The large-scale structure of a universe containing such a cylinder is not asymptotically flat, which differs radically from the universe that we observe.

The Tipler Cylinder

This is a finite version of the van Stockum cylinder of dust -- that is it is a rigidly rotating cylinder of *finite* length. Although it would be exceedingly difficult to find an exact solution to Einstein's equation for this case, Tipler speculated that the cylinder would likely generate CTCs for a sufficiently rapid rotation. In 1992 Stephen Hawking proved that such a time machine occupying a finite region of spacetime could not form CTCs without violating the weak energy condition. Tipler's time machine, then, like wormholes, requires exotic matter. Unlike the van Stockum's cylinder, the large-scale structure of the spacetime hosting a Tipler cylinder would be compatible with that of the observable (asymptotically flat) universe. However, it is still the case that no known form of conventional matter could survive the stupendously high rotation required to produce CTCs.

The Spinning Infinite Cosmic String

A cosmic string is a theoretical type of unconventional matter. There are extremely dense, perhaps light years long, but are no thicker than a subatomic particle. Imagine an infinitely long and straight cosmic string whose axis of rotation was along the length of the string. Its rotation, then, could not be detected visually, even if one could somehow magnify the string enough to see it. Its arrangement is similar to an ultra-skinny van Stockum cylinder, except that it is not made of dust. Unlike dust the string material possesses an internal pressure. This pressure is parallel to the length of the string. If the string spins fast enough, CTCs will form. As in the cases of van Stockum and Tipler cylinders, anyone who travels in a loop around the spinning string will go back in time. Like the van Stockum cylinder but unlike that of Tipler, the CTCs are global, affecting the whole spacetime. It should come as no surprise that physicists also consider the spinning cosmic string to be unphysical. Although it does not require exotic matter, its length is infinite. Its composition remains speculative. Despite occasional claims to the contrary, the consensus view is that no cosmic strings have been observed.

The Gott Cosmic String Pair

In 1991 physicist J. Richard Gott of Princeton's Institute for Advanced Study solved Einstein's equations to discover a cosmic-string time machine with a surprising property. The cosmic strings in the Gott time machine are not spinning. Instead of depending on spin angular momentum, this time machine depends on *orbital* angular momentum. To better understand orbital angular momentum, imagine that you are standing in the middle of a two-way street, right on the double yellow line dividing the lanes of opposing traffic. In one lane, cars travel north, in the other south. Suppose further that there are only two cars on the street. One of them is a kilometer north of you and travels southward toward you. The other is a kilometer south of you and travels northward toward you. They travel at the same speed and remain in their respective traffic lanes. You stretch out your arms so that one arm is in the northbound lane, and the other is in the southbound lane. When the cars reach you, your hands each come into contact with the windshield of a car. One of your hands is pushed northward, the other southward. This causes you to spin, that is, to acquire angular momentum transferred to you from the cars. The moral: Even though the cars are moving in straight lines and are not spinning, they nevertheless possess a type of angular momentum – called orbital angular momentum -- about the point on which you stand.

Now imagine that each car has an ultra-thin pole of infinite length jutting vertically out of its roof. These poles – our cosmic strings – would also have angular momentum about the point on which you stand. As cosmic strings moving toward each other in this way near their point of closest approach, their combined angular momentum density is not dissimilar to that of the spinning infinite van Stockum cylinder or the spinning infinite cosmic string. A traveler whose spaceship takes him in a loop about this central point, while the cosmic strings approach it, will find it possible to return to his starting point before he started -- to travel backward in time. As with the all angular-momentum-dependent time machines, closed timelike curves are only possible if the angular momentum *density* is sufficiently high. For spinning cylinders, spinning black holes, or a universe full of spinning fluid, the angular momentum density is constant. For the approaching and separating cosmic strings it changes. The angular momentum density of a system is the ratio of the angular momentum of the system to the volume of the smallest box into which the system fits. As the cosmic strings approach, the volume of this box (i.e. of any arbitrarily tall but finite box) decreases. Hence, the angular momentum density rises. As the strings pass each other and separate, this volume increases, and the angular momentum density falls. Time travel is only possible, when the angular momentum density exceeds a certain value, i.e. only when the cosmic strings are sufficiently close.

===/===

Why High Angular Momentum Densities Create Time Machines

Each of these angular-momentum-dependent time machines applies a realization that emerges from Einstein's theory of special relativity. If an object can somehow travel faster than light, it can travel backward in time. Although special relativity forbids objects from exceeding the speed of light, there's a loophole. The space in which the object exists can itself induce motion. Just as a tailwind allows an airplane to effectively exceed its maximum speed, so can "currents" in space allow a spacecraft to exceed the speed of light. A familiar example of this is the expansion of the universe. Because of this expansion, the rate of separation between our galaxy and any other galaxy increases in proportion to their separation. For sufficiently large separations, this rate – the speed at which each galaxy moves relative to the other – exceeds the speed of light.

In these time machines, the "current" or motion of space itself is not due to expansion, but to rotation. Just as the water swirling around a drain causes a toy boat to circle the drain faster than its little motor would normally allow, so does the space swirling around a rapidly rotating object boost the top speed of an orbiting spacecraft. Any matter possessing a sufficiently high concentration of angular momentum will induce a detectable swirling of this sort. This will allow a spaceship to whip around the rotating source -- the black hole, cylinder, cosmic string, etc. – at speeds effectively exceeding that of light.

Why Faster-than-Light Travel Permits Time Travel

But how is it that an object able to exceed the speed of light can travel backward in time? To understand this, recall the properties of Minkowski space – the description of spacetime with a rectilinear coordinate system in which the vertical axis corresponds to time and the horizontal axis to space. Objects moving at subluminal (slower-than-light) speeds have timelike world lines, those tangents point vertically more than horizontally. Objects moving at the speed of light have lightlike world lines, those tangents point exactly as much vertically as horizontally (i.e. straight lines at 45 degrees to either axis). Objects that move at superluminal (faster-than-light) speeds have spacelike world lines, those whose tangents point horizontally more than vertically.

A spacelike world line corresponding to an object traveling at superluminal speed must have a *positive* displacement in time. This makes sense. A trip to Alpha Centauri, even at superluminal speed, would certainly require *some* positive displacement in time, wouldn't it? According to stationary observers on Earth, it certainly would. However, to certain other observers moving with respect to Earth, it would not. *Spacelike world lines describing a positive time displacement in a particular reference frame, describe a negative such displacement in certain others.* Imagine that Alice pilots her spaceship on a trip at superluminal speed from Earth to Bob's spaceship, which is heading away from earth at a high but *sub*luminal speed. The dashed line connecting events A and B in Figure 10.4 depicts this leg of her trip. Figure 10.4 shows the relevant part of Bob's coordinate system as determined by the rules of special relativity. According to observers on Earth, Alice's rendezvous with Bob involved a positive displacement in time ($t_B > t_A$). But according to Bob, Alice's trip involved a negative displacement in time ($t'_B < t'_A$). Moreover, Alice's return trip (dashed line connecting events B and C in Figure 10.4), appears to observers on Earth to require a negative time displacement ($t_C < t_B$). But to Bob it is a superluminal trip with a positive displacement in time ($t'_C > t'_B$). Alice's superluminal trip away from Bob and back to Earth cannot -- according to Einstein's Principle of Relativity -- be forbidden. According to

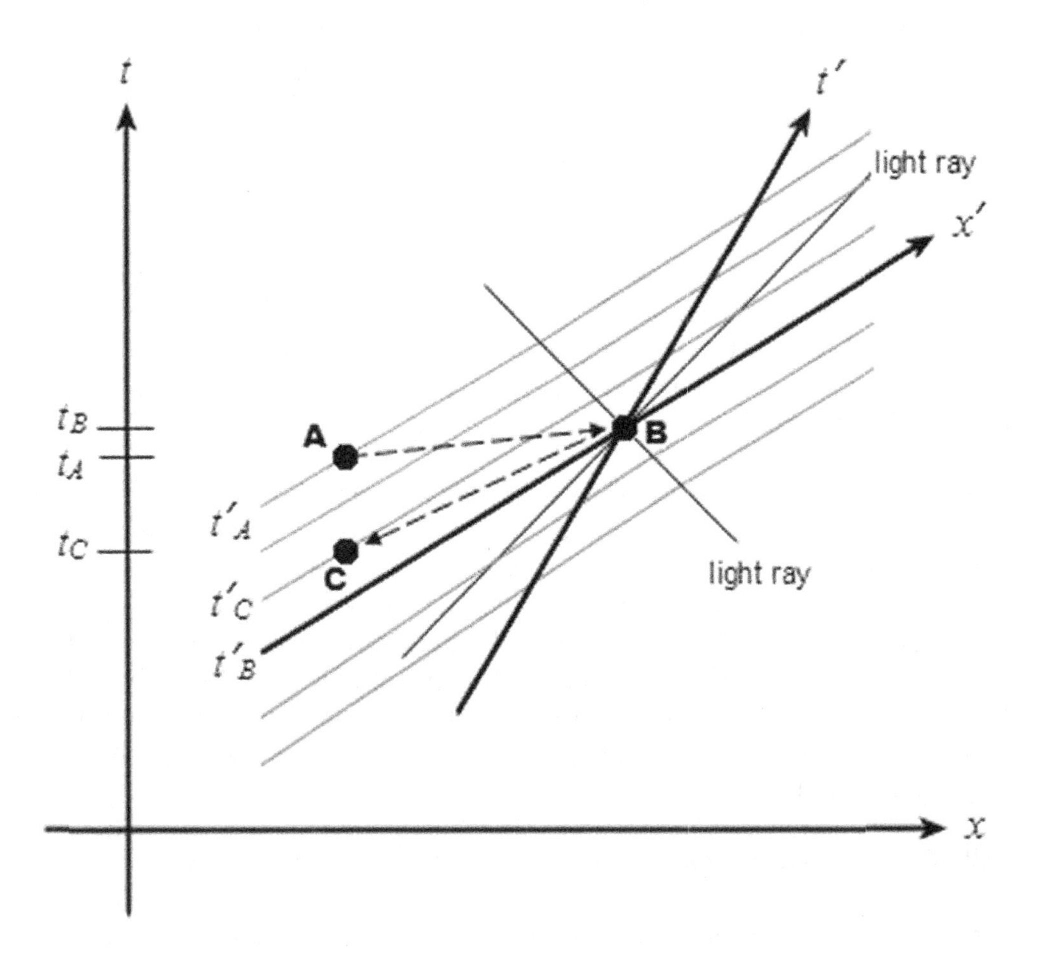

Figure 10.4. Faster-than-light travel implies time travel. According to Einstein's special theory of relativity, observers in relative motion see a different chronological order of the same events. Observers on Earth use the (t, x) coordinate system. Observers on Bob's spaceship speeding a way from Earth at a large but ***sub***luminal speed use the (t', x') coordinate system. Gray lines of equal time are shown for the (t', x') coordinate system. For the (t, x) coordinate system these are just horizontal lines (not shown). Superluminal travel occurs along spacelike intervals. In the (t, x) system these are any lines whose angle with respect to the x axis is less than that of light rays (45 degrees). In the (t', x') system these are any lines whose angle with respect to the x' axis is less than that of light rays shown. Three events in Alice's superluminal journey (dashed lines) are shown:

Event A: Alice leaves Earth at superluminal speed

Event B: Alice arrives at Bob's spaceship, which is moving rapidly away from Earth at subluminal speed, and she immediately heads back toward Earth at superluminal speed.

Event C: Alice arrives back on Earth before she left.

As viewed from Earth, A precedes B ($t_A < t_B$). As viewed from Bob's spaceship, B precedes A ($t'_B < t'_A$) and C precedes A ($t'_C < t'_A$). Superluminal travel from event B to event C appears to require a backward time shift in Earth's reference frame, but it only involves a forward time shift according to observers in Bob's spaceship. The principle of relativity states that the laws of physics are the same in all reference frames. If superluminal travel as seen from Earth is allowed, superluminal travel as seen from Bob's moving spaceship must also be allowed, even if this allows Alice to travel backward into her own past.

this principle, the laws of physics are the same in all inertial reference frames. If superluminal speeds are allowed in Earth's reference frame, they must be allowed in all reference frames, including Bob's. The combined effect of both leg's of Alice's trip is to take her backward into *her own* past ($t_C < t_A$). This is genuine backward time travel.

Hence any means of superluminal transportation permits time travel. For example, the "warp drive" proposed in 1994 by Mexican physicist Miguel Alcubierre will do so. It uses exotic matter to contract the space in front of a spaceship and to expand that behind it in order to propel the ship at superluminal speed. It is another means of exploiting a current in space itself – this time a superluminal bubble containing the ship – in order to effectively violate the cosmic speed limit. A round trip journey enabled by the Alcubierre drive could (even in the absence of angular momentum) carry a voyager into his own past. Similar remarks hold for the Krasnikov tube mentioned in Chapter 1.

P89. *There is no restriction on the speed at which space itself moves.*

P90. *The motion of space itself (e.g. expansion, rotation, bubble wave, or tubular flow) can be exploited to move an object at superluminal speed.*

P91. *Round-trip travel at effectively superluminal speed permits backward time travel.*

Spacetime-Surgery-Dependent Time Machines

Time machines that depend on spacetime surgery are those whose theoretical construction depends on cutting and joining pieces of spacetime. They are not a result of a particular distribution of mass and energy. Rather they require a particular specification of the geometry or topology of the spacetime manifold. Although Einstein's equations are satisfied throughout, these time machines are less realistic than others in that there is no apparent procedure, such as creating suitable concentration of spinning matter, through which they could be constructed. These are chiefly applied as thought experiments that probe for inconsistencies in our understanding of physics. Two examples follow.

The Minkowski Tube World

If we ignore two of the spatial dimensions, we can visualize Minkowski space as an infinite sheet of paper. The time dimension is vertical; space is horizontal. Suppose that we cut this sheet along two horizontal lines. Let the first be the x axis, $t = 0$. Let the second be a parallel line above the x axis, $t = T$. We now have an infinitely wide sheet of paper of finite height T. This completes the cutting part of our surgery. Now we'll do the joining. Join the bottom of this sheet with the top of it. That is, we "identify" $t = 0$ with $t = T$. We now have a tube. We saw this process before in Figure 5.23.

Any observers in a Minkowski tube world live in endless time loops. Some of them cannot distinguish their future from their past. They encounter in general numerous younger and older versions of themselves. These meetings lead to the time travel paradox that we'll consider in the next section.

The Deutsch-Politzer Room

In 1991 Oxford's David Deutsch considered a particular surgical alteration of spacetime that was independently considered by Caltech's Nobel-prize-winning physicist H. David Politzer in 1992. Suppose that we have a special room with an open door at either end. If you enter the room between 12:00 pm and 12:01 pm, you will be caught in a time loop. When the clock in the room strikes 12:01, it becomes 12:00 again. You are transported backward in time by 1 minute. This endlessly repeats. Just as in a Minkowski tube world, you will in general encounter numerous older and younger versions of yourself. Anyone, however, who enters the room before 12:00 or after 12:01 experiences no such loops. If you enter before 12:00 pm, you are, when the clock strikes noon, instantaneously transported to 12:01 pm -- the moment at which the room ceases to operate as a time machine.

There are two particularly interesting features about this time machine. The first is that it shows that unpredictability can seemingly arise from predictability. In the absence of quantum mechanics the future is precisely determined by the past. This is called "classical determinism". The Deutsch-Politzer room appears to fall completely within the purview of classical physics. Yet when this time machine activates at 12:00 pm, classical determinism vanishes.

Here's how. Suppose you walk into the room one second after 12:00 pm, when the room is functioning as a time machine. Let's say that it's a large room. It takes you a full 2 minutes to walk to the opposite door and leave the room. One possible outcome is that you walk through the room without incident. However, another possibility is just as likely. You might enter, walk a few meters, and collide with a slightly older version of yourself, who was standing stationary in the room. As a result of this collision, two things happen. You come to a halt. And the slightly older version of you is jostled into forward motion across the room. As you continue to stand in place, the clock strikes 12:01 pm, you go back in time by 1 minute, and shortly thereafter become the slightly older version of yourself that caused the initial collision. After the collision, you are jostled into proceeding toward the other door, and you exit the room.

Because both of these scenarios (and numerous others) are equally valid, the presence of the time machine destroys classical determinism. Note that this loss had nothing to do with human consciousness. Had a bowling ball attempted to traverse the room instead of a human being, it too would have collided with a slightly older version of itself *or* not. Predictability would still have been lost. All that remains is a relatively jejune form of "quantum determinism".

The second feature, which is more clearly evident with the Deutsch-Politzer room than with the Minkowski tube world, is that there is a clear division of spacetime into regions where classical determinism holds and those where it does not. Recall that the boundaries between these regions are called "Cauchy horizons". Once the time machine activates, it emanates a sphere of unpredictability that spreads outward at the speed of light. Closed timelike curves (time loops) within the time machine will beget CTCs within this growing sphere. These CTCs would in turn beget others. Every point in the entire universe will eventually fall within the Cauchy horizon and thereby become infected with unpredictability. Activating a time machine, even for a single minute, would have repercussion for the entire universe and the rest of time.

Time Travel Paradoxes

If one assumes that time travel to the past is possible, two types of paradoxes arise. I shall refer to them as the paradox of the "self-preventing timeline" and that of the "creator-less artifact". Both have figured prominently in science fiction literature and popular entertainment. Their importance to wormhole physics follows from

P92. *Because traversable wormholes are in general time machines, wormhole physics is incomplete until all time travel paradoxes are resolved.*

The Self-Preventing Timeline

This is the more familiar of the time travel paradoxes. It is commonly known as either the "grandfather paradox", the "grandmother paradox", or the "consistency paradox". Imagine that you go back in time and kill your grandfather before he fathers any children, thus preventing your own birth. If you were never born, how could you have gone back in time to have done this? The paradox does not, of course, depend upon killing your grandparents. The same paradox would arise if you went back in time to the moment five minutes before you entered the time machine. If at that moment you were to kill your earlier self and commit suicide, how could you have returned from the future to have committed these acts? This paradox will occur whenever events resulting from backward time travel prevent the backward time travel from occurring.

This paradox is not limited to conscious actors possessing free will. It applies equally well to simple objects. As string theorist Joseph Polchinski once pointed out to Kip Thorne, a billiard ball could enter a wormhole, go back in time, collide with its earlier self, and thus prevent itself from entering the wormhole.

The Creator-less Artifact

The existence of creator-less artifacts is not a true paradox in that it does not define a contradiction. It is instead an unpalatable condition that allows entities to exist without their having been created. Such entities have been dubbed "jinns" by physicist Igor Novikov. Pseudo-paradoxes of this type are also known as bootstrap paradoxes. They are named for Robert Heinlein's 1941 story, "By His Bootstraps", in which this pseudo-paradox plays a pivotal role.

Suppose that you take your copy of *Moby Dick* back to1840 and hand it to Herman Melville. Suppose further that instead of writing the novel, he copies it word for word and submits it to his publisher. A century and a half later, you purchase the copy that you will later transport to Melville. Who wrote the novel? The existence of the time machine allows the novel to exist without having a creator.

This pseudo-paradox need not be limited to creative works. It pertains as well to physical objects. Suppose that I have retained possession of a silver dollar that was given to me by a mysterious stranger, when I was ten yeas old. Suppose that as an adult I take this silver dollar back in time and become the mysterious stranger, who gave it to me when I was ten. The silver dollar exists in a time loop, but it was never created.

How to Resolve Time Travel Paradoxes

There are three major approaches to eliminating time travel paradoxes: 1) forbid time travel, 2) forbid classical inconsistency, or 3) forbid quantum inconsistency.

Forbid Time Travel

There are two ways in which time travel can be banned from physics. The first is to regard as unphysical any solution to Einstein's equations that contains closed timelike curves (paths that permit backward time travel). This would mean banning the Kerr solution, which, as you recall, describes a rotating black hole. While the degree to which the Kerr solution describes the end state of stellar collapse is unknown, the creation of closed timelike lines as a consequence of stellar collapse cannot be ruled out. Banning stellar collapse from physics in order to prevent the creation of closed timelike curves would, of course, be unreasonable. The alternative is to require all closed timelike curves to exist within event horizons. This would further imply that the description of regions of spacetime interior to event horizons is beyond the scope of physics.

The second approach to banning time travel is less drastic. It would allow the existence of traversable wormholes but not that of closed timelike curves. It conjectures that quantum effects prevent wormholes from forming closed timelike curves. This would prevent wormholes from being used as time machines. This, as you will recall, is known as Hawking's Chronology

Protection Conjecture. Attempts to justify it involve using semiclassical methods to calculate the vacuum energy density to which a prospective time traveler would be exposed. There is reason to suppose that this energy density might become arbitrarily high, thus destroying the wormhole and its closed timelike curve.

One might expect this to occur because the closed timelike curve creates a positive feedback loop. If a vacuum fluctuation creates a particle, the closed timelike curve could return that particle to the time and place of its creation. There are now two particles there. These two are also returned by the closed timelike curve, resulting in four particles at the point of origin. The process continues to double the number of particles until the increased energy density caused by them alters spacetime and destroys the wormhole and its associated closed timelike curve.

These calculations have only been carried out to first order in the quantum correction to general relativity. They can hint at but cannot confirm the onset of such a destructive feedback loop. Confirmation would require a theory valid at all orders of the quantum correction -- a full blown theory of quantum gravity. The calculations are vitiated in another way. They implicitly rule out the Many Universes Interpretation of quantum theory. As will become clear below, the Many Universes Interpretation prevents the existences of destructive temporal feedback loops.

Forbid Classical Inconsistency

Forbidding classical inconsistency is the same as forbidding changes to history. In other words, any attempt to return to the past and change it is conjectured to be impossible. If you go back in time and attempt to kill yourself, events will conspire to prevent you. You will slip on a banana peel just as you are taking aim, or your gun will jam, or someone will hit your arm just as you are squeezing the trigger, or you will be thwarted in some other way. Despite your best efforts you will not be able to change the past. The only allowed journeys to the past are those that are perfectly consistent with it. Suppose, for example, that you remember a childhood incident in which someone who looks as you do now played a game of one-on-one basketball with you. You may go back in time and play that game with your childhood self. You will find, however, that any of your attempts to change the game from the way you remember it (assuming that you remember it accurately) will fail. This imposition of classical consistency is the approach championed by Igor Novikov.

Imposing classical consistency eliminates the paradox of the self-preventing timeline. Unlike the ban on time travel, it does not prevent creator-less-artifact pseudo-paradoxes.

P93. *The creator-less-artifact time travel pseudo paradox is not resolved by imposing classical consistency.*

Forbid Quantum Inconsistency

Terminology: Parallel vs. Co-resident Universes

Before considering quantum inconsistency, we must eliminate a source of confusion in our terminology. You should understand that the word "universe" can have at least two distinct meanings. The appropriate meaning depends upon the context. The first definition is the familiar one. A universe is an isolated and self-contained region of space and its contents. The region is isolated in the sense that it has no contact with anything external to it. More than one such region may exist. These regions have no influence whatsoever on each other. They may be described as universes coexisting independently in an unphysical higher-dimensional embedding space. This space is unphysical in the sense that no particles or forces can propagate within it. Distance is not defined within this space. If, however, the Large Hadron Collider ever sees evidence of higher dimensions, it would be safe to suppose instead that this embedding space is indeed physical and that it allows the propagation of gravity. Until such astounding experimental results occur, I will continue to adhere to the conventional denial of the physicality of this space. I will further refer to universes contained with this unphysical space as *co-resident* universes (Figure 10.4).

If two co-resident universes are connected to one another solely by a wormhole, they may be said to be quasi-isolated. Because they are quasi-isolated, we still speak of this situation as consisting of two distinct (but connected) universes, rather than describing it as a single composite universe.

P94. *Multiple co-resident universes may independently exist. When they are connected solely by diffuse concentrations of wormholes, the resulting network is not conventionally regarded as a single composite universe.*

The other meaning of "universe" stems from the Many Universes Interpretation of quantum theory, which is more familiarly known as the Many Worlds Interpretation. In this case universes are defined as coexisting "planes" of reality. One is tempted to say that these planes occupy the same physical space. But this is not quite accurate in that physical space itself can vary from plane to plane. Just as one can imagine co-resident universes all occupying a higher-dimensional embedding space, one can imagine these planes coexisting in an infinite-dimensional metaphysical "reality" space. A point in such a space would specify every aspect of a particular reality – your height, your weight, your income, the number of teacups in your kitchen, the number of insects in Africa, the number of ships on the oceans, the diameter of the earth, its distance from the sun, the sun's mean surface temperature, etc. For each of these planes, one of a typically infinite number of physically possible histories has occurred. These planes of reality are called parallel universes. These are universes normally considered in the context of quantum theory.

P95. *Just as co-resident universes can be regarded as coexisting in a higher-dimensional embedding space, parallel universes can be regarded as coexisting in an infinite-dimensional "reality" space.*

As mentioned above, the higher-dimensional space hosting co-resident universes, while conceptually useful, has traditionally been regarded as physically meaningless. The same can be said of the "reality" space containing all parallel universes. In short, physics has hitherto offered unsatisfying answers to the question of *where* these other universes physically exist. Nor has it explained the influence that parallel universes are known to have on each other.

Irrespective of where parallel or co-resident universes exist, they remain conceptually distinct. A parallel universe exists for each possibility permitted by the laws of physics, and these universes influence each other. Co-resident universes, by contrast, are isolated and governed in general by distinct physical laws. While the Many Universes Interpretation asserts that universes exist that are like ours except that, say, the Soviet Union never fell, there is no compelling reason to believe that co-resident universes of this sort necessarily exist.

Or is there? Suppose that reality only admits a *single* system of physical laws. Suppose further that this system is sufficiently general to include numerous possibilities for the values of the fundamental constants of nature, such as the mass of the electron. Then any co-resident universe – even those with different fundamental constants – is a possibility permitted by this single system of physical laws. This means that every co-resident universe is part of the multiverse defined by these laws. If every co-resident universe is part of the multiverse, then we might *suppose* that every parallel universe of the multiverse is co-resident. In the context of these assumptions, then, we are free to identify co-resident universes with parallel ones. This identification would necessarily endow the embedding space a physical property. The space would have to permit the interference (as opposed to the interaction) known to occur between parallel universes.

Interestingly, the so-called "landscape" of string theory (widely regarded as evidence of the theory's failure to make testable predictions) seems to be an example of the aforementioned single, yet general system of physical laws. Although it is formulated as a single system of physical principles, it engenders a virtually unlimited number of manifestations (valleys in the landscape), each of which is characterized by a distinct set of values for the constants of nature. String theory has, moreover, popularized the view of our universe as a 3-dimensional membrane, or "3-brane", embedded in a higher-dimensional space known as the "bulk". The bulk differs from the unphysical embedding space described above in two ways. Distances are defined within it, and it can be permeated by one of the four forces of nature – gravity. In certain "brane world" scenarios parallel universes are conjectured to be co-resident 3-branes that are literally parallel in the bulk to our universe. These spatially parallel 3-branes are, moreover, only fractions of a millimeter away from each other. You may imagine this collection of parallel yet co-resident (flat) 3-branes as a floating stack of paper in which the sheets are separated by a fraction of a millimeter. Recently, it was shown that higher-dimensional wormholes might exist in the bulk. If so, they would explain how universes that are not adjacent in the stack of 3-branes could nevertheless be connected. Brane worlds, then, might be a step toward answering the questions of where parallel universes exist physically and why they influence each other.

In summary, the concepts of "parallel universe" and "co-resident universe" are not the same. We may only equate them if we assume reality to admit a single fundamental system of physical laws. This allows any co-resident universe – no mater how weird its physics – to be included in the set of all possible universes parallel to the one that we live in. Be that as it may, the universes considered in the remainder of this chapter are parallel universes.

Warning: Some authors use the term "parallel universes" to refer to: 1) what I have called "co-resident universes", 2) regions of the universe that are out of causal contact due to the cosmic expansion (Hubble volumes), 3) "valleys" in the string theory landscape, 4) bubble universes formed by chaotic eternal inflation, or 5) realms supporting distinct sets of mathematical axioms.

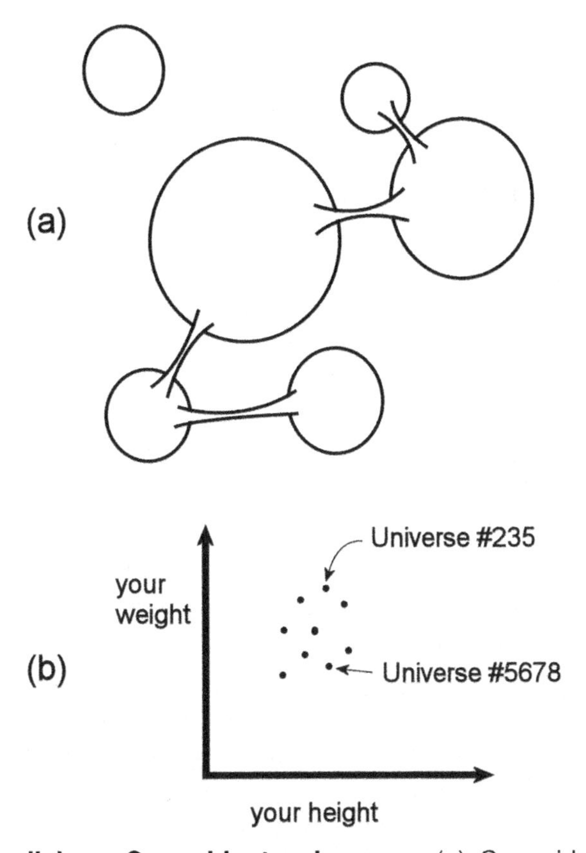

Figure 10.5. Parallel vs. Co-resident universes. (a) Co-resident universes co-existing in a higher-dimensional embedding space can be connected by wormholes. (b). Parallel universes may be thought of as points in an infinite-dimensional "reality" space. Each point specifies a particular reality. The two-dimensional subspace of this reality space that specifies your height and weight, which varies across parallel universes, is shown.

Imposing Quantum Consistency

Like the imposition of classical consistency, imposing quantum consistency prevents the past from being changed. Despite this it will not seem to a time traveler that her efforts to affect the past are being thwarted by an apparent conspiracy of events. Unlike the imposition of classical consistency, it will seem to her that she is free to alter the past however she wishes.

The imposition of quantum consistency relies on the Many Worlds Interpretation of quantum theory. The basic idea, which seems to have originated in science fiction, is that a time traveler bent on changing the past may do so without creating a paradox, because his activities after emerging from the time machine occur not in his universe of origin but in a newly branched parallel universe. This idea was elaborated and formulated rigorously by David Deutsch in late 1990 and published in 1991.

The sense in which the past is protected from change by quantum consistency is that it is now conjectured to consist of a collection of an infinite number of parallel universes. Each of these universes describes one of every possible history. These histories include every possible intervention by me (and any other time travelers for that matter). I can now go back in time and kill my younger self without creating a paradox. This is because the past already contains a universe in which I returned and shot myself, one in which I returned and stabbed myself, and others for every possible form of violence that I could have employed. When I return to the past with the intention of eliminating my younger self, I return to that preexisting universe whose history seamlessly matches my intention. If, for example, I decide to go back in time with a crowbar to bludgeon my younger self to death, I do not return to any of the universes in which I had earlier emerged from the time machine carrying, say, a rifle. I will instead emerge in one of the universes in which I had earlier emerged from the time machine in the same state in which I now enter it – carrying a crowbar, dressed in a particular way, and intent on a particular plan of action (bludgeoning). When I carry out that plan of action, I do not affect my universe of origin. This is the way in which the paradox of self-preventing timelines is resolved. Self-preventing timelines are impossible, because a time traveler cannot in general enter a time machine from a particular universe and visit the past of that universe. The time traveler may visit the past of her universe of origin, only if her path is classically consistent in the sense of the previous section. That is to say, she can visit the past of her own universe only if by so doing she does not change it in any way. This requirement effectively bans time travel to the past of your universe of origin. If you did *not* emerge from a time machine in the past of your universe of origin, you cannot change that. You cannot go back. Even if the history of your universe does record your having emerged from a time machine, it is nevertheless impossible to return to the past of your universe. In order for such a journey to occur every particle of your body would have to be in precisely the same state as they were when you previously emerged from the time machine. If a single electron is out of place when you enter the time machine, you will not return to the past of your own universe but to that of an exceedingly similar parallel universe (Figure 10.6).

To impose quantum consistency, Deutsch uses what's called a "density matrix". This is a means of specifying the state of a quantum system that might be a statistical mixture of pure quantum states. He imposes quantum consistency by requiring the state of the system, specified by its density matrix, to be single-valued (consistent) at all points in spacetime. In other words, when the quantum mechanical descriptor of the system (the density matrix) is evolved from a particular position in spacetime forward in time through a wormhole mouth in the future and back via the wormhole to the original position, its evolved value at that position must match its original value.

The imposition of quantum consistency eliminates self-preventing timelines and their associated paradoxes. Moreover, by effectively forbidding time travel to the past of one's own universe, the creator-less-artifact pseudo paradox no longer occurs. Although entities – be they physical objects or packets of information – can be moved between universes, they cannot do so without first having been created in one of them. Quantum consistency also reduces the cogency of the Chronology Protection Conjecture. Recall that the Conjecture follows in part from semiclassical calculations that indicate that a wormhole time machine would be destroyed by a positive feedback loop caused by virtual particles endlessly reentering the wormhole. But this will no longer occur. Virtual particles entering the wormhole will not emerge in the past of the universe in which they originated but in the pasts of other universes in the multiverse. This breaks the feedback and preserves the wormhole.

A Consequence of Quantum Consistency: "Sliding"

We see, then, that in order to make the Many Universes Interpretation of quantum theory consistent in the presence of closed timelike curves, it must be possible for objects from one universe to enter into the past of another. Nothing in principle prevents us from setting up our wormhole in such a way as to make this time jump to the past as short as we like. This means that our wormhole can, for example, send us a fraction of a second into the past of parallel universes. Such travel between (similar) contemporaneous parallel universes has in recent years acquired a name. A popular American television series that debuted in the mid 1990s called it "sliding". Most physicists, were they to consider it, would regard the notion of sliding as preposterous. After all, most of them so regard the Many Universes Interpretation. I suspect, however, that even those that subscribe to the Many Universes view would consider this possibility to be outrageous. Yet sliding is merely an obvious special case of what I have described above -- a well known means of resolving time travel paradoxes.

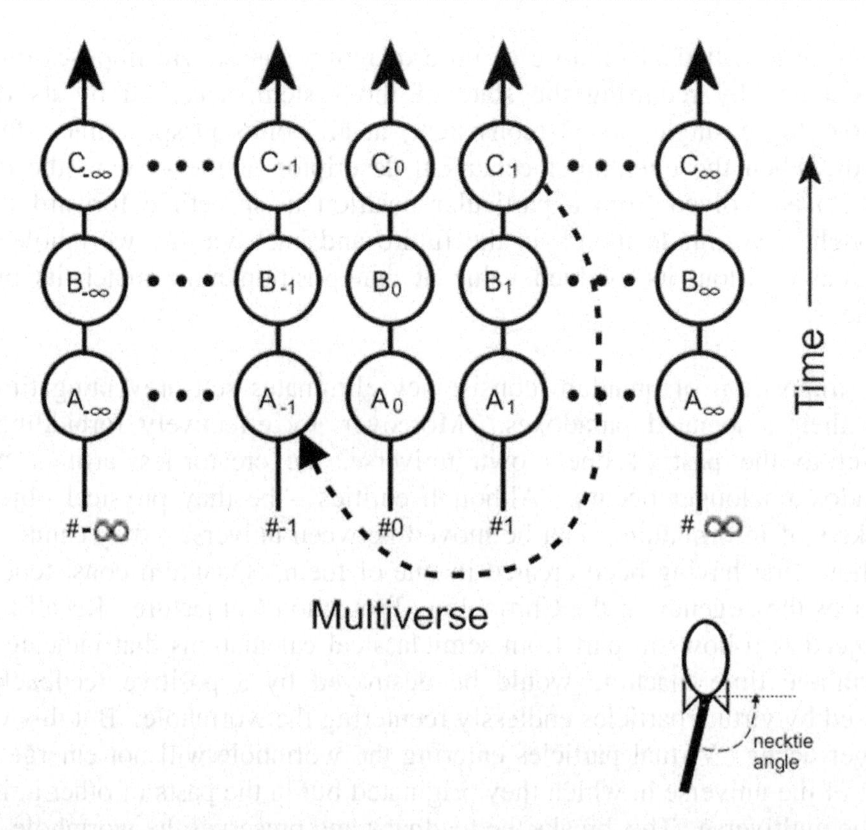

Figure 10.6. Quantum consistency in the Many Worlds Interpretation. Each universe of the multiverse has a history of events (circles) experienced by its occupants. There are an infinite number of these history strings, each of which has a label, e.g. # -27or #482. Labels range from negative infinity ($^-\infty$) to positive infinity (∞) in order to depict the infinite number of history strings. A time traveler from Universe (history string) #1 travels to the past of Universe #-1. Event C_1 is defined as the traveler entering the time machine with his necktie skewed at an angle of 91 degrees. Event A_{-1} is the time traveler emerging from the time machine with his necktie skewed at some angle. Quantum consistency requires this angle to be 91 degrees. The traveler does not return to his own past in Universe #1, because the chances of the necktie angle of Event A_1 being *precisely* equal to that of C_1 is virtually zero. To be concrete, we may define the events of this example as follows:

- C_{-1} – Traveler enters time machine with necktie skewed at 89 degrees.
- C_0 – Traveler enters time machine with necktie skewed at 90 degrees.
- C_1 – Traveler enters time machine with necktie skewed at **91** degrees.
- B_{-1} – Traveler dons necktie skewed at 89 degrees.
- B_0 – Traveler dons necktie skewed at 90 degrees.
- B_1 – Traveler dons necktie skewed at 91 degrees.
- A_{-1} – Traveler observes older self exit time machine with necktie skewed at **91** degrees.
- A_0 – Traveler observes older self exit time machine with necktie skewed at 92 degrees.
- A_1 – Traveler observes older self exit time machine with necktie skewed at 93 degrees.

Events C_1 and A_{-1} are consistent. So traveler goes from Universe # 1 to Universe # -1, where his actions have no effect on his universe of origin (# 1). Of course, there is no time clock external to all universes. Hence, only exceedingly similar universes with nearly identical clocks can be organized as shown.

P96*. The Many Universes Interpretation of quantum theory resolves all time travel paradoxes and pseudo paradoxes.*

In hindsight, it shouldn't have been particularly surprising that wormholes permit sliding or inter-parallel-universe communication. Back in 1991 Joseph Polchinski, then at the University of Texas at Austin, all but predicted as much. Steven Weinberg, a Nobel-prize-winning physicist also at the University of Texas, had earlier formulated a theory of nonlinear quantum mechanics. Polchinski realized that it implied the possibility of communication between distinct parallel universes. Nonlinear quantum mechanics enables a theoretical universe-to-universe communicator that Polchinski dubbed an "Everett phone", after Hugh Everett, who originated the Many Worlds Interpretation in 1957. What does nonlinear quantum mechanics have to do with wormholes? Quantum mechanics formulated in a spacetime containing wormholes is manifestly nonlinear. By connecting through closed timelike curves events at which the quantum state vector of the system is defined, the wormhole forces the state vector to interact with itself. Such self interaction is the very definition of nonlinearity.

P97. *An intra-universe wormhole prepared as a time machine with a slight time jump permits travel (sliding) between contemporaneous parallel universes.*

So a slightly time shifted *intra*-universes wormhole connection permits sliding between parallel universes. While an *inter*-universe wormhole connection permits travel between co-resident universes. One might speculate, then, that parallel universes and co-resident ones are somehow related. And, as we have seen in the discussion in the previous section, they are.

Pseudo Quantum Consistency

An alternative approach to imposing what might seem at first to be a form of quantum consistency proceeds as follows.

1. Impose classical consistency, i.e. forbid all self-preventing paths.

2. Calculate the quantum mechanical probability of particular states of the universe by summing over all possible histories that are consistent with step one, i.e. exclude self-preventing paths from the sum over histories.

This is not a true imposition of quantum consistency in that no intrinsically quantum mechanical object, such as the wave function or density matrix, is restricted. It is merely a restriction of quantum mechanics to a classically consistent domain. As such it imposes no additional constraints on timelines beyond those imposed by classical consistency.

P98*. There are two theoretical means through which all time travel paradoxes and pseudo paradoxes can be resolved: the development of a quantum theory of gravity that renders (backward) time travel impossible, or the embrace of the Many Universes Interpretation of quantum theory.*

The Case against Chronology Protection – The Roman Ring

The strongest indication of the validity of the Chronology Protection Conjecture is the potentially destructive effect of the aforementioned positive feedback loop. Recall that this occurs when vacuum fluctuations are amplified with each passage through a wormhole time machine. The resulting unbounded growth of the local energy density might be enough to distort spacetime sufficiently to destroy the wormhole. We cannot be sure that this occurs, because the rising energy density eventually renders invalid the semiclassical approximation used to calculate the rise. We must make do with a semiclassical approximation, because a complete theory of quantum gravity has yet to be formulated.

It turns out, however, that there is a way to construct a wormhole time machine that circumvents this chronology-protecting feedback loop. Back in 1986 Thomas Roman of Central Connecticut State University pointed out to Morris and Throne that their traversable wormhole -- by virtue of effectively permitting faster-than-light travel -- could be used to construct a time machine. This led Morris and Thorne to include a description of a two-wormhole time machine in their original paper. An interesting feature of Roman's two-wormhole configuration is that neither of the wormholes is itself a time machine. A traveler could go back in time only if she traverses *both* wormholes. Moreover, Maxim Lyutikov, then at Caltech, was later able to show that this configuration arbitrarily weakens the energy-density growth resulting from the positive feedback loop.

Matt Visser extended these results to the case of a time machine constructed from an arbitrary number of wormholes. As in the two-wormhole Roman configuration, none of the wormholes or any subset of them acts as a time machine, only the entire configuration. The wormholes are arranged in a spatial ring that first leads away from the point of origin and then back to its vicinity. The traveler is thus able to go backward into more or less the same spatial location. Visser dubbed this arrangement a "Roman ring".

At this point you might be wondering how it is possible for a collection of wormholes to function as a time machine without any one of them functioning as a time machine. The answer is that it is possible for a wormhole to carry a traveler back in time (as judged by a particular observer) without being considered a time machine. A wormhole is a time machine only when it carries the traveler back into her *own* past -- the region of spacetime consisting of events that might have affected her in the present. This region is called the traveler's *past light cone*.

If, for example, you enter a wormhole that instantaneously takes you to the Alpha Centauri system as it was five minutes ago, this is not considered time travel. The rules of special relativity guarantee that there are observers for whom your traversal through the wormhole seemed to be a journey *forward* in time. The only time that there are no such observers is when you travel into your own past light cone. The Alpha Centauri system as it was five minutes ago is not within this cone. No event in this region of spacetime could affect you here and now, because it would take over four years for any influence (traveling at the maximum permissible speed, i.e. that of light) to reach you.

The trick of the Roman ring is that each of its constituent wormholes takes you back in time, as judged by an observer who is stationary with respect to the wormholes. But for each wormhole, the event of the traveler's exit is never in the past light cone of the event of the traveler's entrance. No wormhole acts as a time machine. However, the ring's chain of wormhole jumps is arranged to return the same spatial location from which the traveler left. The traveler's exit from this location is indubitably in the past light cone of the traveler's entrance from the same location. The ring, then, is a time machine, even though none of its wormholes are.

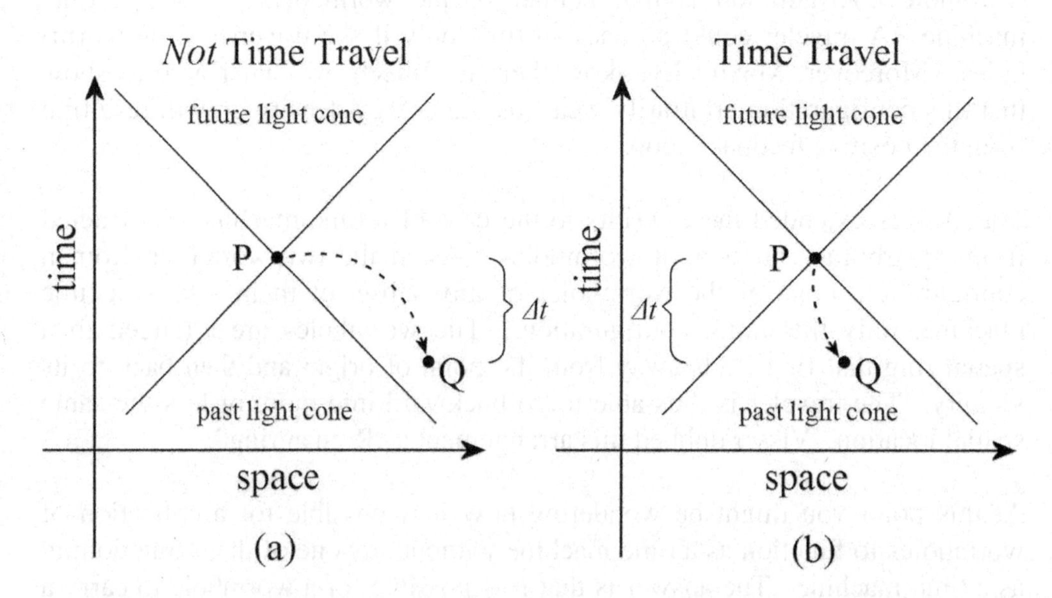

Figure 10.7. A time jump is not necessarily time travel. Consider two events: P – traveler enters wormhole, and Q – same traveler exits wormhole at its other end. (a) An example of a backward time jump Δt that is not time travel, because Q is not unambiguously in the past of P. In our reference frame, travel from P to Q appears to be travel into the past. But for some observers moving relative to us, the trip from P to Q appears to be travel into the future. (b) When Q is in P's past light cone, there are no such observers.

Visser's result, being based on a semiclassical calculation, does not eliminate the possibility of chronology protecting effects. It merely demonstrates that to the extent that a semiclassical calculation may be relied upon, these effects can be made arbitrarily weak. Nevertheless, to the degree that the single-wormhole time machine indicates the validity of the chronology protection conjecture, the Roman ring counter-indicates it.

P99. *Semiclassical calculations suggest that 1) single-wormhole time machines are subject to unbounded energy-density increases due to the feedback of vacuum fluctuations, and 2) multi-wormhole time machines need not be.*

You might have noticed that the recipe for building a wormhole time machine, with which I opened this chapter, fails to provide a crucial instruction – that for obtaining the wormhole. It is to this task that we now turn our attention.

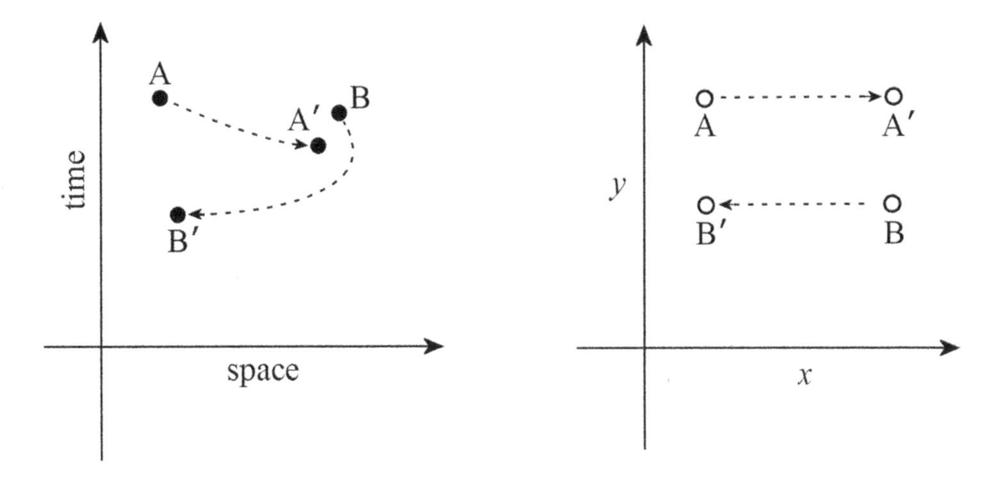

Figure 10.8. Two-wormhole time machine. Event *A* – traveler enters first wormhole. Event *A'* – traveler exits first wormhole. Event *B* – traveler enters second wormhole. Event B' – traveler exits second wormhole. Acting separately, neither wormhole functions as a time machine (See Figure 10.7). Acting together, the wormholes form a system that does. Traveler goes from *A* to *A'* via wormhole, *A'* to *B* via normal space, and *B* to *B'* via wormhole. (a) Spacetime view. (b) Spatial arrangement.

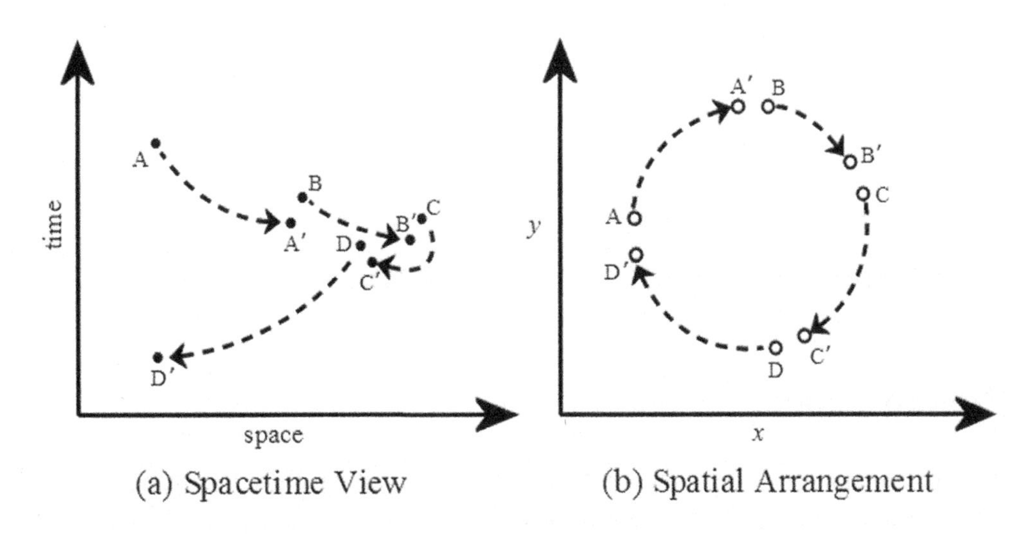

Figure 10.9. A four-wormhole Roman ring. A traveler moves backward in time from event *A* to event *D'* through a series of wormhole jumps (*AA', BB', CC', DD'*). Although in our reference frame, each wormhole appears to send the traveler backward in time, none of them act individually as true time machines. Collectively they do.

11. Wormhole Engineering

How advanced would a civilization have to be to successfully undertake the construction of a wormhole? It would have to be sufficiently advanced to accomplish the following construction steps.

1) Change the macroscopic topology of spacetime.

2) Position the wormhole mouths.

3) Enlarge the wormhole with exotic matter.

4) Stabilize the wormhole.

Let us consider each step.

Changing the Topology of Spacetime

Pseudo Topology Change

As previously discussed, general relativity assumes that the topology of spacetime is fixed. This is not to say that topology changes are necessarily impossible, only that their consideration would counter the conventional assumption that accompanies general relativity. For the moment, let us continue to assume that genuine topology change is impossible. We can nevertheless in effect consider changes in the topology of spacetime. We need only consider pseudo topology changes instead of actual ones. Recall that this was the notion emphasized by Visser in which a spacetime can seem to change its topology by changing its macroscopic topology without changing its actual topology.

There are two likely ways of changing the macroscopic topology of spacetime while retaining its submicroscopic topology. The first is to enlarge a submicroscopic virtual wormhole. Presumably, one would attempt this by flooding a small region with high-intensity, ultra-high-frequency exotic radiation. Several beams of such exotic radiation could, for example, be focused on a single spot. It is perhaps reasonable to suppose that the gravitational vacuum would include wormholes of all masses. The larger its mass, the larger its throat, the shorter is its virtual existence. One could hope that a virtual wormhole large enough to swallow a quantum of exotic radiation would come into existence long enough to do so. Upon ingesting the radiation it would begin to enlarge, which would assist in its ingestion of additional radiation, thus accelerating its enlargement. This process poses a danger, however. Once the wormhole has been enlarged beyond a certain point, there is some chance that it will begin to exponentially inflate. This could jeopardize the experimenter's solar system, not to mention her funding.

The other approach to changing a spacetime's macroscopic topology, while retaining its actual topology, begins with finding a black hole. By converting the black hole into a traversable wormhole, the previously blocked passage to another universe becomes accessible. This changes the spacetime's effective macroscopic topology. Dumping a sufficiently large quantity of exotic matter into the black hole should accomplish this. Unfortunately, this approach requires the black hole to already be connected to another universe or distant region. In other words, it must be described by the black hole sector of the *maximally extended* Schwarzschild, Reissner-Nordstrøm, or Kerr solutions. This is exceedingly unlikely for a naturally occurring black hole that has resulted from stellar collapse. The maximally extended solutions might, however, be found amongst the fresh virtual black holes to be found in the spacetime foam. We might further guess that the instability of the extended solutions is unproblematic, because virtual black holes are ephemeral by definition. Virtual black holes have another advantage over stellar-scale black holes. Converting them into traversable wormholes will require less exotic matter. Moreover, like black hole evaporation through Hawking radiation, the maximum rate of the process would be greatly accelerated over that expected for a stellar-scale black hole.

Actual Topology Change

Let us explore the consequences of discarding the conventional assumption. Let us consider *bona fide* topology changes in general relativity.

When Wheeler first likened the vacuum state of the gravitational field to sea foam, he envisaged actual microscopic topology fluctuations. As he well knew, however, the quantization of a classical theory in which topology change is forbidden will not yield a quantum theory in which such change is allowed. The same is true for dimensionality. Quantizing a classical system in which a particle is restricted to motion in one dimension does not magically yield a quantum system in which the particle can move in any number of dimensions. This issue, then, of topology change must be resolved solely within a classical theory of gravity.

Given the assumption of traversable wormholes and therefore the assumption of closed timelike curves, we cannot use Geroch's theorem to exclude the possibility of topology change. Yet knowing that topology change might be possible is not the same as knowing how to induce it. Such knowledge could be obtained as follows.

> *Step 1:* Specify a dynamic spacetime geometry that describes the desired topology change. [As it is unlikely that a single coordinate system could be defined to describe the entire spacetime, divide the spacetime into overlapping patches on which local coordinates are defined and used to specify the dynamic metric that describes the geometry.]

> *Step 2:* Insert this metric into Einstein's equations and obtain the dynamic stress-energy tensor – the densities and pressures of matter required to induce the topology change described by the spacetime's dynamic geometry.

Were these steps carried out for several topology-changing spacetimes of interest, we might discover that singularities in the density or pressure distributions would be required each time. In that case we would conjecture that these topology changes are unlikely to be achieved, or that they could only be achieved within the cloaks of the event horizons formed by the matter singularities. If, instead, the density and pressure distributions were well behaved, we would then have a recipe for inducing the desired topology change.

Back in the 1970s Frank Tipler considered this type of program. It was again considered in the 1990s by Borde. They concluded that the topology-changing spacetimes are forbidden, because the matter that sources them would have to violate a reasonable energy condition -- a liberalized version of the WEC in which the lower bound of zero is replaced by a negative constant. In other words, the possibility of topology change turns on whether the energy density of matter, as measured by any observer, cannot be less than some constant negative value.

If it is true that *classical* violations of the original WEC are possible, which appears to be the case for the scalar field, then there can be no such lower bound on the degree to which the WEC is violated. This is because for any given violation -- i.e. any measurement of negative energy density -- there is always an observer moving in such a way as to see a greater violation. In other words, the energy condition ceases to hold as a physical principle. It could not, therefore, be used to justify the prohibition of topology change.

Tipler also showed that if the WEC holds, then singularities must exist in the density or pressure distributions of any topology-change-inducing matter. Of course, if we are free to discard the WEC, such singularities are no longer guaranteed.

If classical WEC violations turn out to be impossible, one might be able to appeal to the quantum inequalities to exclude the possibility of topology change. It might be straight forward to show that any matter required to induce such a change must violate the inequalities and may therefore be rejected as unphysical. To the best of my knowledge this has yet to be proved. Another possible way in which quantum theory can be enlisted to prevent topology change follows from the work of A. Anderson and Bryce DeWitt carried out at the University of Texas at Austin in the mid 1980s. They showed that a quantum scalar field defined on a topology-changing spacetime would become explosive at the onset of the topology change. This turns out not to be a general result, however, as Gary Horowitz of the University of California at Santa Barbara has subsequently shown. Currently, then, quantum theory does not preclude the possibility of topology change. And the fact remains that we have not been able to rule out the possibility of arbitrarily large classical violations of the energy conditions. Hence, I will in this and subsequent sections regard genuine topology change as an option.

The immediate consequence of taking this position is that one is required to answer the following question: If topology change is possible, why do we not observe it? Four possible answers are: 1) Large-scale topology change is dynamically disfavored by the classical theory. As we saw in the chapter on wormhole dynamics, wormholes do not grow unless supplied with sufficient negative energy. 2) Large-scale topology change is dynamically disfavored by the semi-classical theory. For example, it would appear to be the case that the Casimir force works to constrict wormholes. 3) Large-scale topology change is dynamically disfavored by an as yet unknown generalization of the classical theory. One could imagine supplementing general relativity (its action) with a term that somehow suppresses large-scale topology change and whose value in the absence of such change is zero[‡‡]. 4) Large-scale topology change is dynamically disfavored for the same reason that large-scale pseudo topology changes are. Conventional general relativity on a spacetime pregnant with vast numbers of preexisting microscopic wormholes, would probably be dynamically similar on the large scale to a version of general relativity that would allow such wormholes to arise *ex nihilo*.

Returning to the topic of engineering, it is clear that traversable wormholes resulting from topology change will contain the expected energy-condition-violating matter. So it's a safe bet that this means of inducing topology change, if it is possible, will require the concentrated delivery of exotic matter.

All three methods, then – enlarging virtual wormholes, converting black holes, and creating wormholes *ex nihilo* through topology change -- rely on the ability to produce exotic matter. To ease our task we would seek to create the smallest detectable wormhole. This requires our exotic matter to affect spacetime at tiny length scales, which places an upper limit on the exotic matter's wavelength. Whether the exotic matter is delivered as targeted intersecting beams of negatively massed particles or as beams massless radiation, we have

P100*. Inducing macroscopic topology change through any of 1) the enlargement of virtual wormholes, 2) the conversion of virtual black holes, or 3) the creation of wormholes* ex nihilo *seems likely to require high-frequency exotic matter.*

[‡‡] A possible candidate would be the aggregate 4-volume of causality-violating regions.

Positioning the Wormhole Mouths

Suppose we had changed the macroscopic -- by which I mean "detectable" -- topology of spacetime by enlarging a virtual wormhole to the easily detectable size of one Angstrom – about the size of an atom. Because we enlarged a virtual wormhole or created one from scratch, we might hope its mouths to be in close proximity, as opposed to being separated by a few billion light years. If they *are* so separated, we would be denied the luxury of positioning both mouths. To position one of the mouths, we would first charge it. Bombarding it with a high-intensity electron beam should suffice. We need to be careful to ensure that we charge only one of the mouths, or at least that we charge one mouth more than the other. Once a wormhole mouth has been thus charged, we can use its electrostatic attraction to an oppositely charged spaceship to haul the mouth to a desired location.

The ease with which our charged spaceship can tow the wormhole mouth depends on the mouth's effective mass. This mass -- which can be positive, negative, or zero -- is determined by the geometry of the spacetime exterior to the mouth. This in turn is determined by the distribution of matter within the wormhole. If, for example, we were somehow able to create a thin-shell wormhole, the mass of its mouths could be zero (because the spacetime exterior to the mouths could in this case be flat). Towing one of its mouths, then, would be effortless. If, as would be more likely, our atom-sized wormhole instead resembled that of Morris and Thorne (specifically the third example in the appendix of their first paper), the mass of each mouth would be about 10^{15} kilograms. This is about the mass of a medium-sized (~ 10 kilometer diameter) asteroid. Towing such a mass over a distance of several light years should pose no problem for an advanced civilization.

One possible impediment to a successful towing operation is that the *length* of the wormhole's throat might need to increase slightly as its mouths separate. To prevent the wormhole from constricting and pinching off, it might be necessary to periodically inject exotic matter into the mouths as we tow.

Enlarging the Wormhole

Once we have satisfactorily positioned the mouths of our atom-sized wormhole, our next step is to enlarge them to a useful diameter. This would require massive injections of exotic and perhaps normal matter. This matter would be injected so as to join the wormhole's supporting matter, as opposed to merely traversing the wormhole.

Injecting any normal matter required will be no problem. It is the creation of the required quantity of exotic matter that prior to 2003 was thought to pose a challenge. We could have estimated the needed amount of exotic matter by using the aforementioned Morris and Thorne (MT) example. We could have used Visser's formula for a thin-shell wormhole of similar size as a lower bound. MT's wormhole example – the most benign of them save for the example that they called "absurdly benign" -- has a mouth radius that is 600 times the distance between the earth and the sun. It turns out to require between about 10^7 and 10^8 solar masses of exotic matter. This corresponds to about one thousandth of the stellar mass of our galaxy or to that of about one thousand globular clusters.

We see, then, that a small galaxy's worth of exotic matter could be required to a create wormhole. Needless to say, this would only be within the purview of an exceedingly advanced civilization. Such a civilization would presumably favor any possible classical methods, because these are not hampered by quantum inequalities. These methods assume the existence of a fundamental (non-composite) scalar field. While no such field has actually been detected, the standard model of particle physics, which has thus far survived all attempts at experimental refutation, predicts the existence of one. An astoundingly advanced civilization should therefore be able to create an exotic matter beam and inflate our atomic wormhole by injecting into it over time a small galaxy's worth of exotic matter.

P101*. The construction of the most realistically benign of the Morris-Thorne wormhole examples requires an enormous quantity of exotic matter – at least* 10^7 *solar masses.*

Should it turn out that classical violations of the Null Energy Condition are impossible, the wormhole construction crew would face two additional obstacles. The rate at which they could beam negative energy into the wormhole would be severely restricted by the Quantum Inequalities. Moreover, they would be forced by these inequalities to restrict the exotic matter content of the wormhole to a microscopically thin shell at the wormhole's throat. Hence, their astoundingly advanced civilization would have to be even more advanced than originally supposed.

In 2003 the task of constructing a traversable wormhole seemed to have become less daunting. Visser, Kar, and Dadhich showed that a wormhole traversable by light (and potentially traversable by humans) could be constructed using an *arbitrarily small* quantity of exotic matter. They even weakened the definition of exotic matter by requiring only that the matter violate the *Averaged* Null Energy Condition (ANEC). Within a year Peter Kuhfittig showed that, using an arbitrarily small quantity of such exotic matter, it is possible to construct a wormhole that is not only traversable by light but by humans as well.

How is this possible? How could the quantity of exotic matter required to construct a human-traversable wormhole go from 10^8 solar masses to less than a single gram? If such a tiny quantity of exotic matter were the sole content of a wormhole's throat, we would expect the wormhole to flare out very, very gradually (Figure 11.1b). It would flare out so gradually that the distance between the space stations at the upper and lower rims of the wormhole would be immense. The time, as measure by stationary observers, required to

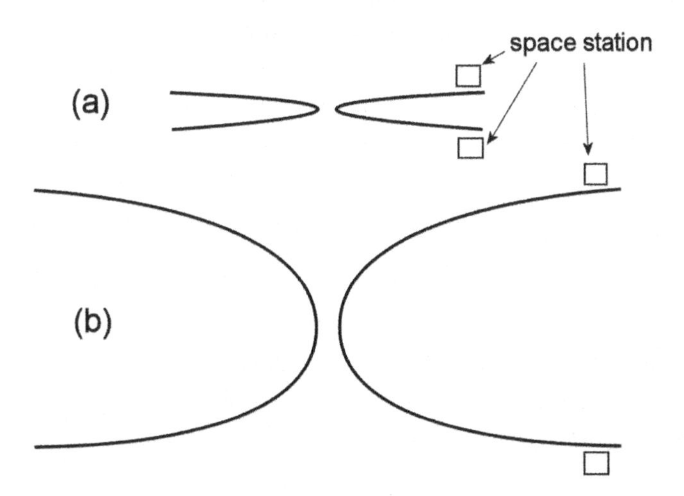

Figure 11.1. Degree of flare out increases with quantity of exotic matter. (a) A large quantity of exotic matter at a wormhole's throat results in a large flare out and a small distance between space stations. (b) A small quantity of exotic matter results in a small flare out and large distances between space stations. The space stations are located on spherical shells surrounding the throat at which the acceleration of gravity matches that of Earth. These two shells (one in each universe) define the wormhole's mouths.

traverse such a distance would be expected to exceed the human life span. This wormhole would, by definition, not be traversable.

How, then, can the exotic matter content of a wormhole be reduced arbitrarily, while its traversability is maintained? The answer, unfortunately, appears to be that it cannot be. There is no free lunch in this regard. Kuhfittig's human-traversable wormhole suffers from the requirement that it be supported by a mass distribution of infinite spatial extent. In other words, there is no cutoff radius from his wormhole's throat within which all of his wormhole's normal matter is contained. If one attempts to impose such a cutoff radius by grafting his wormhole spacetime onto an empty Schwarzschild exterior spacetime, one discovers that this cannot be accomplished without violating one of the assumptions of his solution. This means that an advanced civilization interested in constructing a Kuhfittig wormhole would have to carefully distribute normal matter throughout the whole of space, so that its density diminishes with distance from the throat according to a particular rule. One might hope that a Kuhfittig-like solution in which the matter density is set to zero beyond a certain distance from the throat might exist. However, in light of Birkhoff's Theorem and the apparent impossibility of grafting such a solution onto a Schwarzshild exterior, this seems unlikely.

It should not be surprising that a wormhole traversable by light, but failing to meet all of the conditions for human traversability, could be constructed using an arbitrarily small quantity of exotic matter. As mentioned above, such a wormhole would simply require an unacceptably long time to traverse. Beyond this it would have another drawback – extreme instability. Were the amount of exotic matter used to construct the wormhole comparable to that of a spacecraft attempting to traverse it, the wormhole would be destroyed. If even smaller quantities of ANEC-violating matter were used, the wormhole would be unstable against stray radiation entering its throat.

***P102**. It is possible to construct, using an arbitrarily small quantity of ANEC-violating matter, a wormhole that is traversable by light. If, however, the quantity of exotic matter used is too small, the wormhole will not meet the conditions for human traversability.*

Stabilizing the Wormhole

Once the wormhole has been expanded to its desired size, it must be stabilized. Travelers passing through it or stray matter ingested by it might cause the wormhole to begin to constrict. Given that an advanced civilization possesses an exotic matter beam, it could easily compensate for the constricting perturbation with a suitable injection of exotic matter. In the event that too much matter was inadvertently injected, the wormhole stabilizer must also be prepared to inject a corrective quantity of normal matter.

In order for the stabilizer to function properly, it will need to sense changes in the diameter of the wormhole's throat. Such a sensor might even be within the technological capability of our own civilization. Imagine a buoy positioned at the wormhole's throat. It periodically sends light pulses that travel around the throat and return to the buoy. The buoy measures the time it took for the pulse to return and thus knows the throat's circumference. It transmits this information to the wormhole stabilizer stationed nearby. For the MT wormhole example considered above, the circumference of its throat is ten thousand times smaller than that of its mouth. The time required for a light pulse to circumnavigate this throat is about 3 minutes. This means that the stabilizer would learn about a perturbation 3 minutes after it occurred. For perturbations resulting in small changes in the wormhole's geometry over this time, it should be a sufficiently timely warning for the stabilizer to take corrective action. If it is not, the single-buoy sensor could, for example, be replaced by a triad of buoys that measure local curvature more rapidly by summing the angles of the vertices of a triangle formed by laser beams connecting the buoys.

=/=

P103. *Constructing a Morris-Thorne style wormhole requires a civilization advanced enough to create and deliver exotic matter in quantities equal to 0.1% of the stellar mass of a typical galaxy. Such a civilization would presumably be sufficiently advanced to perform all other steps required to construct a wormhole.*

Reducing the Exotic Matter Requirement

Thin-Shells

As a means of minimizing the use of exotic matter, Morris and Thorne recommended that this matter be concentrated in a thin layer centered about the wormhole's throat. For a given quantity of exotic matter this results in a design that more closely resembles a thin-shell wormhole. Its mouth circumference would be smaller than that of a wormhole in which the same matter is more diffusely distributed. The time to traverse it would be reduced correspondingly. They were also aware that the presence of normal matter in the wormhole required the use of a compensating quantity of exotic matter. Reducing the wormhole's normal matter content, then, would reduce its exotic matter requirement. The reduction is dramatic. A calculation by Visser demonstrates that a thin shell design reduces the exotic matter requirement by ten orders of magnitude below the most benign of the realistic (i.e. not "absurdly benign") examples put forward by Morris and Thorne.

Rotation

In Chapter 5 we encountered another way of economizing on exotic matter. Recall that Peter Kuhfittig had discovered that a spherically symmetric rotating wormhole requires less exotic matter than one that does not rotate. Inducing the shell of exotic matter to spin about an axis, while somehow preventing it from becoming oblate, might further reduce the amount of exotic matter that the wormhole need contain.

The Gravitational Faraday Cage

The reason that a Morris-Thorne wormhole can be so large – a mouth circumference on the order of tens of millions of kilometers – is that it was designed to protect human travelers from harmful tidal and motion-induced accelerations, while retaining the shape that a naturally occurring wormhole would be expected to have. Suppose, however, that instead of modifying the geometry of the entire wormhole, the human traveler could be protected by modifying the gravitational field in his immediate vicinity only. A civilization advanced enough to master the creation of exotic matter could more easily protect the traveler by surrounding him with a gravitational Faraday cage.

A Faraday cage is an enclosure of solid metal or metal mesh that prevents exterior electrical fields from penetrating into the enclosure's interior. You may have noticed that your cell phone does not work in steel elevators, or that you cannot receive AM radio stations when your car enters a steel reinforced tunnel. The electric fields in the waves transmitted by a cell phone tower or a radio station induce currents in the walls of the elevator or rebar in the tunnel. These currents create new fields that act to cancel the fields that created them. This cancellation prevents the exterior field from reaching the interior of the Faraday cage, be it an elevator or a tunnel. [You are able to receive FM radio stations within a tunnel, because the wavelengths of FM transmissions are much smaller than the separation between the rebar rods in tunnels. This allows the waves to pass through the tunnel's metal mesh much as visible light enters a jail cell.]

A gravitational Faraday cage would be an enclosure that prevents exterior gravitational fields from influencing the gravitational field within the enclosure. Currents and static distributions of exotic and normal matter could be setup on the exterior boundary of the cage. These could be arranged to ensure that the traveler standing within the cage would enjoy a uniform gravitational field equivalent to what she would experience standing on the surface of earth. Because it is unclear whether it is possible to create a gravitational conductor analogous to an electrical conductor, the exotic matter currents and density distributions surrounding the cage might have to be actively created. This would be similar to the operation of active noise cancellation headphones.

Were the traveler to be contained within a compact capsule that acts as a gravitational Faraday cage, she would be protected from deadly tidal forces or crushing accelerations. This would allow her traverse a wormhole whose tidal forces would otherwise have killed her and travel at accelerations that would otherwise have crushed her.

Unfortunately, there is a chance that the gravitational Faraday cage will fail as a means of reducing the exotic matter requirements for a traversable wormhole. Its own exotic matter requirements may well exceed any savings in exotic matter that result from eliminating the constraint on accelerations and tidal forces.

The Stargate

Another possible means of reducing the amount of exotic matter requirement is to construct a special type of thin-shell wormhole, one whose throat is flat. Instead of a spherical or cubical wormhole mouth and throat, we could use instead a segment of a plane, a disk or a square for example. This plane segment would be mathematically identified with an identical plane segment in a distant part of the universe or in another (co-resident) universe. As the traveler walks through one plane segment, he emerges from another plane segment, thereby traversing great distances in a single step.

Krasnikov calls such a wormhole a "portal". Visser has described similar wormholes, degenerate variants of his thin-shell wormholes. His "dihedral wormholes" are confined to a plane, while the throats of his "loop-based" wormhole form more general two-dimensional surfaces. I shall refer to these and any other wormholes with planar or quasi-planar throats specifically as *stargates* -- in addition to retaining the general application of this term to any synthetically controlled wormhole.

While stargates do require significantly less exotic matter than do Morris-Thorne wormholes, their requirements are not small. Krasnikov estimates that his stargate design scaled for human travelers will require 0.01 solar masses. Visser's estimate is similar. His dihedral stargate will require about one tenth of this amount – about the mass of Jupiter.

The problem with stargates might at first seem to be the same as that with thin-shell wormholes. They require a seemingly artificial modification of geometry of spacetime. If, however, spacetime foam is like the vacuum of any other quantum field theory, in that it allows the ephemeral existence of *any* possible field configuration, there must be virtual stargates. Indeed, if their energy requirements are in some sense lower than those of the more natural seeming Morris-Thorne-style wormholes, virtual stargates might actually dominate the spacetime foam. If this is true, a civilization that concentrates an exotic radiation beam, as described above, might in general find itself enlarging virtual stargates. As usual, such speculation only becomes fully meaningful within the context of a quantum theory of gravity, which has, of course, yet to be devised.

If, instead, the virtual stargates negligibly contribute to the contents of the spacetime foam, we can still resort to an attempt to create a stargate *ex nihilo*. We simply would follow the two steps described above in the section on changing the topology of spacetime. In this procedure we would use a time-dependent geometry that begins with flat space and ends with flat space containing a tiny stargate.

P104. *Traversable wormholes with planar throats (stargates) require less exotic matter than wormholes with throats that are not confined to a plane.*

Stargate Construction

Let us suppose that all of our previous attempts to mine Wheeler wormholes from the quantum vacuum has only yielded spherically symmetric wormholes of the Morris-Thorne variety. Tiny wormholes of this type are undesirable for our purpose, because they would require us to invest enormous amounts of exotic matter to enlarge them to a useful size. Even for our exceedingly advanced, wormhole-building civilization, exotic matter is dear. We seek to conserve it as best we can. We favor, then, a stargate-style design. But how can we achieve it?

Stargates from Thick-Shell Wormholes

We must begin our stargate construction with the wormholes available to us – the atom-sized, Morris-Thorne, thick-shell wormholes that we have extracted from the vacuum. We must convert these into thin-shell wormholes. Fortunately, this can be accomplished. All we need do is to change the distribution of exotic matter. In the Morris-Thorne-style thick-shell wormhole the exotic matter is confined to a finitely thick spherical shell. In the stargate-style thin-shell wormhole, by contrast, this matter is confined to an infinitesimally thin circular ring. We need, then, to gradually transform a thick shell into a thin ring.

If we are fortunate, our exotic matter is in the form of a fluid or a concentration of dust. We could then easily arrange it in the desired distribution. To accomplish this, we would first compress the thick spherical shell into a thin shell. We would then punch holes in this thin spherical shell at opposite poles. Lastly, we would expand the size of these two holes -- while keeping the distribution thin -- until the holes nearly meet at the "equator" -- the location of the resultant thin ring of exotic matter.

It is, of course, possible that such an operation might destroy the wormhole – causing it to collapse and become a black hole or explode into an inflationary expansion. We can prevent this by exploiting the Einstein equations. To do so we would write down an expression for the desired evolution of the wormhole's spatial geometry. Our expression would be one that describes a gradual morphing of a Morris-Thorne wormhole into a stargate. Inserting this morph-describing metric into Einstein equations will tell us exactly how we need to vary the exotic matter distribution to ensure that the morph succeeds. The instructions thus obtained would probably require us to remove (or even add) exotic matter in particular ways, or it might require us to induce certain currents in the exotic matter. These tasks are the price we must pay to guarantee the wormhole's successful conversion into a stargate.

P105. *Thick-shell wormhole can be converted into thin-shell wormholes.*

Stargates Created *Ex Nihilo*

As we have seen, the usual arguments for excluding topology change from general relativity are not ironclad. It is not unreasonable to suppose, therefore, that Wheeler's intuition was correct. The gravitational vacuum state is a superposition of all possible spatial geometries. The crux of the matter is whether these geometries are limited to a single topology or not.

Limiting spatial geometries to a single topology is likely[§§] tantamount to forever fixing the total number of wormholes to some value *N*. That there seems to be no natural means of selecting *N*, darkens the plausibility of this limitation. We cannot, for example, obtain a constant *N* by dividing the volume of the universe by the Planck volume. The universe's volume is not constant.

It seems odd, moreover, that a manifestly four-dimensional theory would demand a restriction on *three*-dimensional submanifolds. General relativity is not a 3+1 dimensional theory; it is not a theory of three-dimensional space evolving in time. It is a theory of four-dimensional spacetime. We cast it in the 3+1 form for the convenience of our calculations. In so doing, we tacitly assume the impossibility of topology change. Nature, however, is unconcerned with the convenience of our calculations. Nor has she instructed us to make this assumption.

[§§] In the absence of a theorem for classifying three-dimensional topologies analogous to that for classifying two-dimensional topologies, it is difficult to be sure.

It would also seem odd for the quantization of a classical theory to enjoy degrees of freedom unknown to the classical theory. If the quantization of a one-particle system allows the particle to occupy positions in three dimensions of space, so must the corresponding classical system. If, as is widely supposed, Wheeler's quantum foam exists in the quantization of gravity, if it allows spatial geometries to possess any topology, then so must the classical theory of gravity to which it corresponds.

If we take Einstein's equations seriously, they inform us that a topology change requires a particular evolution of stress energy. We have, by assuming the existence of wormholes, abandoned virtually all restrictions on classical matter. The only restrictions that might remain are those demanding that stress-energy densities be finite. But in general relativity, as in classical electrodynamics or any other physical theory, infinite quantities are not taken literally to be infinite. Physics is not mathematics. The appearance of infinities in an otherwise well performing theory is normally taken to be a signal that at least one of the theory's underlying assumptions is not strictly true in the regime of interest. Resolving the matter usually involves couching the theory within another theory of greater scope, such as quantum theory. The consequence of such a resolution is that quantities previously seeming to be infinite become merely "large" in some sense and finite. Should Einstein's equations, then, predict a particular topology change to require an infinite density of exotic matter at some location, this should not be taken literally. Rather, it is better taken to mean that the density there must be "large" – in a sense to be determined precisely by a quantum theory of gravity, once we have it.

It is not unreasonable to suppose that Einstein's equations mean what they say. Extreme and peculiar distributions of stress energy will change the topology of space. Needless to say, such a change will not affect the topology of space-*time*, which, due to its inclusion of time, has no basis against which it can change. This fixed nature of the topology of the whole of spacetime will not, however, prevent an exceedingly advanced civilization from exploiting Einstein's equations to create stargates from nothing.

Exotic Matter Confinement

An aspect of wormhole engineering that is often overlooked concerns the means through which exotic matter is to be confined at wormhole throats. It is normally assumed that gravitationally repulsive exotic matter somehow remains at the throat even in dynamic situations, when it is clear that it would flee. For example, stability analyses routinely assume that exotic matter will remain in place at the throat even as it constricts. This assumption permits the matter's density to rise, which for certain equations of state, increases the matter's gravitational repulsiveness enough to reverse the wormhole's contraction and thus prevent collapse. It might happen this way. It seems at least as likely, however, that self repulsive exotic matter will vacate the throat and allow it to effectively pinch off.

The methods by which we can confine exotic matter are also of interest, because they will allow us to vary the matter's density and thus allow us some measure of control over the wormhole's operation.

P106. *Exotic matter will not in general remain at a wormhole's throat unless it is artificially confined there.*

Holding Matter at the Throat

Let us define exotic matter as matter that violates the pointwise null energy condition. Such matter can be classified as dust or fluid. The dust is merely a collection of discrete particles. Dust, unlike fluid, is by definition without pressure. The only way, then, for dust to be exotic is for its density to be negative in its rest frame. Any two particles of exotic dust repel each other. Exotic dust repels normal matter. Normal matter attracts exotic dust. [An incidental feature of such hypothetical particles is a sort of perpetual motion. An exotic particle with mass *–m* would be attracted to a normal particle of mass *+m*. The latter would be equally repelled by the former. As a consequence the normal particle would flee the exotic particle, while the exotic particle would pursue the normal one. Trapped in an endless chase, the pair would ceaselessly accelerate away.]

Exotic fluid possesses a pressure. One normally assumes that this pressure is the same in all three directions and that the fluid contains no shearing forces or significant currents. Unlike dust, fluid may be exotic without its density being negative. If its density is positive, however, its pressure must be sufficiently negative to compensate for this. If the fluid's pressure is positive, exoticity requires its density to be sufficiently negative to compensate. Fluid exotic matter, even that with positive density, is gravitationally repulsive. Like dust, it not only repels normal matter, it repels itself.

Self-repulsive exotic matter will not remain stationary at a wormhole throat or anywhere else unless it is compelled to do so. In the case of the first Morris-Thorne wormhole example of Chapter 5, the "Infinite-Exotic-Region Wormhole", that compulsion results from the presence of exotic matter throughout space. The exotic matter can be reasonably supposed to remain at the throat, because there is no region free of exotic matter to which it can flee. In the second and the third Morris-Thorne wormholes presented in Chapter 5, the "Large-Exotic-Region Wormhole" and the "Medium-Exotic-Region Wormhole", exotic matter can be supposed to remain at the throat because of the presence of physical barriers. In both of these examples a spherical shell of normal matter surrounds the region containing the exotic matter. For this shell to be able to confine the exotic matter, we would have to assume that the nature of the interaction of exotic and normal matter permits this. For all we know, exotic and normal matter react explosively. Or exotic matter, having only a gravitational interaction with ordinary matter, might easily penetrate it. Beyond these designs that possibly confine exotic matter with a physical barrier, there are those that provide no method of confinement whatever. These include those wormhole designs that economize on exotic matter -- such as the "Small-Exotic-Region Wormhole" of Chapter 5, stargates and other thin-shell wormholes.

How might exotic matter in such wormholes be confined to their throats? Given that confinement by physical barriers remains questionable, we shall seek other means. Accordingly, we will assume henceforth that exotic matter possesses or can be induced to possess an electric (or magnetic) charge. Charging exotic matter would worsen the problem of confining. It would add electrostatic self repulsion to the existing gravitational self repulsion. Fortunately, external electrostatic forces can be brought to bear. Because electromagnetism is 10^{38} times stronger than gravity, these external forces can easily overwhelm exotic matter's inherent anti-gravitism.

Consider a spherically symmetric wormhole. By surrounding both of its mouths with a conducting sphere, we can concentrate or dilute charged exotic matter at its throat simply by increasing or decreasing the charge on the sphere. In order to preserve the wormhole's essential structure as a solution to the Einstein equations, the mass of this sphere would have to be very small compared to the mass of the wormhole's exotic matter. In order to lighten the sphere, we could replace it with a wire mesh. We could, for example, erect a geodesic sphere – the extension of Buckminster Fuller's famous dome. Not only would its mass be negligible, it would permit the passage of spaceships small enough to fit within its facets.

Electrostatic confinement could also improve the aforementioned wormhole designs that feature an external shell of ordinary matter, whose effectiveness at containing exotic matter is unknown. These shells would simply be charged. This would protect them from interacting with charged exotic matter by confining such matter to the vicinity of the throat. As above, it would moreover permit fine tuning of the density of exotic matter.

For spherically symmetric wormholes a confinement field will be effective only if it is applied at both ends of the wormhole. If it is not, the one-sided confinement field will simply push the exotic matter through the wormhole, causing the matter to emerge from the other side, while the wormhole collapses behind it.

Wormhole designs lacking in spherical symmetry would require a modified approach. In a stargate, for example, the formerly spherical throat has been flattened to a disk. The formerly spherically symmetric distribution of exotic matter has been concentrated into a thin ring surrounding this disk. In order to electrostatically confine charged exotic matter to this ring, we would surround it with a torus-shaped wire mesh. Suitably charging this toroidal mesh would ensure any desired density of exotic matter in the ring. We could, moreover, hedge our bets by coiling the mesh with a conducting cable. An electric current through this cable would create a magnetic field parallel to the exotic matter ring. Any exotic matter that attempts to leave the ring would be deflected by the magnetic field into a tight, inward-spiraling orbit with the effect of confining the matter as desired.

As in the spherically symmetric case, the mesh or coil generating a stargate's confinement field would have to exist on both sides – in both universes – connected by the wormhole.

These confinement schemes suffer from a potential drawback. The charge added to the exotic matter and the electromagnetic energy of the confinement fields might be great enough to prevent the entire system – charged exotic matter cum confinement field – to be a wormhole solution of the Einstein equations. This seems unlikely, however, given the existence of charged wormholes qualitatively similar to their uncharged counterparts and the quasi-exoticity of the electromagnetic field.

P107. *Exotic matter is gravitationally self-repulsive.*

P108. *The density of exotic dust is negative, its pressure zero.*

P109. *The density of exotic fluid is normally taken to be positive, its pressure negative.*

P110. *Charged exotic matter can be confined by an electrostatic field.*

P111. *Charged exotic matter in stargates can be confined by magnetic fields.*

P112. *Exotic matter confinement fields are ineffective unless applied from both ends of a wormhole.*

Regulating Wormhole Operation

The advantage of using electric or magnetic fields to confine exotic matter is that it allows us to vary the density of matter in the vicinity of the throat however we like. We can in this way control the operation of the wormhole. By arbitrarily lowering the exotic matter density at the throat, we would arbitrarily reduce the size of the throat. By constricting the throat until its size is, say, that of an atom, we would render the wormhole unfit for human travel. We would in effect have turned the wormhole "off". Turning the wormhole back "on" would simply be a matter of applying charge to a mesh or current to coils. Keeping the wormhole activated would consume power. Charge leaking from the electric-field-producing mesh would have to be replaced. Current flowing through the magnetic-field-producing coils would have to be maintained. In the absence of a charge on the mesh or a current through the coils, the self repulsiveness of the exotic matter would lower its density at the throat. The matter would continue to spread until a low-level confinement field stops it. This low-level field would thus impose a minimum exotic matter density and, consequently, a minimum throat radius. Maintaining this condition, the "off" state mentioned above, would therefore require less power than keeping the wormhole turned "on".

Confinement fields would have other margin effects. They could to some degree alter the time required to traverse the wormhole. This time decreases with increasing exotic matter density. It should also be possible to use asymmetric confinement fields to impart angular momentum to the exotic matter, thus allowing another means through which the wormhole's behavior could be fine tuned.

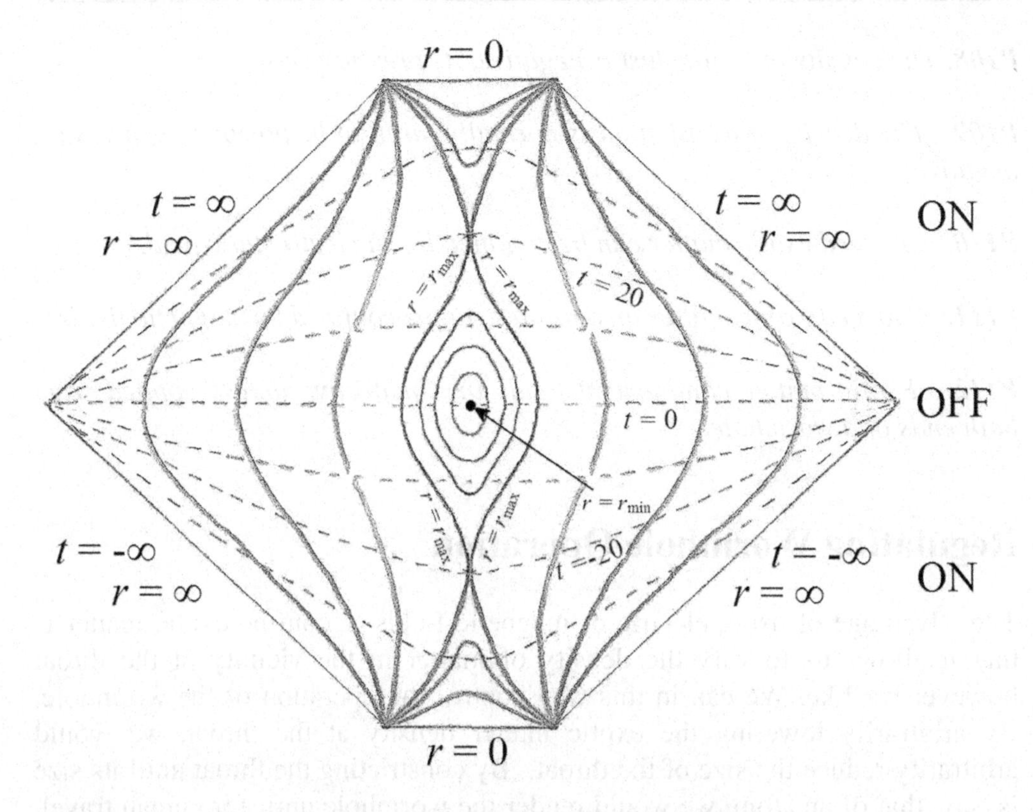

Figure 11.2. Turning a Wormhole Off and On. Penrose diagram shows throat radius to be zero in the distant past. Throat expands as usual for a dynamic wormhole. Expansion decelerates to stop at t = -20, when throat is fully ON – i.e. quasi static with radius r_{max}. Confinement field required for quasi staticity is relaxed, causing throat to contract. Contraction decelerates to stop at t = 0, when throat is OFF – i.e. quasi static with radius r_{min}. Confinement field is strengthened, causing throat to re-expand. Expansion decelerates to stop at t = 20, when throat is ON again. Deactivation of confinement field allows throat to collapse again to zero, as wormhole effectively pinches off, the usual end state for a dynamic wormhole.

P113. *The radius of a wormhole's throat varies directly with the density of the exotic matter there.*

P114. *Confinement fields can be used to effectively activate or deactivate a wormhole.*

P118. *An activated wormhole consumes more power than a deactivated one.*

One-Way Wormholes

An advanced civilization wanting to strictly control the direction in which interstellar traffic flows might choose to construct a one-way wormhole. The entrance to the wormhole would appear to be a black hole in that it would be marked by the presence of an event horizon. The wormhole's exit, by contrast would be horizon-free. Travelers entering from the black hole side will successfully traverse the wormhole, as long as they leave while the wormhole is sufficiently young. Unlike those who enter one-way Reissner-Nordstrøm or Kerr wormholes, these travelers need not risk their lives penetrating an internal blue sheet singularity. However, as shown in Figure 11.3, travelers who cross the event horizon too late in wormhole's history are doomed to die in its spacelike singularity. Such a death is invariably the fate of *any* traveler foolish enough to enter the wormhole from the opposite direction. Adequate signage could reduce such carnage.

The one-way wormhole evinces an important feature of wormhole physics that we have thus far overlooked. The solutions to Einstein's equations at the separate sides of a wormhole are independent. The only condition that they need satisfy is that they must match at the geometric throat. A wormhole mouth in one region of space could bear little resemblance to its associated mouth in another region.

P119. *The mouths of a wormhole need not resemble each other. As solutions to Einstein's equations, they may be asymmetric across the geometric throat as long as they satisfy matching conditions there.*

Wormhole Networks

A wormhole through which one can travel to a single destination tends to lose its allure after a while. It would be much improved, if it could somehow render multiple destinations accessible. If these multiple destinations were accessible to each other, things would be better still. How could we construct such a wormhole network?

To illustrate one answer to this question, let us consider the construction a network of three locations directly connected by wormholes. Our objective is

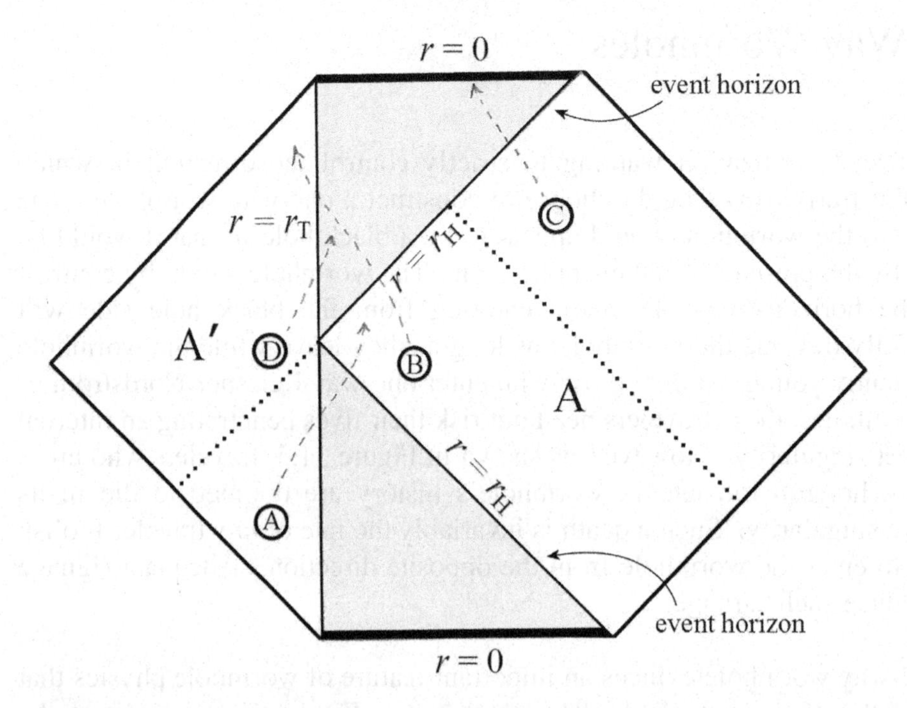

Figure 11.3. Example of a One-Way Wormhole. A wormhole is horizon-free in universe A' and appears to be a black hole in universe A. It functions as a one-way conduit for much of its history. Traveler (A) traverses successfully, but only avoids the singularity if he continues to rocket away from the event horizon with sufficient speed. Traveler (D) is destined to die in the singularity, unlike traveler (B), who traverses successfully. Traveler (C) waited too long to attempt traversal and pays with her life. Attempted traversal trajectories beginning above the dotted lines in A or A' will end in the singularity.

to enable access through a single mouth at each location to both of the other locations. We would proceed as follows.

1. Find two wormholes: #1 and #2.

2. Move a mouth of #2 into the mouth of #1 until the mouth of #2 is at the throat of #1.

This completes the construction of our network. We could continue the procedure. We could in principle create a wormhole network connecting *N*+2 locations by moving *N* different mouths into the throat of one of the wormholes. Would the Einstein equations permit such a monstrosity? Of course they would. We merely need to specify the desired geometry, insert it into the equations, and obtain the required distribution of stress energy. Like any arbitrarily chosen wormhole geometry, it will be unstable. Stabilizing it would at the least require each mouth to retain the same stabilization unit and confinement field generator that presumably preserved its structure before its associated wormhole was incorporated into the network.

The throat of the original host wormhole would of course need to be expanded to accommodate *N* mouths. A judicious injection of exotic matter would likely accomplish this. The resulting nexus of mouths would be difficult to navigate. Like all networks the wormhole network would require a routing system. This could consist of a computer stationed at the nexus that would receive transmitted destination codes of incoming spaceships as they emerge into the nexus from their respective mouths. It would then direct the ships to the local coordinates of the particular mouth that will send them to their proper destinations.

One could further fantasize that it might be possible to construct the network all at once instead of incrementally. After all, if wormhole creation *ex nihilo* is possible – and I have argued that it is not entirely absurd to suppose that it might be – then so is wormhole network creation.

P120. *A network of wormholes can be formed by sequentially moving the mouths of wormholes leading to desired locations into the throat of a designated host wormhole.*

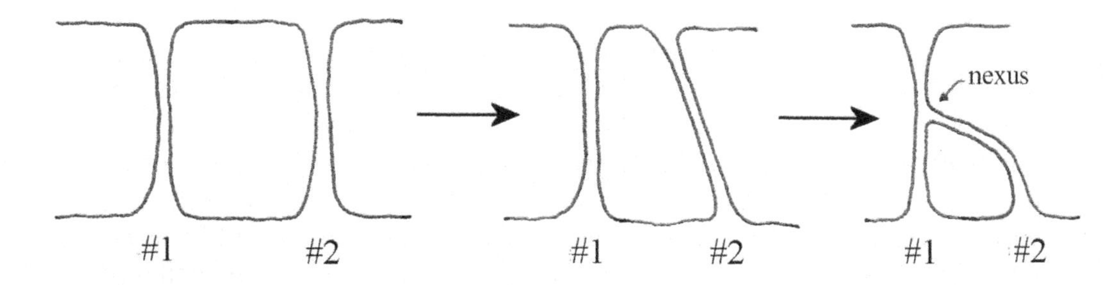

Figure 11.4. Making a 3-location Wormhole Network. The network is formed by moving a mouth of wormhole # 2 to the throat of wormhole # 1. The only advantage of the network is that it gives travelers at each location a single embarkation point for interstellar travel.

12. Wormhole Mysteries

One aspect of my education in physics, of which I was never particularly fond, was what seemed to me to be an under-emphasis on the unknown. A detailed awareness of one's ignorance is not only the first step toward greater understanding, it also engenders appreciation for the hard-won knowledge that we as a civilization do possess. So, in contrast to the Introduction, I have listed here questions in wormhole physics for which definitive answers are currently unknown.

Does classical exotic matter exist?
Unless it does, the prospect for finding wormholes or creating them is dim. They would require quantum effects to hold them open. The Ford-Roman constraints would apply. Quantum exotic material would be confined to an impossibly thin sliver about the wormhole's throat. Construction would be complicated immeasurably, and natural evolution would seem exceedingly unlikely. In short, there would be bad news for wormhole fans. There are three ways that I see that this might not be the actual state of affairs. 1) A non-minimally-coupled scalar field (Higgs field?) exists, and its potential exoticity does not violate the generalized second law of thermodynamics (or the GSL is wrong). 2) Brane world cosmologies, M-Theory-inspired constructs featuring negative tension, apply to our universe. 3) The dark energy believed to pervade the universe turns out to be exotic.

Can classical violations of the weak energy condition be generated by a non-scalar field?
Currently, the answer appears to be no. It seems strange, however, that of all known matter fields only the lowly (and fundamentally nonexistent) non-minimally coupled scalar field can generate classical WEC violations.

What is the equation of state of exotic matter?
The question of wormhole stability cannot be fully settled without an answer to this question. Unfortunately, this cannot be answered through astrophysical observations. Such observations can only determine the equation of state of the *diffuse* form cosmic exotic matter – the type of dark energy better known as phantom energy or superquintessence. To determine the equation state of dense exotic matter, on which wormhole stability depends, it will be necessary to collect a suitable quantity of it. This would require a civilization capable of sending collector ships through great expanses of intergalactic space. Alternatively, it might be possible to create exotic matter artificially and determine the equation of state in local experiments. A further complication stems from the realization that although most forms of matter do adhere to some sort of equation of state, matter is by no means required to do so.

Is it possible to create a valid energy condition?
If is not, then geometries can no longer be excluded from physics on the grounds that they correspond to unphysical stress-energy. This would mean that anything goes. In addition to accepting wormholes and other causality-violating spacetimes, topology-changing geometries might have to be deemed physical. While it might be possible to rule them out on other grounds, they could no longer be excluded on the basis of their weirdly behaving stress-energy tensors. Recent attempts to find an inviolable energy condition have not succeeded.

Does Hawking radiation necessarily result in black hole evaporation?
We know that it is possible to convert an untraversable Schwarzschild wormhole into a traversable wormhole by suitably infusing the corresponding black hole with negative mass. The Hawking process effectively performs a negative mass infusion. It is clear that one possible result of this infusion is a black hole of reduced mass. But is it clear that this is the only possible result? The effective distribution of negative energy within the black hole due to the Hawking process could be such as to render Birkhoff's Theorem inapplicable there. If so, could Hawking radiation turn the black hole sector of the maximally extended Schwarzschild solution into a dynamic traversable wormhole?

Do quantum inequalities rule out topology-changing spacetimes?
It appears that the quantum inequalities are the closest thing we have to valid energy conditions. One argument against topology change was that it seemed to require violations of energy conditions. Now that the energy conditions are defunct, can we argue that topology change somehow requires violations of the quantum inequalities?

A wormhole with negative energy would presumably have negative temperature. Is this consistent with laws of thermodynamics?
One might be justified in blithely applying the laws of thermodynamics to systems containing negative energy. Nevertheless, an explicit incorporation of negative energy into thermodynamics would be welcomed. It would in particular be nice to see an elaboration (along the lines of Hayward 1994) of the view that there are laws of wormhole thermodynamics for which the laws of black hole thermodynamics are a special case.

Can any of the laws of wormhole thermodynamics be derived from string theory?
One of the successes of string theory is its derivation of the expression for black hole entropy in a particular case. Can this be repeated to obtain the corresponding expression for any particular class of wormholes?

Can wormholes shed light on the black hole information paradox?
When an object is tossed into a black hole the information it contains is, after complete evaporation of the black hole, loss. But the laws of physics, as currently understood, forbid information lost. A paradox. Useful strategies for solving this problem could involve wormholes. What happens when the wormhole is a contracting dynamic wormhole (a red hole) whose throat is arbitrarily close to being an event horizon? It might be availing to solve the problem for such a traversable wormhole (which should be straight forward, because there would be no paradox) and take the limit of this solution as the wormhole becomes a black hole. One might also speculate that quantum gravitational effects in the proximity of the black hole's singularity could induce changes in topology. These could include the formation of quantum wormholes, through which the information ingested by the black hole might be expelled back outward past its event horizon.

Can a simple wormhole be formulated in the context of loop quantum gravity?
Comparisons to a quantized thin-shell wormhole might serve as a sanity check on the theory.

What is to be learned by reformulating the physics of wormholes in the language of quantum computation?
Physical systems can be regarded as computers that perform a particular calculation – converting its initial state to its final state. The laws of physics, then, can be recast as programs running on hardware with assumed properties. This approach applied to closed timelike curves -- such as those due to worm-

hole time machines -- lead to a demonstration of the consistency of CTCs with quantum theory. Will recasting wormhole physics in this vain yield further insight unavailable through the conventional approach?

Is there any relationship between the universes of the Many Worlds Interpretation and the universes that wormholes connect?
They seem to have nothing to do with each other. Yet it seems strange that multiple universes would just happen to pop up independently in gravity and quantum mechanics. Given that these theories must ultimately be unified, this coincidence seems like some sort of clue. Moreover, wormhole time-travel paradoxes can be resolved by supposing that the wormhole traveler emerges in a parallel universe. Taken seriously, this seems to imply, as I have argued in previous chapters, that wormholes can be adapted to explore the Multiverse. Is there another theoretical means – besides time-travel paradox resolution – through which this conclusion can be reached?

Can a charged or rotating wormhole retain its access to an infinite number of universes after it becomes traversable?
The Reissner-Nordstrøm and Kerr (collectively known as Kerr-Newman) black hole solutions feature access (in principle) to an infinite number of universes. When these black holes are converted into traversable wormholes, this access is lost. The wormhole instead connects our universe to one universe only. Is traversability necessarily inconsistent with access to an infinite number of universes?

Are planar wormholes really possible?
Image a flat metal plate with an arbitrarily great surface area. Now rapidly accelerate it in a direction perpendicular to its surface. If it accelerates fast enough, an event horizon will form above its surface. It becomes a planar black hole. Just as a spherical black hole can be converted into a traversable wormhole, so it must be with its planar (plane symmetric) counterpart. If two such plane symmetric traversable wormhole are simultaneously (in their center of mass frame) brought into existence, will they connect? If so, why? If not, why not?

Are Cauchy horizons within wormholes traversable?
Blue sheet (mass inflation) singularities appear at the inner horizons of charged and rotating black holes. They have been conventionally regarded as impassable instabilities. It has been argued that the shock encountered when traversing these horizons is of exceedingly short duration and therefore survivable by humans. What arguments can be adduced to refute this?

Would the existence of wormholes imply the reality of topology change?
A standard argument against topology change rested on a 1966 theorem by Geroch. This theorem stated that absence of closed timelike curves implied the absence of topology change. But if wormholes exist, so do closed timelike curves. This argument, then, ceases to apply. Another argument observed that topology-changing spacetimes produced badly behaving stress-energy tensors. But if there are no longer any energy conditions to impose restrictions on stress-energy, there is no longer any such thing as bad behavior.

Where are the wormhole dynamics movies?
Back in the 1970s computational physicist Larry Smarr pushed the computing envelope by producing a movie of colliding black holes. Today, any undergraduate carries hundreds of times the computing power required to accomplish this in her backpack. Animating wormhole evolution, interaction, collapse, inflation, and reaction to traversal attempts would be a worthwhile student project.

Can supergravity quantum wormholes form the basis for a model of elementary particles?
Attempts to realize Wheeler's idea by modeling elementary particles with pure-gravity quantum minisuperspace wormholes have not succeeded. The wormhole's mass is invariably of the Planck scale – far too large for an elementary particle. Moreover, it is not clear how a minisuperspace quantization of the Kerr-Newman black hole will result in spin quantization to say nothing of charge quantization. Work in the early 1990s on supersymmetric quantum wormholes seemed to suggest that this approach might shed light on these problems. Is it perhaps time for an interesting alternative to string-theoretic particle models?

Can alternative theories of gravity allow wormholes to exist in the absence of exotic matter?
This question has usually been asked with regard to "*f*(*R*)" theories (those in which the Lagrangian density of general relativity, *R*, is modified by replacing *R* with a function of *f* of *R*, e.g. $R \rightarrow R^2$). The answer was "no", except when the parameters in the alternative gravity theory were improbably "fined-tuned". Today, this question is asked with regard to the "brane worlds" concept that emerges from M Theory (superstrings).

Is a spherical wormhole more likely to be destabilized by spherically symmetric or spherically asymmetric perturbations?
If would be interesting to know whether the minimum mean amplitude required to destabilize a wormhole depends on the symmetry of the perturbing

matter wave. Asymmetric perturbations are scarcely mentioned in the literature.

Does a mindless generalization of the Kerr solution lead to a rotating traversable wormhole?

Teo, in developing his rotating traversable wormhole, was careful to begin with the most general stationary axisymmetric metric. Following Morris and Thorne, he constrained it so that it would be free of horizons. What if he had not been careful? What results from the most naïve generalization of the metric – replacing the black hole mass with a shape function and requiring the temporal component to be finite?

What are the consequences of entanglement through a wormhole?

Consider two particles that are entangled. Send one of them through a wormhole to a distant part of the universe. Now bring the particles in close proximity by moving them toward each other through normal space the long way. What is the consequence of the entanglement path being thousands of light years long, when the entangled particles are only inches apart? What happens when the wormhole is closed? What is the consequence of entanglement through a wormhole acting as a time machine?

Is Hawking's chronology protection conjecture valid?

As the time-shifted mouths of a single wormhole are moved toward each other, semiclassical calculations show that the vacuum stress-energy begins to diverge just as the mouths reach the point where the wormhole would begin operating as a time machine. Of course, a semiclassical calculation ceases to be valid in the midst of a divergent vacuum energy density. It nonetheless argues that a single wormhole at least *resists* becoming a time machine. Interestingly, the same does not seem to apply to a Roman ring of wormholes. This arrangement can come arbitrarily close to functioning as a time machine without the vacuum stress-energy diverging. Unlike the single-wormhole case, there is no resistance. We need a tie breaker.

Can the quantum inequalities be used to rescue the singularity, topological-censorship and other lost theorems of general relativity?

The invalidation of the energy conditions by quantum effects rendered otiose certain cherished theorems of general relativity. The quantum inequalities are the only remaining restrictions on the stress-energy of matter. It might be interesting to explore the dim possibility that they might be used (in a manner analogous to Ehrenfest's Theorem) to place restrictions on classical matter and thus recover at least one of the lost theorems in a weakened form.

What happens when entangled systems are in separate universes?
Imagine two entangled particles. One of them traverses a wormhole into another universe. The wormhole pinches off. Is entanglement lost? What would that mean? If it is lost, does this occur before the diameter of the contracting throat reaches zero? If entanglement is not lost, does this mean that quantum teleportation through a closed wormhole hole is possible? What happens if the wormhole reopens? Suppose the wormhole forms an intra-universe connection. When one of the entangled particles passes through, and the wormhole pinches off, is the entanglement preserved because both particles remain in the same universe? What if the wormhole is acting as a time machine?

Can a wormhole be used to see the interior of a black hole?
As first noted by Frolov and Novikov in the early 1990s, there appears to be nothing to prevent a wormhole connection from crossing an event horizon. An external observer would then be able to obtain information from within the horizon. A traveler who had fallen into the black hole could use the wormhole to escape his impending death. Timelike singularities cloaked by the event horizon would be visible through the wormhole to external observers. This would seem to imply that existence of traversable wormholes is inconsistent with the cosmic censorship hypothesis. It would be interesting to work out the physics of a tiny wormhole mouth approaching the horizon of a large black hole. Does a semiclassical treatment "resist" the mouth's crossing of the horizon? In other words, as the mouth approaches the horizon, does any observable quantity diverge?

What is to be learned from the theory of wormhole computers?
Black hole computers have been described in the literature. They arise in attempts to understand the black hole information paradox. Wormhole computers would be interesting, because one would naively expect a negative-energy wormhole to have a negative temperature. Does the physics of computation need to be modified to properly describe such a system?

Are Wheeler's geons possible?
A geon is a tiny wormhole threaded by lines of an electric or magnetic field. It mimics an electric or magnetic point charge in both universes connected by the wormhole, even though no charge is actually present. As mentioned in Chapter 2, geon solutions have recently been found, but they assume that the electromagnetic field couples directly to the curvature of spacetime. What evidence or arguments support the existence (or nonexistence) of such a nonminimal coupling?

Is the universe itself a giant expanding wormhole?

What arguments would rule out the possibility that we are sitting at the throat of the mother of all wormholes? Such a wormhole could as easily accommodate an inflation-inducing matter field as do the standard cosmological models. Moreover, it is not clear that anisotropy arguments -- that observations of the sky would depend on the direction in which we looked – would suffice to eliminate the possibility of this wormhole cosmology. In this scenario, the Big Bang would simply have been the usual past singularity of a dynamic wormhole. Dispelling this notion could be an enlightening exercise.

13. The Central Enigma of Wormhole Physics

Exotic Matter Matters

We have learned that the existence of traversable wormholes depends on the existence of exotic matter. If cosmic exotic matter exists in abundance, naturally occurring macroscopic wormholes cannot be ruled out. If such exotic matter is rare or nonexistent, as the latest experimental evidence suggests, naturally occurring wormholes can only be of the infinitesimally small quantum variety. Such quantum wormholes can be synthetically enlarged, provided that one can create or harvest the requisite quantity of exotic matter. If physics permits topology change, wormholes could be synthetically created from scratch, but only if the required quantity of exotic matter can be brought to bear.

There is no reason to believe that synthetic concentrations of exotic matter at two widely separated locations in the universe will result in the formation of a wormhole shortcut between these locations. Deliberate topology change, even if possible, could only be used to affect local geometry. The most that we could expect would be the creation of a new quantum wormhole with mouths in close proximity. These mouths would require enlargement and separation just as in the case of wormholes mined from spacetime foam. Wormholes whose mouths must be dragged to their intended locations, although useful, cannot address the problem elucidated in the introduction. Beyond this, it seems far more likely that a synthetic concentration of exotic matter would only succeed in inflating a tiny bubble universe – a little pocket of our universe connected to the greater universe by a newly formed wormhole. Such a bubble might be a useful place in which to hide precious objects, store toxic waste, or banish dangerous criminals. But its associated wormhole entrance would be worthless as a means of superluminal communication or travel.

Irrespective of whether exotic matter exists in nature it appears currently that it can in principle be created. This is not an allusion to the Casimir Effect, in which the negative energy that forms between parallel conducting plates is more than cancelled out by the larger positive mass of the plates themselves. Rather, it is the conclusion of theoretical research conducted within the last decade. As I mentioned in the chapter on wormhole traversability, if fundamental scalar fields exist, you can violate the averaged null energy condition as much as you please. These violations may be quantum – requiring the scalar field to be in a superposition of particular quantum states, or classical – requiring only the usual (conformal) interaction between the scalar field and the curvature of spacetime. The question of the existence of exotic matter, then, turns on that of the existence of a fundamental scalar field. The conventional expectation among physicists is that such fields do exist. The yet-to-be discovered Higgs boson predicted by the highly successful Standard Model of particle physics is a fundamental scalar field, as is the so-called inflaton that cosmologists use to explain the rapid inflation of the early universe. Given that a fundamental scalar field is likely to exist, that the average null energy condition can therefore be violated, and that, consequently, the existence of a form of exotic matter is probable, we can suppose with Morris and Thorne that a sufficiently advanced civilization will be able to produce it.

Stargates and Survival

Recall that in the introduction I pointed out that humanity's distant descendants – or whatever intelligence comes to dominate the universe -- would need wormholes to compensate for the expansion of the universe. They would do so by using them to effectively annihilate distance. This annihilation would prevent the growth in the speed at which humanity's intellectual heirs can collectively think from continually slowing down. This slow down would result in part from their computational resources becoming increasingly dispersed throughout space due to the expansion. As this expansion would increase the physical dimensions of their group brain, so would it increase the time required for signals to traverse it. Part of that brain would be forever lost by expansion past the cosmological horizon. This would result in an effective mental retardation. Distance-nullifying stargates would prevent this retardation by creating shortcuts for signals and thus neutralize the effects of the expansion. Humanity's descendents could then keep pace with the growing complexity of the universe and thereby model it -- that is, understand it --

arbitrarily well. Were they to lack this understanding, they would be unable to defend themselves from a growing number of threats that they would find unfathomable. Eventually, perhaps after the passage of countless eons, one of these newly encountered threats would destroy them.

Recall also that stargates could provide our descendants with the option of avoiding the Big Freeze by importing energy from inaccessible regions of the universe or from parallel universes, including those that correspond to their past. In so eliminating the limit on their supply of energy, they would eliminate the limit on the number of computations that they could perform, the number of thoughts they could think, and the subjective time that their civilization could endure. Without so freeing themselves of these limits, they would – even if successful against all emerging threats – surely die out.

Recall further that these sorts of deliberations are required to determine the truth of the strong form of Thales' doctrine of rationalism -- the idea that humanity can understand the universe arbitrarily well. This doctrine asserts in effect that reality is free of supernatural entities.

The Microgeometry of the Universe: Is it Conducive to Our Survival?

If humanity's descendants are to survive, they cannot do so if the mouths of the wormholes that they create must be dragged great distances to their desired locations. As a result of the accelerating expansion, a growing number of locations in the universe will be receding from each other at effective speeds exceeding that of light. But in moving the mouths our descendants cannot exceed the speed of light. This limitation will prevent them from placing each mouth where they will need it to increase maximally the rate at which they process information.

In short, the only apparent hope for humanity's heirs lay in the nature of the primordial Planck-scale wormholes of the quantum foam. If their mouths are sufficiently separated to permit communication anywhere in the universe, including regions beyond the small section in which we are in casual contact, we may take heart. Such wormholes would annihilate distance to the greatest possible degree. If, by contrast, the wormhole mouths are near each other, their capacity for distance annihilation will be reduced in proportion to their proximity. The mouths of these Planck-scale wormholes could, for example, be separated by Planck-scale distances of about 10^{-33} cm. In this case they would be worthless until separated at some cost to the degree possible given the aforementioned limitation. Although such a wormhole would be useful for, say, communication within a single galaxy, it would not serve to annihilate the distances required to ensure the survival of humanity's descendants. It would merely help to forestall their demise. Nor, given the cost of separating their mouths, does it seem likely that such wormholes could be created in sufficient quantities.

Superbeings or Supernatural?

If we assume that our descendants or members of some other advanced civilization can obtain exotic matter in sufficient quantities, there are two possibilities of interest relevant to their long-term survival. Either they can create an intricate network of wormholes able to compensate for the expansion of the universe, or they cannot. In the latter case, they are, as mentioned above, doomed by their inability to understand new threats to their existence that will undoubtedly arise as the eons pass. For them, reality will possess a supernatural sector containing elements that they will never fathom irrespective of how hard they try. They just won't have the needed speed to process information. The complexity of the universe will rise faster than the ever declining increase in their processing speed will allow them to formulate theories about it.

In the case that they can create an expansion-compensating wormhole network, they could survive by using wormholes to annihilate distance and thus increase their effective speed of collective thought. This would only be possible if the network of primordial wormholes that they exploit interconnects the entire universe as opposed to some small fraction of it. As a consequence of their ability to think fast enough to keep up with the growth of complexity in the universe, they would regard reality as being free of supernatural entities. Interestingly, they would nonetheless be affected by the influence of god-like superbeings. Recall that wormholes permit communication between parallel universes. The dissimilarity between the parallel universes thus brought into communication is likely dependent on the spacetime separation of the mouths of the facilitating wormhole. In this case we have assumed a universe-spanning network of wormholes. Such numerous and widely separated mouths likely ensure a wide sampling of the multiverse. The network will open its users to influence by arbitrarily advanced civilizations in parallel universes.

We may summarize these cases as follows. In the absence of a universe-spanning wormhole network, humanity's descendants will be aware of the existence of a supernatural sector of reality and will be ultimately destroyed by it. In the presence of such a network, humanity's descendants will know a reality free of a supernatural sector, but will be open to interactions – possibly fatal -- with god-like beings belonging to arbitrarily advanced civilizations in parallel universes.

In which situation will our descendants fare better? That depends on the relative probability of being killed by unfathomable complexity (e.g. diseases or environmental threats beyond comprehension) verses being killed by contact with an arbitrarily advanced civilization. If it is true that civilizations of arbitrarily great power and knowledge tend to respect knowledge-acquiring life, we can conclude that contact with these is the lesser threat and quite possibly a boon.

The Central Enigma

Which of these two possibilities represents the human destiny, or, to be more precise, the destiny of the intellectual descendants of whichever intelligence comes to dominate the universe? No stargates together with the growing risk of growing ignorance? Or stargate-enabled knowledge growth together with the risks and benefits of contact with advanced beings in parallel universes? The answer turns on what I call the central enigma of wormhole physics. It follows from the paradoxical role of exotic matter. The enigma is this: The presence of exotic matter implies the existence of wormholes, which permit the importation of energy and neutralize the expansion-induced reduction in the growth of information processing speed, which removes limits to understanding, which implies that humanity's descendants will survive indefinitely. But the presence of exotic matter also implies a rapidly accelerating expansion of the universe, which implies a Big Rip, which implies a time limit on the universe, which implies limits on understanding, which implies that humanity's descendants will *not* survive indefinitely. In other words, exotic matter is both our savior and our nemesis. Or in epistemological terms, Thales' doctrine of rationalism is both true and false.

The solution to this enigma will be a matter of cosmology. There are two ways to solve it that are consistent with Thales' doctrine. The first requires us to show that a wormhole-populated cosmological model exists in which the time limit imposed by a looming Big Rip does not limit what intelligent life can understand. Alternatively, we would need to find a cosmological model in which the exotic matter required to support a profusion of wormholes somehow fails to induce a Big Rip and thereby imposes no limit on the knowledge acquired by such life. In other words, either humanity's successors will exist for an effective eternity despite the presence of a Big Rip, or they will exist eternally because of the absence of a Big Rip. In both cases Thales' doctrine requires our successor civilizations to exist for a subjective eternity. Otherwise, there would be no guarantee that they would have sufficient time to comprehend *every* aspect of reality as required by the doctrine. Cosmological models meeting either of these criteria have yet to be found.

The other solution to the central enigma of wormhole physics is to simply discard Thales' doctrine. This would eliminate the need for wormholes and exotic matter required to support them. As mentioned in the introduction, this is tantamount to assuming the ultimate destruction of all intelligent life in the universe. Without stargates to nullify the retarding effects of the expansion of the universe on their collective mind, our intellectual descendants will be unable to keep pace with the ever growing complexity of reality. Their knowledge will grow, but their ignorance will grow faster. The sector of reality forever beyond their conceptual reach – an effectively supernatural sector – will eventually spell their doom. This will follow from the old adage, "What you don't know could kill you." Sooner or later a deadly threat to their existence will emerge from this supernatural sector. Their inability to comprehend it will prevent them from surviving it.

Our Three Options

What, then, do we conclude? Was Thales right or wrong? Are our distant descendants doomed or not? The answer is that we do not yet know. Given our ignorance we face three choices: We may believe the strong form of Thales' doctrine to be false. We may believe it to be true. Or we may commit to neither view. Let's consider these options in turn.

Assuming that Thales was wrong might seem advantageous to those of us who would find comfort in the presumed existence of a supernatural sector of reality. We might suppose that we are free to populate this sector with our favorite deities and supporting supernatural entities. Unfortunately, knowing that such a sector exists is distinct from knowing its contents. Even if supernatural entities do exist -- entities about which we cannot increase our knowledge irrespective of how hard we try -- we have no idea what they are. We are not therefore justified in assigning to the supernatural sector any particular set of comforting characteristics.

If we assume Thales to have been wrong, we face the disadvantage of having to explain the answers to certain questions. Why does Thales' doctrine *seem* to be true? How is it that all aspects of reality to which we have thus far been exposed seem susceptible to human attempts at comprehension? If our understanding of reality is truly limited, how precisely is this limit to be specified? Where precisely is the boundary between the comprehensible natural world and the incomprehensible supernatural one? If the supernatural sector of reality actually interacts with the natural sector, how then is such a boundary even possible?

Lastly, there are the disadvantageous emotional effects that would accompany our disbelief in Thales' doctrine. Our confidence in our ability to solve scientific problems – the single most important factor in our being successful against them – would be shaken. We could never be sure that the problem before us that seems merely difficult is in fact impossible as a consequence of its connection to the supernatural. We would, moreover, experience to some degree the debilitating pessimism that would result from our having to believe that the human project is ultimately doomed.

Instead of believing Thales to be wrong, we may choose to be agnostic – refusing to make up our minds one way or the other. This might seem to be the most rational position, given our lack of knowledge. Recall, however, that our objective is to understand reality as best we can. This means finding the simplest explanation for it. Agnosticism about Thales' principle is like agnosticism regarding the existence of a civilization in the center of the earth. The latter would complicate our geophysical theories by requiring them to accommodate the *possibility* of such a civilization. The possible falsity of Thales' doctrine would similarly require us to answer the aforementioned questions. In addition to the complexity that this would introduce, agnosticism would -- like outright disbelief but to a lesser extent – also reduce our confidence in our ability to solve scientific problems and increase our pessimism. The milder case of the latter would result from our belief in the possibility, rather than the certainty, that humanity is a dead end.

Advice from the 17th Century

We are led, then, to a secular version of Pascal's wager. Pascal, you will recall, was the seventeenth-century French mathematician who argued that in the face of uncertainty about the existence of God, we are better off acting as though He exists. The cost, he reasoned, of living a sinful life, if God exists (eternal damnation), well exceeds the cost of living a virtuous life, if He does not (missed pleasures). Here we can conclude that the cost of disbelieving Thales' doctrine, if it is true, well exceeds the cost of believing Thales' doctrine, if it is false. If Thales' doctrine is true, but we do not believe it, our confidence in our ability to solve our problems through mental effort would weaken. We would be more likely to succumb to mysticism. We would direct less of our brain power toward overcoming threats to our survival. And the risk of our demise would thus progressively increase. The cost, then, of our loss of belief would be the needless extinction of humanity. If, by contrast, Thales' doctrine is false, but we believe it nonetheless, we merely pay the cost of mental effort wasted in futile attempts to understand aspects of reality that are in fact beyond comprehension.

Our predicament is this. If the expansion of the universe is in fact accelerating and Thales was right, then we must develop stargates or other distance-nullifying technology, lest our distant descendants be destroyed. If such expansion is occurring and Thales was wrong, our efforts in this regard will come to naught. Humanity, its intellectual descendants, or that of whatever intelligent life comes to dominate the universe, are already doomed to ultimate extinction.

In view of this predicament we should in effect take Pascal's advice. We should fully embrace the Greek half of our cultural heritage by living and acting as though Thales was right. One manifestation of this renewed faith in our ability to comprehend the whole of reality will be the attention that we pay to the physics of wormholes. Just as Thales' ancient musings are of relevance today, so may we hope that our study of wormholes might ensure sufficient understanding by some distant tomorrow. For it may well be that our distant intellectual heirs will thank us. Our current attention to wormhole physics may well ensure their timely possession of the knowledge they will require to one day save themselves from imminent doom, to build the instrument of their salvation – the stargate.

Glossary

Here are definitions of terms that you are likely to encounter in reading further on the topic of wormholes and related physics.

acausal -- spacelike.

acausal region – a region of spacetime containing no events separated by timelike or lightlike intervals.

achronal – spacelike or lightlike, i.e. not timelike.

achronal region -- a region of spacetime containing no events separated by a timelike interval.

AdS spacetime – see "anti-de Sitter spacetime".

AdS/CFT correspondence – short for "anti-deSitter/Conformal Field Theory correspondence" -- the equivalence discovered in 1997 between two types of theories: 1) a string theory that describes gravity within a region of anti-de Sitter space and 2) a conformally invariant field theory that does not describe gravity and is defined on the boundary of this region. It is of interest because it makes it likely that established perturbative techniques of quantum field theory may be successfully applied to gravity and thus advance the effort to quantize it.

action – essentially the time-averaged value of the difference between the kinetic and potential energies of a physical system, which the motion of the system seeks to minimize. A value used to weight the history of any particular universe in a sum that will determine the prevalence of that history in the multiverse.

amplitude – a complex number the square of whose magnitude is proportional to the prevalence of universes in a particular state (a.k.a. the probability of the state).

affine – having a coordinate transformation rule that is not purely multiplicative (as it is for tensors) but also requires the addition of a term (e.g. $v' = Tv + a$).

affine parameter – a parameter whose value specifies a particular position along a path.

ANEC – see "averaged null energy condition".

anomaly – a symmetry of a classical field theory that ceases to exist in the corresponding quantum field theory.

anisotropy -- a lack of uniformity of observed properties of a system (normally taken to be the universe) with respect to the direction in which the system is observed.

anthropic principle – the idea that the laws of physics must be consistent with the evolution of intelligent life in the universe.

anti-de Sitter spacetime – a spacetime with a constant negative curvature that is driven to contract at an exponentially decreasing rate by a negative cosmological constant. Also known as "anti-de Sitter space".

antihorizon – the horizon that surrounds a white hole that functions as a time-reversed version of the event horizon of a black hole. Nothing exterior to it can enter its interior; everything in its interior must pass to its exterior.

anti-trapped surface – a past trapped surface. See "past trapped surface".

apparent horizon – a surface defined by local properties that is likely to become an event horizon in the future.

asymptotically flat – having curvature that progressively decreases with spacelike distance from a specified point, becoming arbitrarily close to zero at arbitrarily large distances.

averaged null energy condition – the requirement that the null energy condition hold on average along any null path.

averaged weak energy condition – the requirement that the weak energy condition hold on average along any timelike path.

AWEC – see "averaged weak energy condition".

axion -- a hypothetical spin-0 particle of low mass predicted by supergravity and string theory and represented mathematically by a rank-3 tensor field.

Ashtekar variables – variables that specify geometry in terms of the Cartanian connection instead of the Riemannian metric, that can be used to express general relativity in a form that resembles a gauge theory familiar to particle physics, and that have engendered the development of the "loop quantum gravity" approach.

baby universe -- a universe created from the spacetime of a parent universe typically through a vacuum fluctuation.

back reaction – the effect of a perturbation on the cause of the perturbation.

base state – one of a set of quantum states of a superstition that can be used collectively to specify a general quantum state.

Bianchi types – a classification of spatially homogenous generally anisotropic cosmology models.

Big Rip, The – the infinite expansion of the universe within a finite time that would result in the ripping apart of all bound states of matter including galaxies, solar systems, molecules, atoms, and nuclei.

Birkhoff's Theorem -- any spherically-symmetric solution to the Einstein field equations in empty space is a piece of the Schwarzschild solution.

black hole -- a compact region of high curvature culminating a singularity and surrounded by an event horizon. The contracting phase of an untraversable wormhole.

blue hole -- a term used only in this book for the expanding phase of a wormhole with a dynamic geometric throat. The expansion results in light escaping the interior region being blue shifted in the sense that its gravitational redshift is reduced. In the limit that its geometric throat expands rapidly enough to prevent light from entering its interior, it becomes a white hole.

blue sheet singularity – the concentration of infalling, blue-shifted radiation energy that occurs on the anti-horizons of white holes and the Cauchy horizons of certain black hole solutions. Also known as a mass inflation singularity.

boson -- a particle with integral spin (e.g. 0, 1, 2, …) that serves as a means of propagating a force and is not restricted from occupying any quantum state already occupied by another boson.

boson star – a massive, dense, gravitationally bound concentration of particles with integral spin at zero temperature, whose density is just below that required to form a black hole.

Boulware vacuum – the quantum vacuum state of a scalar field in a Schwarzschild spacetime. The vacuum is singular at the event horizon and violates all pointwise energy conditions.

brane – the mathematical objects of varying dimensionality used to model elementary particles. The traditional point particle is a 0-brane; particles modeled as strings are 1-branes; particles modeled as membranes are 2-branes, etc.

brane world – an 11-dimensional string-theoretic model in which the universe is taken to be a 4-dimensional subspace – a 3-brane evolving in time -- of a typically 5-dimensional space (the "bulk") with the remaining spatial dimensions compactified. The properties of the universe are determined from its motion in the 5-dimensional space and from projections onto the 4-dimensional subspace of fields defined in the 5-dimensional space.

Casimir effect – the attractive force between parallel conducting plates due to the dependence on plate separation of the negative energy of the quantum vacuum between the plates.

Cauchy data -- the value of a field at each point on a particular spacelike hypersurface together with the value at each point of the rate of change of the field in a direction perpendicular to the hypersurface.

Cauchy horizon -- The boundary between 1) the part of a spacetime whose geometry is determined by the evolution of Einstein's equations from initial data, and 2) the part of the spacetime whose geometry cannot be so determined, i.e. the boundary between a region of spacetime in which physics is predictable and a region in which it is not.

Cauchy surface – a spacelike hypersurface that is intersected precisely once by every non-spacelike curve.

causal curve – a non-spacelike curve in spacetime.

causality – that property of a spacetime that ensures that events that can influence any particular event are distinct from the events that that particular event can influence. The absence of closed causal curves.

causality horizon – the boundary between 1) the set of events that are connected via a non-spacelike curve to an event in a causality-violating region and 2) a region of spacetime containing no such events.

causality-violating region – the region of a spacetime consisting of the set of all events that are on a closed non-spacelike curve.

causality violation – the presence in a spacetime of closed non-spacelike curves.

CGHS model – a two-dimensional model proposed in 1992 by Callan, Giddings, Harvey, and Strominger in which matter is coupled to gravity in such a way as to describe the formation of an evaporating black hole.

CFT – see "Conformal Field Theory".

chirality – that property of being either left handed or right handed

chronal curve – a timelike curve in spacetime.

chronology horizon – the boundary between 1) the set of events that are connected via a timelike curve to an event in a chronology-violating region and 2) a region of spacetime containing no such events.

chronology-preserving region –a region spacetime in which the events that can influence any particular event through a timelike signal are distinct from the events that that particular event can influence through a timelike signal. A region of spacetime containing no closed timelike curves.

chronology-preserving spacetime – a spacetime containing no closed timelike curves.

chronology-violating region – the region of a spacetime consisting of the set of all events that are on a closed timelike curve.

chronology violation -- the presence in a spacetime of closed timelike curves.

Chaplygin gas – a substance with the equation of state $p = -A/\rho$, where A is a positive constant. Considered as a gas model by Chalplygin over one hundred years ago, it today serves as a joint model of cosmic dark matter and dark energy, having the characteristics of the former at early times in the history of the universe and of the latter at late times.

classical limit – the description, suitable for macroscopic phenomena, that a quantum mechanical description of a system approaches as Planck's constant is set to zero.

closed timelike curve -- a curve that corresponds to the world line of a traveler whose future precisely matches his past. Also known as "closed timelike lines".

cosmic censorship hypothesis – the idea that all singularities exist within event horizons, i.e. naked singularities, those not cloaked by an event horizon, do not exist. In its strongest form, it is the view that experimenters exterior to any event horizons cannot observe a singularity directly or indirectly (by accessing recorded observational data).

coherence -- that condition of a quantum state in which it is a superposition of base quantum states.

collapse of the wave function – the transition triggered by a measurement, through which a system seems to change from a superposition of quantum base states to a single base state.

compact manifold – a manifold on which a metric is defined for which there is a finite maximum distance between any two of its points.

compactified dimensions – dimensions whose spatial extent is so small as to escape detection through normal means (as, for example, drastically shrinking the circumference of a drinking straw would make the straw seem to be a one-dimensional object, even though it is still 2-dimensional).

complexity – that property of an object quantified by the length of the shortest set of instructions required for a "universal constructor" (a hypothetical machine capable of constructing any object) to construct an indistinguishable copy of the object from scratch.

conformal field theory – a quantum field theory that is symmetric under angle-preserving coordinate transformations called conformal transformations.

conformal transformation – an angle-preserving coordinate transformation whose effect is 1) a change of scale (akin to "zooming in" or "zooming out"), or 2) a special scale change, followed by a displacement (akin to "panning"), followed by another special change of scale, or 3) a Poincaré transformation (akin to a displacement and an instantaneous acceleration).

conformally flat spacetime – a spacetime whose metric g is a product of the flat spacetime metric η and a positive scaling function Ω, i.e. $g_{\mu\nu} = \Omega\eta_{\mu\nu}$. The term is used when this is locally true about every point requiring in general many distinct scaling functions, and it is also used when this is globally true for a single scaling function.

conformal coupling – a coupling between the massless scalar field and the gravitational field which is unchanged by conformal transformations thus ensuring scale-invariant physics.

consciousness – the awareness of a program, such as a mind, of its internal state in addition to its awareness of the state of its external environment.

cosmological constant – a coefficient that appears in the Einstein field equation that causes the universe to expand or contract depending on whether the constant is positive or negative. It may be interpreted as the vacuum energy density of the matter of universe.

cosmological constant problem – the value estimated for the cosmological constant (as the vacuum expectation value of the stress-energy tensor of matter fields) is many orders of magnitude larger than the observed value.

cosmological horizon -- for a given point in the universe the boundary between those parts of the universe that will at some time in the future be in causal contact with that point and those parts that will never be. Such horizons can result from the expansion of the universe, which causes regions to separate from each other at speeds effectively exceeding the speed of light.

cosmological principle – the idea that averaged over large scales, the locally observable properties of the universe are the same everywhere – there is no privileged location. In other words, the universe is spatially homogeneous and isotropic.

covariant – retaining the same form in any coordinate system.

covariant derivative -- an extension to curved space of the flat space concept of the derivative of a tensor field that compensates for changes in the tensor field that are due solely to the curvature of the underlying space.

CTC – see "closed timelike curve".

consistent histories – an interpretation of quantum mechanics in which histories that do not obey the rules of probability are excluded from consideration.

curvature scalar -- a single number that quantifies the curvature at a given point in spacetime, and whose average over spacetime the evolution of the gravitational field seeks to minimize. A contraction of the Ricci tensor, which is in turn a contraction of the Riemann curvature tensor.

dark energy -- the substance presumed to be responsible for the accelerating expansion of the universe and is believed to account for 70% of its matter. Its equation of state is $p = \omega\rho$, where $\omega < -1/3$.

dark matter – low-temperature, low-pressure, non-luminous matter whose non-gravitational interaction with ordinary matter is weak and is thought to comprise about 25% of the matter in the universe.

DEC – see "dominant energy condition".

decoherence – the collapse of the wave function induced by subtle and ubiquitous interactions with the environment in advance of deliberate measurements.

density matrix – a matrix formed from the outer product of the state vector of a system in a pure state or the weighted sum of outer products of the state vectors of a system in a mixed state. The trace of its matrix product with the matrix representation of a quantum operator yields the expectation value of the operator.

de Sitter spacetime – a spacetime with constant positive curvature that is driven to expand at an exponentially increasing rate by a positive cosmological constant. Also known as "de Sitter space".

de Sitter temperature – the temperature characterizing the Hawking radiation emitted by a cosmological horizon, i.e. an event horizon due to the expansion of the universe. Also known as the "Hawking temperature" of a cosmological horizon.

diffeomorphism – a one-to-one mapping from one manifold to another (or the same) manifold that is continuous and differentiable, and whose inverse is similarly characterized.

dilaton – a scalar field that was originally the 5th component, $A_5(x)$, of the vector potential in Kaluza-Klein theory. In string theory it is the trace of the graviton.

Dirac equation – equation of motion of a relativistic fermionic field (such as the electron).

dominant energy condition -- a defunct energy condition requiring the local energy density be non-negative and the local energy-flux vector to be non-spacelike.

dS – see "de Sitter spacetime".

dual – having equivalent physical meaning despite apparent fundamental differences in formulation.

duality – the property of being dual.

Ehrenfest's theorem – a classical equation of motion can be obtained from a quantum equation of motion through the application of a simple technique (which essentially replaces operators with their expectation values).

Einstein frame – the form of the Lagrangian of a scalar-tensor theory of gravity in which the scalar field is minimally coupled to the curvature scalar and which is obtained from the Lagrangian in the Jordan frame by a conformal rescaling of the metric tensor.

embedding – the insertion of a geometry of dimension *n* into a space of dimension greater than *n* in which the curvature of the geometry is manifested as a distention into the additional dimension.

energy condition -- a restriction on the stress-energy tensor conjectured to hold for all physically reasonable matter.

energy-momentum tensor – see "stress-energy tensor".

entanglement – the connection between elements of a physical system in a particular quantum state that persists irrespective of the distance by which the elements are separated.

entropy – a measure of the number of ways in which a system can be in a particular macroscopic state. A measure of disorder. [e.g. there are many ways in which a kitchen can be messy, but comparatively few in which it can be neat. Hence, a messy kitchen has higher entropy than a neat one.]

equation of state – the equation that expresses the pressure of a substance as a function of its density and temperature.

ergobelt -- the narrow region straddling the equatorial plane of a rotating wormhole from which work may be extracted. Also called an ergo region.

Euclidean signature – having the time coordinate preceded by a positive sign in the generalized Pythagorean expression for distance: $s^2 = t^2 + x^2 + y^2 + z^2$. Denoted by (+ + + +).

Euclidean geometry -- geometry for which the expression of any infinitesimal distance has a Euclidean signature.

event – a place and a time, i.e. a point in spacetime.

event horizon – the boundary between the events of spacetime from which emitted light can reach infinity and those from which it cannot.

exotic matter – matter that violates the null or weak energy conditions.

extremal black hole – a black hole whose mass is the smallest value allowed by its charge and angular momentum.

extrinsic curvature – a measure of the dissimilarity of the subsequent space-like hypersurfaces in a foliation of spacetime. It is zero when the hypersurfaces are embedded in spacetime in a manner that results in their normal vectors at each point being parallel.

factor ordering problem – a problem that arises in the canonical quantization of a classical system when it is unclear how to order factors that commute in the classical theory but no longer commute in the corresponding quantum theory.

fermion – a particle with half-integral spin (e.g. 1/2, 3/2, 5/2, ...) that does not serve as a means of propagating a force and is restricted from occupying any quantum state already occupied by another fermion.

flare out – that feature of a wormhole geometry that describes the transition between its throat and the asymptotically flat region that contains it.

Ford-Roman constraints -- see "quantum inequalities".

FRW spacetime – see "Friedman-Robertson-Walker" spacetime.

Friedman-Robertson-Walker spacetime -- a uniquely homogenous and isotropic solution to Einstein's equations corresponding to a dynamic space with either positive, zero, or negative curvature that formed the basis of one of the earliest relativistic cosmologies.

foliation – a decomposition of spacetime into a stack of temporally ordered spacelike hypersurfaces.

functional – a function of whose domain is a set of (closely related) functions and whose range is (typically) a set of numbers. A function of functions.

functional throat – a surface at which the expansion of a perpendicularly incident bundle of light rays is momentarily zero before becoming positive.

future trapped surface – a closed surface from which simultaneous emissions of outward (and inward) light pulses from every of its points produces a wave front whose area is less than that of the surface. In the context of Schwarzschild geometry, it is a surface for which all intersecting word lines will terminate in a singularity.

geodesic – the shortest distance between any two events in spacetime. When timelike, the world line of free falling particles.

geodesically complete -- that property of a spacetime requiring it to contain no events at which a geodesic is undefined. Singularity-free.

Geroch's Theorem – name used only in this book to label Geroch's 1966 theorem: If a compact spacetime manifold is time orientable and free of closed timelike curves, then its spatial topology cannot change.

gauge – any of an infinite number of mathematically distinct yet physically equivalent means of eliminating the ambiguity in the solutions to equations of motion of a field theory describing a physical wave or particle.

gauge theory – a field theory that is designed to be invariant under a particular group of position-dependent symmetry transformations (that each correspond to a change in gauge).

generalized second law – a generalization of the second law of thermodynamics to include the entropy of black holes. A black hole's entropy is defined as one fourth of its surface area. The law requires the total entropy in the universe, including that from all black holes, never to decrease.

geometric throat – a hypersurface of minimal area in a wormhole spacetime. A surface of minimal area in a spacelike hypersurface of wormhole spacetime.

ghost particle – particle with negative energy.

ghost radiation -- massless particles whose kinetic energy is negative.

globally hyperbolic spacetime -- a spacetime on which Cauchy data defined on one hypersurface determines the entire spacetime through the operation of the Einstein equations.

gravastar – a hypothetical end state of stellar collapse in which black hole formation is prevented by a sort of phase change in spacetime, induced by quantum effects, that results in a dense, impenetrable, non-singular, non-luminous object equal in size to a black hole of the same mass.

graviton – the hypothetical, massless, spin-2 quantum of the gravitational field.

GSL – see "generalized second law".

Gyr – a gigayear. One billion years.

Hamiltonian constraint – the requirement that the Hamiltonian of the gravitational field, which may in some sense be regarded as its total energy, be zero. This follows immediately from insensitivity of gravity to the particular the choice of a time coordinate.

Hartle-Hawking vacuum – the quantum vacuum state of a scalar field in the vicinity of a black hole that is in thermal equilibrium with its environment. Unlike the Boulware vacuum, it is defined in a coordinates system that prevents it from being singular at the event horizon. It violates all pointwise energy conditions near the event horizon.

Hausdorf spacetime – a spacetime that nowhere splits (in manner analogous to partially splitting a hair).

Hawking radiation – the emission of particles near the surface of a black hole that results from the gravitationally induced splitting of virtual particle-antiparticle pairs and the concomitant consumption by the black hole of one member of these pairs.

Hawking temperature – the temperature characterizing the Hawking radiation emitted at an event horizon. If the horizon surrounds a black hole, this temperature is proportional the mass of the black hole. If the horizon is due to the expansion of the universe, this temperature is proportional to the Hubble constant and is also known as the "de Sitter temperature".

hindered surface – a term used only in this book to describe 1) a surface within a red hole from which particles may escape to the exterior only by exceeding a certain outward speed, or 2) a surface within a blue hole that can be reached from the exterior only by particles that exceed a certain inward speed.

holographic principle – the idea that the spatial universe is fundamentally two-dimensional in that the patterns that describe it fully may be realized on a two-dimensional surface and projected by the laws of physics to create our three-dimensional spatial reality in a manner analogous to the way in which a two-dimensional hologram is by laser illumination projected to create the illusion of a three-dimensional scene.

homogeneity – uniformity of the local characteristics of the universe at all locations. Perfect homogeneity implies isotropy. A universe can be homogeneous above a certain length scale and anisotropic below it.

hoop conjecture – an object will only collapse to form a black hole, if its longest dimension fits within a circular hoop whose radius is that of the event horizon of a nonrotating uncharged black hole possessing the mass of the object.

Hubble constant – the constant of proportionality quantifying the dependence of the speed at which an object recedes from an observer and the separation of the object and the observer: (recession speed) = (Hubble constant) x (separation).

inertial reference frame – the frame of reference of an observer who feels no forces caused by acceleration or gravitation.

inflation – the exponentially accelerated expansion of the universe conjectured to have occurred briefly a short time after its creation.

inflaton -- a hypothetical scalar field that upon seeking its minimum potential energy caused the universe to inflate.

instanton – a classically forbidden, imaginary-momentum solution to field equations in Euclidean space that is interpreted as a "tunneling" transition between allowed quantum states of the field in Lorentzian space.

intrinsic curvature – curvature of a space that does not depend upon how the space is embedded in space of higher dimension. What is normally meant by "curvature".

isometry – a mapping of spacetime into itself (e.g. from one region to another) that leaves the metric unchanged.

isotropy – the uniformity in the appearance of the universe in each direction of observation from a given point. Isotropy at all points implies homogeneity.

Jordan frame – the native form of the Lagrangian of a scalar-tensor theory of gravity in which the scalar field is coupled directly to the curvature scalar.

junction conditions – the requirement that in crossing any surface certain components of spacetime curvature be continuous and certain other components be discontinuous to a degree determined by the stress energy tensor at the surface.

Kaluza-Klein model -- any theory of gravity featuring extra dimensions (beyond the usual four) that are taken to be compactified.

Killing vector – a vector field that defines directions in which the metric tensor does not change.

Killing horizon – a null hypersurface in which the Killing vector field is null.

Klein-Gordon equation – equation of motion of a relativistic scalar field.

Kantowski-Sachs spacetime – an anisotropic homogeneous spacetime equivalent in the case of zero and negative curvature to Bianchi type I and Bianchi type III cosmologies respectively and recently shown be among the vacuum states of some string models.

Lagrangian -- the difference between the kinetic and potential energies of a system whose minimization or maximization determines the motion of the system.

landscape, the – a means of coping with M-Theory's infinite plethora of possible vacuum states and consequent lack of predictive power by regarding its various vacua as low-energy valleys in a landscape of possibilities from which that corresponding to our universe is to be selected by the anthropic principle.

lapse and shift functions – functions used to specify how sequential hypersurfaces of a foliation are to be stacked.

Lens-Thirring effect – the dragging of inertial reference frames in the vicinity of a rapidly spinning massive object.

lightcone – that region of spacetime defined with respect to a particular event *P* that consists of (a) all events that can be reached by a light pulse emitted from *P* (future light cone) and (b) all events from which an emitted light pulse could have reached *P* (past light cone).

lightlike – associated with the motion of particles that are constrained to travel at the speed of light.

lightlike curve – a curve in spacetime that specifies a trajectory of a particle whose speed is that of light.

lightlike vector – a vector that is proportional to a tangent vector of a lightlike curve. Having a negative length as calculated using the spacetime's metric to form an inner product

line element – the expression in terms of the metric tensor of the distance between two points whose coordinates are infinitesimally close.

Loop Quantum Gravity – a background-independent, non-perturbative approach to quantizing gravity that relies on the quantization of a type of coarse grained means of specifying geometry (spin network) and its time evolution (spin foam).

Lorentzian signature – having the time coordinate preceded by a negative sign in the generalized Pythagorean expression for distance: $s^2 = -t^2 + x^2 + y^2 + z^2$. Denoted by (— + + +).

Lorentzian geometry – geometry for which the expression of any infinitesimal distance has a Lorentzian signature.

M-Theory -- an overarching framework that seeks to unify string theory by showing that the five original string theories as well as supergravity are actually special cases of a general, 11-dimensional theory. [There is no consensus on what 'M' stands for, as this was never specified by string theorist Edward Witten, who coined the term.]

microscopic wormhole – a wormhole that is not macroscopic. See "macroscopic wormhole".

MACHO – MAssive Compact Halo Object. A dark cosmic object that can be detected by the deflection of starlight by its gravity that produces an apparent halo around the object when viewed from Earth.

macroscopic topology -- the topology that can be detected using geometry-probing instruments with the highest resolution available.

macroscopic wormhole – a wormhole whose throat is sufficiently large to be detected by an instrument probing at the maximum energy that the human race can bring to bear.

manifold – an infinite set of points for which a well-behaved coordinate system can be defined. A space: that mathematical entity on which distances between points can be defined and to which curvature can be ascribed. [e.g. Spacetime in general relativity is a manifold.]

marginal surface – a surface at which the divergence of light is momentarily zero as it transitions from converging to diverging or vice versa.

metaphysics – the branch of philosophy that seeks to determine the fundamental constituents of reality (ontology) and their origin (cosmogony).

metric tensor – a means of specifying the distances between nearby points with known coordinates and the angles between coordinates axes.

microlensing – small deflection (on the order of milliarcseconds) of stellar light by an object that is too small to allow the separate resolution of the star and the object.

minimal coupling – the weakest possible interaction between fields of a field theory.

minimally coupled scalar field – a scalar field that interacts with the curvature of spacetime in the weakest possible way. [The coupling term in the Lagrangian density between the curvature scalar $R(x,t)$ and scalar field $\varphi(x,t)$ is $\alpha\ \varphi(x,t)R(x,t)$. Minimal coupling occurs when the coupling constant α is zero.]

minisuperspace – a tiny, finite-dimensional subset of superspace consisting only of certain chosen families of geometries whose members are specified by a finite set of parameters. See "superspace".

Minkowski space -- a flat four-dimensional spacetime with a Lorentzian signature.

mixed state – a weighted sum of pure states in which the numerical weights in the sum contain no phase information, which prevents the amplitude for the system to be in particular base states from being known precisely.

modulus – 1. the square root of the product of a complex number and its complex conjugate. 2. A field whose value at a particular point in space specifies which member of family of closely related manifolds describes the compactified subspace of extra dimensions there.

MT – Morris-Thorne.

multiverse -- the whole of reality consisting of a collection of an infinite number of separate non-interacting universes of which ours is only one.

NEC – see "null energy condition".

null – lightlike.

null energy condition – the condition that the pressure p and density ρ of matter be restricted as follows: $p + \rho c^2 \geq 0$, where c is the speed of light. The requirement that matter should never have an anti-gravitating effect on light.

null geodesic – a path in spacetime taken by light.

off shell – not satisfying a system's classical equations of motion. Virtual as opposed to real.

on shell – satisfying a system's classical equations of motion. Real as opposed to virtual.

ontological – pertaining to the essence of reality.

orientable spacetime – a spacetime that at every point permits a distinction between 1) past and future (time orientable), or 2) left and right (space orientable).

partial Cauchy surface – a spacelike hypersurface that is intersected by every non-spacelike curves no more than once (i.e. once or not at all).

past trapped surface – a closed surface from which simultaneous emission of inward (and outward) light pulses from every of its points produces a wave front whose area is greater than that of the surface. In the context of Schwarzschild geometry, it is a surface for which all intersecting word lines began in a singularity.

path integral -- an integral that is a sum of the values of a functional evaluated over a range of functions (paths), as opposed to the usual integral that is a sum of the values of a function evaluated over a range of numbers. First entered physics through the formulation of quantum mechanics in terms of the principle of least action.

penetration speed – the term used in this book to denote the minimum speed, determined by the rate of contraction or expansion of a wormhole's geometric throat, at which a particle must travel to escape the interior of a red hole or reach the interior of a blue hole.

phantom energy – a gravitationally repulsive form of dark energy whose equation of state is $p = \omega\rho$, where $\omega < -1$. A type of cosmic exotic matter.

photon -- the massless, spin-1 quantum of the electromagnetic field.

Planck mass -- the energy scale at which classical gravity fully ceases to be valid -- approximately 2×10^{-8} kilograms.

Planck length – the length scale at which classical gravity fully ceases to be valid – approximately 2×10^{-35} meters. A Planck-Wheeler length.

pointwise – applicable to a single point of spacetime (such as an energy condition that applies at such a point).

positive mass theorem -- any asymptotically flat spacetime free of violations of the dominant energy condition has a nonnegative total mass.

proper time -- the length, as determined by the Lorentzian metric, of an interval of an observer's word line which corresponds to the elapse of time measured by his clocks during the interval.

pseudo topology change -- a change in the geometry of spacetime that changes its macroscopic topology without changing its actual topology.

pure state – a quantum state for which the amplitude to be in each base state is known precisely.

quantum computation – a computation (i.e. a formulaic generation of outputs from inputs) that relies on the rules of quantum theory.

quantum cosmology – cosmology that treats the entire universe as a quantum mechanical system.

quantum field theory – a theory that describes the dynamics of a field, i.e. a function of position and time, in which the configuration of the field and the rate at which this configuration changes cannot, unlike a classical field theory, simultaneously be specified with arbitrarily high precision. The theory restricts the field to certain so called states that correspond to the presence of various quantities of an associated elementary particle.

quantum gravity -- a theory that would unite quantum theory with a theory of gravity that has yet to be devised despite great effort.

quantum interest -- the additional magnitude of the positive energy to which an observer must be exposed, according to the quantum inequalities, as consequence of delaying his exposure to compensating positive energy after an exposure to negative energy.

quantum inequalities – a discovery in quantum field theory that states that any observer's exposure to negative energy must be followed by a compensating of exposure to positive energy, whose magnitude and duration exceeds that of the negative energy exposure.

quantum teleportation – a means of exploiting the quantum entanglement between a local particle A and a distant particle B together with a classical communication channel between the local and distant locations to transfer the quantum state of a local particle X to the distant particle B at the cost of the destruction of the original state of the local particle X.

quintessence -- a hypothetical type of dark energy that could cause the accelerating expansion of the universe and is manifested as a scalar field whose energy is gradually moving towards its minimum possible value. Its equation of state is $p = \omega\rho$, where $-1 < \omega < -1/3$,

Randall-Sundrum construction -- a brane world that models the universe as a 4-brane embedded in a 5-dimensional anti-DeSitter spacetime with physics confined to the brane by the constricting effect of the large negative cosmological constant of the anti-DeSitter bulk.

Raychaudhuri equation – an equation that gives the degree to which light pulses moving in parallel will diverge or converge from each other as a function of the curvature and other properties of the spacetime through which they move.

red hole -- a term used only in this book for the contracting phase of a wormhole with a dynamic geometric throat. The contraction induces an additional redshift in light escaping from the interior region. In the limit that its geometric throat contracts rapidly enough to prevent light from leaving its interior, it becomes a black hole.

Regge-Wheeler equation -- an equation that determines the perturbations in the metric that can occur in the vicinity of a black hole.

renormalization -- a collection of techniques used to remove infinities that arise in calculation of physical observables in quantum field theory.

reparametrization invariance – invariance of the laws of physics with respect to the choice of a time parameter (coordinate). May also refer to invariance of these laws with respect to any reparametrization (i.e. choice of coordinate system).

Riemannian geometry – a configuration of a smooth manifold on which a symmetric, non-degenerate metric tensor is defined.

Rindler vacuum – the quantum vacuum state of a scalar field in a uniformly accelerating reference frame. It is characterized by the presence of radiation whose temperature is proportional to the acceleration of the reference frame.

ringhole -- A wormhole whose throat and mouths have the topology of a torus. Any solution to Einstein's equations containing a compact region of non-simple topology that is bounded by a region of topology S^1xS^1 (a torus).

Roman ring -- a ring-shaped arrangement of a set of wormholes in which no subset of wormholes is close to operating as a time machine even as the entire arrangement becomes arbitrarily close to operating in this way.

S-duality – the equivalence of a theory whose particles have a fundamental charge (coupling constant) g with another theory whose particles have a fundamental charge 1/g. [The nature of the equivalence between the Type I string and the SO(32) heterotic string.]

scalar – a quantity whose calculation in any coordinates system produces the same value.

scalar field – a spinless bosonic field represented classically by a function of spacetime that retains its value under arbitrary coordinate transformations.

Schrödinger equation -- the fundamental equation of quantum mechanics, obtained from the classical equation of energy conservation, whose solutions are wave functions that specify the state of a quantum system.

Schwarzschild radius – the radius of the spherical event horizon surrounding a nonrotating and uncharged black hole.

SEC -- see "strong energy condition".

second law of thermodynamics -- entropy of the matter in the universe cannot decrease.

self-consistent solution – a complete solution to coupled field equations describing a system of interacting fields, as opposed to a partial solution consisting of a solution to the equation of motion of one of the fields that assumes that the other fields are fixed.

singularity -- a location in spacetime at which geodesics end or begin and around which curvature increases arbitrarily as the location is approached.

soliton -- a solution to field equations that corresponds to a localized concentration of energy that persists for an extended period of time.

spacelike – associated with the motion of hypothetical particles that are constrained to travel at speeds that always exceed that of light. Not timelike or lightlike.

spacelike curve – a curve in spacetime that specifies a trajectory of a particle whose speed never falls below that of light. A curve that is nowhere timelike or lightlike.

spacelike hypersurface – a three-dimensional subspace that contains only spacelike curves.

spacelike vector – a vector that is proportional to a tangent vector of a spacelike curve. Having a positive length as calculated using the spacetime's metric to form an inner product.

spacetime – a geometric object consisting of an infinite set of events, each specified by a time and a location, on which physics unfolds. A manifold used to describe physics.

stargate -- a wormhole whose throat is a flat surface that does not appear to enclose a volume (e.g. a disk as opposed to a sphere) and whose exotic matter requirements are reduced as a consequence. In science fiction, a synthetically controlled wormhole.

state vector -- a vector whose components are the amplitude of the system being in each of the possible base states.

stress-energy -- short hand for that property of matter, fully specified in the stress energy tensor, responsible for the curvature of spacetime.

stress-energy tensor – a tensor that characterizes the density, energy-fluxes, stresses, and pressures within a matter distribution and serves as a source of spacetime curvature in the Einstein gravitational field equations.

string theory – a research program in theoretical physics founded on the assumption that the fundamental building blocks of nature are not restricted 0-dimensional points but may also include tiny 1-dimensional strings.

strong energy condition -- a defunct energy condition on matter that requires the sum of its density and its pressure in each direction to be non-negative and the sum of its density with the sum of its pressures to be non-negative, i.e. if the pressure p in each direction is the same: $\rho + p \geq 0$ and $\rho + 3p \geq 0$.

subluminal – slower than light.

supergravity -- a field theory that results when supersymmetry is taken to be a local symmetry, i.e. it is the gauge theory associated with supersymmetry. The superstring theory in the limit of a zero-length string. In eleven dimensions, the low-energy limit of M-theory.

superluminal – faster than light.

supernatural – belonging to a hypothetical part of reality about which knowledge cannot be increased irrespective of the quantity of effort expended to do so.

superposition – a pure quantum state that is a linear combination (a weighted sum) of more than one base state.

superquintessence -- quintessence for which the equation of state is $p = \omega\rho$, where $\omega < -1$. Also known as phantom energy, it is a type of exotic matter in that it violates the weak energy condition.

superspace – 1) the space of all three-dimensional geometries. 2) the extension of spacetime to include an N-dimensional anticommuting sector on which supergravity is defined.

superstring theory – string theory incorporating supersymmetry. The usual formulation of string theory.

supersymmetry – a symmetry transformation through which every boson is shown to be related to a hypothetical fermion “super partner” and vice versa.

susy – see “supersymmetry”.

T-duality -- the equivalence of 1) a string theory at the nth mode of momentum excitation due to a compact space of radius R about which the string is wrapped m times, and 2) another string theory at the mth mode of momentum excitation due to a compact space of radius $1/R$ about which the string is wrapped n times. [The nature of the equivalence between Type IIA and Type IIB strings, and between E8xE8 and SO(32) heterotic strings.]

tensor – an m-dimensional array of numbers each of which assume particular values in particular coordinate systems according to a special rule. This rule (requiring, for example, that physical quantities such as length remain constant under changes in the coordinate system) ensures that physical laws expressed in terms of tensors retain the same form in all coordinate systems. m is called the rank of the tensor. A rank-0 tensor is called a scalar. A rank-1 tensor is called a vector.

tensor field – a tensor that is a function of location in spacetime.

thin-shell wormhole – a wormhole whose geometric throat is characterized by an infinite spike in curvature and an infinitesimally thin concentration of exotic matter.

throat -- see "geometric throat" or "functional throat".

timelike – associated with the motion of particles that are constrained to travel at speeds that never reach or exceed that of light.

timelike curve – a curve in spacetime that specifies the trajectory of a particle whose speed never reaches or exceeds that of light.

timelike vector – a vector that is proportional to a tangent vector of a timelike curve. Having a negative length as calculated using the spacetime's metric to form an inner product

topology – a set of shapes that are continuously deformable into one another without cutting, pasting, adding holes, or removing them. The branch of mathematics devoted to the study of such sets. Example: the topology of a sphere is the same as that of a pyramid but differs from that of a torus.

topological censorship theorem – In a "reasonable" spacetime (one that is fully determined by evolving the Einstein equations from initial data) for which the ANEC holds for all causal curves from past lightlike infinity to future lightlike infinity, any such curve can be deformed into a rudimentary causal curve – one that goes from past to future infinity in a topologically simple region of spacetime near spatial infinity. This required topologically equivalence of all causal curves from infinity to infinity is tantamount to the absence or censorship of non-simple topologies.

torus – a donut-shaped surface whose topology is specified as $S^1 \times S^1$ (the Cartesian product of two 1-dimensional spheres, i.e. circles)

trapped surface – normally refers to a future trapped surface, but technically refers as well to a past trapped surface. See "future trapped surface" and "past trapped surface".

trapping horizon – a region of spacetime, of which an event horizon is a special case, at which the expansion of light due to gravity reaches a minimum or maximum value.

traversable wormhole -- a wormhole through which a healthy human traveler can pass in either direction within reasonable time, usually taken to be one year as measured by the traveler or distant stationary observers, without substantial risk of death or injury.

tunneling -- the apparent instantaneous passage, as a consequence of quantum theory, of a particle through a barrier that it is classically forbidden to penetrate.

type A region – a term used only in this book to label the region exterior to a wormhole that corresponds to our universe.

type A' region – a term used only in this book to label the region exterior to a wormhole that a traveler attempts to reach in traversing the wormhole. The "other universe" or the distant asymptotically flat region of our universe.

type B region – a term used only in this book to label the black hole interior region encountered immediately upon passing through the event horizon.

type B' region – a term used only in this book to label the white hole interior region separated from the wormhole exterior by an anti-horizon.

type C region – a term used only in this book to label the region between the Cauchy horizons of the maximally extended Reissner-Nordstrøm and Kerr solutions.

Unruh vacuum – the quantum vacuum of a scalar field in the vicinity of a black hole that is not in thermal equilibrium with its environment. Sometimes described as a hybrid of the Boulware and Hartle-Hawking vacua, it violates all pointwise energy conditions.

vacuum – a configuration of a classical field for which the energy is a local minimum in the space of all possible configurations. The corresponding state of the quantum field theory derived from the classical field.

Vaidya spacetime – spacetime containing the spherically symmetric ingoing and outgoing massless radiation on a Schwarzschild background.

vector – a rank-1 tensor. See "tensor".

vector field – a vector whose value depends on its location in spacetime.

virtual particle – a particle whose ephemeral existence is consistent with the time-energy uncertainty principle.

virtual wormhole – see "Wheeler wormhole".

wave function – a state vector whose components (base states) correspond to a particular configuration of the system. A function whose squared modulus is traditionally interpreted as the probability of the system being in the configuration specified by the argument of the function.

weak energy condition -- a condition on the stress-energy tensor of matter that requires in most cases that the density ρ and the pressure p be constrained by: $\rho + p \geq 0$ and $\rho \geq 0$.

WEC -- see "weak energy condition".

Wheeler-DeWitt equation – an equation conceptually similar to the Schrödinger equation that is obtained by quantizing the Hamiltonian constraint of general relativity. Its solution determines the amplitude of the universe possessing any particular three-dimensional geometry.

Wheeler wormhole – a microscopic wormhole that exists in any of the spatial geometries that comprise the vacuum state of the gravitational field. A virtual wormhole.

Wick rotation -- an analytic continuation of a mathematically ill-defined expression in Lorentzian spacetime to a well defined counterpart in Euclidean spacetime through the replacement of real time with imaginary time.

wisdom – counterintuitive knowledge, typically about the means of achieving desirable outcomes, that is normally obtained through experience and is seldom understood explicitly.

WKB approximation – the Wentzel-Kramers-Brillouin method for obtaining approximate solutions to the Schrödinger equation which is also useful in solving the Wheeler DeWitt equation in minisuperspace.

Wilkinson Microwave Anisotropy Probe – a NASA satellite launched in 2001 that has created high-resolution maps of the angular distribution of cosmic microwave background radiation and whose data is used to construct cosmological models.

WMAP – see "Wilkinson Microwave Anisotropy Probe".

world line – the path of an observer in spacetime especially when displayed as a graph in Minkowski space.

wormhole -- a spacetime for which a foliation exists that contains a spacelike hypersurface in which there is a closed surface of locally minimal, non-zero area. A tunnel to another universe or to other regions (usually assumed to be distant) of our universe.

Sources

Miguel Alcubierre, "The Warp Drive: Hyper-Fast Travel within General Relativity", Class.Quant.Grav. **11,** L73 (1994)

C. Armendáriz-Picón, "On a class of stable, traversable Lorentzian wormholes in classical general relativity", Phys. Rev. D **65**, 104010 (2002)

L. J. Alty, P. D. D'Eath and H. F. Dowker, "Quantum wormholes states and local supersymmetry", Phys. Rev. D **46** (1992)

Alexander B. Balakin, José P. S. Lemos, Alexei E. Zayats, "Nonminimal coupling for the gravitational and electromagnetic fields: Traversable electric wormholes", Phys. Rev. D. **81**, 084015 (2010)

Carlos Barceló and Matt Visser, "Scalar fields, energy conditions, and traversable wormholes", Class. Quantum Grav. **17**, 3843 (2000)

Carlos Barceló and Matt Visser, "Twilight for the Energy Conditions", Int. J. Mod. Phys. **11**, 1 (2002).

Arvind Borde, "Topology change in classical general relativity", gr-qc/9406053 (1994)

S.E. Perez Bergliaffa and K.E. Hibberd, "On the stress-energy tensor of a rotating wormhole", gr-qc/0006041 (2000)

Biplab Bhawal and Sayan Kar, "Lorentzian wormholes in Einstein-Gauss-Bonet theory", Phys. Rev. D **46**, 2464 (1992)

Patrick R. Brady, Ian G. Moss, and Robert C. Myers, "Cosmic censorship: as strong as ever", Phys. Rev. Lett. **80**, 3432 (1998)

K. A. Bronnikov and S. Grinyok, "Instability of wormholes with a nonminimally coupled scalar field", Grav.Cosmol. **7**, 297 (2001)

K. Bronnikov and S. Grinyok, "Charged wormholes with non-minimally coupled scalar fields. Existence and stability", gr-qc/0205131 (2002)

K. A. Bronnikov and Sung-Won Kim, "Possible wormholes in a brane world", Phys. Rev. D **67**, 0604027 (2003)

Robert R. Caldwell, Marc Kamionkowski, and Nevin N, Weinberg, "Phantom Energy and Cosmic Doomsday", Phys. Rev. Lett **91**, 071301 (2003)

[Sean M. Carroll, Mark Hoffman, and Mark Trodden, "Can the dark energy equation-of-state parameter w be less than -1?", Phys. Rev. D **68**, 023509 (2003)

Mauricio Cataldo, Patricio Salgado, and Paul Minning, "Self-dual Lorentzian wormholes in n-dimensional Einstein gravity", Phys. Rev. D **66**, 124008 (2002)

Celine Cattoen and Matt Visser, "Cosmological milestones and energy conditions", J.Phys.Conf.Ser.**68**, 012011 (2007)

Marco Cavaglià, Vittorio de Alfaro, and Fernando de Felice, "Anisotropic wormhole: Tunneling in time and space", Phys. Rev. D **49**, 6493 (1994)

Chiang-Mei Chen, "Dyonic wormholes in five-dimensional Kaluza-Klein theory", Class. Quantum Grav. **18**, 4179 (2001)

Sidney Coleman, "Why is there nothing rather than something", Nucl. Phys **B310**, 643 (1988)

D. H. Coule, "Comment: No wormholes with real minimally coupled scalar fields", Phys. Rev. D **55**, 6606 (1995)

John G. Cramer, Robert L. Forward, et. al, "Natural wormholes as gravitational lenses", Phys.Rev. D **51,** 3117 (1995) | astro-ph/9409051

A. DeBenedictis and A. Das, "On a general class of wormhole geometries", Class. Quantum Grav. **18**, 1187 (2001)

David Deutsch, "Quantum mechanics near closed timelike lines", Phys. Rev. D **44**, 3197 (1991)

David Deutsch and Michael Lockwood, "The quantum physics of time travel", Sci. Am. **270**, 68 (1994)

David Deutsch, *The Fabric of Reality*, Penguin, London (1997)

David Deutsch, Proc. Roy. Soc. A455, 3129 (1999)

B.S. DeWitt, "Quantum theory of gravity I. – The canonical theory", Phys. Rev. **160**, 1113 (1967)

B.S. DeWitt, "Quantum theory of gravity II. – The manifestly covariant theory", Phys. Rev. **162**, 1195 (1967)

B.S. DeWitt, "Quantum theory of gravity III. – Applications of the covariant theory", Phys. Rev. **162**, 1239 (1967)

Bryce S. DeWitt and Neill Graham, ed., *The Many-Worlds Interpretation of Quantum Mechanics*, Princeton University Press, Princeton, 1973

Homer G. Ellis, "Ether flow through a drainhole: A particle model in general relativity", J. Math. Phys. **14**, 104 (1973)

V. Faraoni and W. Israel, "Dark energy, wormholes, and the Big Rip", Phys. Rev. D 71, 064017 (2005)

Bo Feng, Xiulian Wang, and Xinmin Zhang, "Dark energy constraints from the cosmic age and supernova", Phys.Lett. B **607**, 35 (2005)

Christopher J. Fewster and Thomas A. Roman, "On wormholes with arbitrarily small quantities of exotic matter", Phys.Rev. D **72**, 044023 (2005)

Ludwig Flamm, "Beiträge zur Einsteinschen Gravitationstheorie.", Physik. Zeitschr. XVII, 448 (1916)

L. H. Ford and Thomas A. Roman, "Quantum field theory constraints traversable wormhole geometries", Phys. Rev. D **53**, 5496 (1996)

Lawerence H. Ford and Thomas A. Roman, "Negative Energy, Wormholes and Warp Drive", Sci. Am. **282**, 46 (2000)

Katherine Freese and William.H. Kinney, "Ultimate Fate of Life in an Accelerating Universe", Phys.Lett. B **558**, 1 (2003)

John Friedman, Michael S Morris, Igor D. Novikov, Fernando Echeverria, Gunnar Klinkhammer, Kip S. Thorne and Ulvi Yurtsever, "Cauchy problem in spacetimes with closed timelike curves", Phys. Rev. D **42**, 1915 (1990)

John L. Friedman, Nicolas J. Papastamatiou, and Jonathan Z. Simon, "Failure of unitarity for interacting fields on spacetimes with closed timelike curves", Phys. Rev. D **46**, 4456 (1992)

N. Furey and A. DeBenedictis, "Wormhole throats in R^m gravity", Class. Quantum Grav. 22 313, (2005)

Sijie Gao and José P. S. Lemos, "The Tolman-Bondi-Vaidya spacetime: Matching timelike dust to null dust", Phys.Rev. D **71** 084022, (2005)

Luis J. Garay, "Quantum state of wormholes and path integral", Phys. Rev. D **44**, 1059 (1991)

Robert P. Geroch, "Topology in General Relativity", J. Math. Phys. **8**, 782 (1967)

Kazuo Ghoroku and Teruhiko Soma, "Lorentzian wormholes in higher-derivative gravity and the weak energy condition", Phys. Rev. D **46**, 1507 (1992)

Steven B. Giddings and Andrew Strominger, "Axion-induced topological change in quantum gravity and string theory", Nucl. Phys **B306**, 890 (1988)

Pedro F. González-Díaz, "Ringholes and closed timelike curves", Phys. Rev. D **54**, 6122 (1996)

Pedro F. González-Díaz, "Wormholes and ringholes in a dark-energy universe", Phys. Rev. D **68**, 084016 (2003)

Noah Graham and Ken D. Olum, "Achronal averaged null energy condition". Phys. Rev. D **76**, 064001 (2007)

Brian Greene, *The Elegant Universe: Superstrings, Hidden Dimensions, and the Quest for the Ultimate Theory*, Norton, New York, 1999

Brian Greene, *The Fabric of the Cosmos: Space, Time, and the Texture of Reality*, Knopf, New York, 2004

John Gribbon, *In Search of the Edge of Time: Black Holes, White Holes, Wormholes,* Penguin, London 1992.

J. Richard Gott, *Time Travel in Einstein's Universe: The Physical Possibilities of Travel Through Time*, Mariner, Boston, 2001

Michael Gutperle and Wafic Sabra, "Instantons and wormholes in Minkowski and (A)dS spaces", hep-th/0206153 (2002)

Paul Halpern, *Cosmic Wormholes – The Search for Interstellar Shortcuts*, Dutton Books, New York, 1992

Steen Hannestad and Edvard Mortsell, "Probing the dark side: Constraints on the dark energy equation of state from CMB, large scale structure and Type Ia supernovae", Phys. Rev. D **66**, 063508 (2002)

S. W. Hawking and G. F. R. Ellis, *The Large Scale Structure of Spacetime*, Cambridge University Press, Cambridge, 1973

S.W. Hawking, "Spacetime foam", Nuclear Physics **B144**, 349 (1978)

S. W. Hawking, "Quantum coherence down the wormhole", Phys. Lett. B 195, 337 (1987)

S. W. Hawking, "Wormholes in spacetime", Phys. Rev. D **37**, 904 (1988)

S.W. Hawking, "Chronology protection conjecture", Phys. Rev. D **46**, 603 (1992)

S. W. Hawking, "Information loss in black holes", Phys. Rev. D **72**, 084013 (2005)

Sean A Hayward, "General laws of black-hole dynamics", Phys. Rev. D **49,** 6467 (1994)

Sean A. Hayward, "Dynamic wormholes", Int. J. Mod. Phys. **D8**, 373 (1999)

Sean A. Hayward, Sung-Won Kim, and Hyunjoo Lee, "Dilatonic wormholes and black holes", J. Korean Physical Soc. **42**, S31 (2003)

Sean A. Hayward, Sung-Won Kim, and Hyundoo Lee, "Dilatonic wormholes: construction, operation, maintenance and collapse to black holes", Phys. Rev. D **65**, 064003 (2002)

Sean A. Hayward, "Black holes and traversible wormholes: a synthesis", gr-qc/0203051 (2002)

Sean A. Hayward, "Recent progress in wormhole dynamics", gr-qc/0306051 (2003)

Sean A. Hayward and Hiroko Koyama, "How to make a traversable wormhole from a Schwarzschild black hole", Phys.Rev. **D70**, 101502 (2004)

David Hochberg, "Lorentzian wormholes in higher order gravity theories", Phys. Lett B **251**, 349 (1990)

David Hochberg, "Quantum-mechanical Lorentzian wormholes in cosmological backgrounds", Phys. Rev. D **52**, 6846 (1995)

David Hochberg, Arkadiy Popov, Sergey V. Sushkov, "Self-consistent wormhole solutions of semiclassical gravity", Phys. Rev. Lett. **78**, 2050 (1997)

David Hochberg and Matt Visser, "Geometric structure of the generic static traversable wormhole throat", Phys. Rev. D **56**, 4745 (1997)

David Hochberg and Matt Visser, "Null energy condition in dynamic wormholes", Phys. Rev. Lett. **81**, 746 (1998)

David Hochberg and Matt Visser, "Dynamic wormholes, antitrapped surfaces, and energy conditions", Phys. Rev. D **58**, 044021

Shahar Hod and Tsvi Piran, "The inner structure of black holes", Gen. Rel. Grav. **30**, 1555 (1998) | gr-qc/9902008

Soon-Tae Hong and Sung-Won Kim, "Can wormholes have negative temperatures?", gr-qc/0303059 (2003)

Gary T. Horowitz, "Topology Change in General Relativity", hep-th/9109030 (1991)

Daisuke Ida and Sean A. Hayward, "How much negative energy does a wormhole need?", Phys. Lett. A **260**,175 (1999)

Jim Al-Khalili, *Black Holes Wormholes & Time Machines*, Institute of Physics, Bristol, U.K., 1999

Sayan Kar, Naresh Dadhich, and Matt Visser, "Quantifying energy condition violations in traversable wormholes", Pramana **63**, 859 (2004)

Nail R. Khustnutdinov and Sergey V. Sushkov, "Ground state energy in a wormhole space-time", Phys. Rev. D **65**, 084028 (2002)

Sung-Won Kim, "Exact solutions of charged wormhole", Phys. Rev. D **63**, 064014 (2001)

Sung-Won Kim, "Rotating wormhole and scalar perturbation", Nuovo Cim. **120B**, 1235 (2005)

Hiroko Koyama and Sean A. Hayward, "Construction and enlargement of traversable wormholes from Schwarzschild black holes", Phys. Rev. D **70**, 084001 (2004)

Hiroko Koyama, Sean A. Hayward, and Sung-Won Kim, "Construction and enlargement of dilatonic wormholes by impulsive radiation", Phys. Rev. D **67**, 084008 (2003)

S. Krasnikov, "Traversable wormhole", Phys. Rev. D **62**, 084028 (2000)

S. Krasnikov, "The quantum inequalities do not forbid spacetime shortcuts", Phys.Rev. D **67**, 104013 (2003)

Peter K. F. Kuhfittig, "Static and dynamic traversable wormhole geometries satisfying the Ford-Roman constraints", Phys. Rev. D **66**, 024015 (2002)

Peter K. F Kuhfittig, "Axially symmetric rotating traversable wormholes", Phys. Rev. D **67**, 064015 (2003)

Peter K. F. Kuhfittig, "Can a wormhole supported by only small amounts of exotic matter really be traversable?", Phys. Rev. D **68**, 067502 (2003)

Peter K. F. Kuhfittig, "Wormholes supported by small amounts of exotic matter: Some corrections", gr-qc/0508060 (2005)

José P. S. Lemos, Francisco S. N. Lobo, and Sérgio Quinet de Oliveira, "Morris-Thorne wormhole wormholes with a cosmological constant", Phys. Rev. D **68**, 064004 (2003)

José Lemos and Francisco S. N. Lobo "Plane symmetric traversable wormhole in anti-de Sitter background", Phys. Rev. D **69**, 104007 (2004)

Li-Xin Li "Two open universes connected by a wormhole: exact solutions", J. Geometry and Phys., **40**, 154 (2001)

Seth Lloyd and Y. Jack Ng, "Black hole computers", Sci. Am. **291**, 52 (2004)

Francisco S. N. Lobo, "Energy conditions, traversable wormholes, and dust shells", Gen. Rel. Grav. **37**, 2023 (2005)

Francisco S. N. Lobo, "Phantom energy traversable wormholes", Phys.Rev. D **71**, 084011 (2005)

Francisco S. N. Lobo, "Stability of phantom wormholes", Phys. Rev. D **71**, 124022 (2005)

Mark S. Madsen, "A note on the equation of state of a scalar field", Astrophys. Space Sci. **113**, 205 (1985)

Elisabetta Majerotto, Domenico Sapone, and Luca Amendola, "Supernovae type Ia data favour coupled phantom energy", gr-qc/0410543 (2004)

Charles W. Misner, Kip S. Thorne, and John Archibald Wheeler, *Gravitation*, W. H. Freeman, San Francisco, 1973

Michael S. Morris and Kip S. Thorne, "Wormholes in spacetime a their use for interstellar travel: A tool for teaching general relativity", Am. J. Phys. **56**, 395 (1988)

Michael S. Morris, Kip S. Thorne, and Ulvi Yurtsever, "Wormholes, time machines, and the weak energy condition", Phys. Rev. Lett. **61**, 1446 (1988)

Kamal Kanti Nandi, Yuan-Zhong Zhang, and K. B. Vijaya Kumar, "Volume integral theorem for exotic matter", Phys. Rev. D **70**, 127503 (2004)

Y. Jack Ng, "From computation to black holes and space-time foam", Phys. Rev. Lett. 86, 2946 (2001)

Hermann Nicolai, Kasper Peeters and Marija Zamaklar, "Loop quantum gravity: an outside view", Class.Quant.Grav.**22**, R193 (2005)

I.D. Novikov, “Developments in general relativity: Black hole singularity and beyond”, gr-qc/0304052 (2003)

Roland Omnès, *Understanding Quantum Mechanics*, Princeton University Press Princeton, 1999

Amos Ori, “Inner structure of a charged black hole: An exact mass-inflation solution”, Phys. Rev. Lett. **67**, 789 (1991)

S. Perlmutter, “Discovery of a supernova explosion at half the age of the Universe”, Nature **391**, 51 (1998)

Eric Poisson, “Black-hole interiors and strong cosmic censorship”, gr-qc/9709022 (1997)

Joseph Polchinski, “Weinberg’s nonlinear quantum mechanics and the Einstein-Podolsky-Rosen paradox”, Phys. Rev. Lett. **66**, 397 (1991)

H. David Politzer, “Simple quantum systems in spacetimes with closed timelike curves”, Phys. Rev. D **46**, 4470 (1992)

Ian H. Redmount and Wai-Mo Suen, “Quantum dynamics of Lorentzian spacetime foam”, Phys. Rev. D **49**, 5199 (1994)

Adam G. Riess *et al.*, “Observational evidence from supernovae for an accelerating universe and a cosmological constant”, Astron. J **116**, 1009 (1998)

Enrico Rodrigo, “Higher-dimensional bulk wormholes and their manifestations in brane worlds”, Phys. Rev. D **74**, 104025 (2006)

Enrico Rodrigo, “Wormholes, Void Bubbles, and Vacuum Energy Suppression”, Classical and Quantum Gravity, **24**, 3221 (2007)

Enrico Rodrigo, “Denouement of a Wormhole-Brane Encounter”, International Journal of Modern Physics, **18**, 1809 (2009)

Thomas Roman, “Inflating Lorentzian wormholes”, Phys. Rev. D **47**, 1370 (1993)

Thomas A. Roman, “Some thoughts on energy conditions and wormholes”, gr-qc/0409090 (2004)

Barbara Ryden, *Introduction to Cosmology*, Addison Wesley, San Francisco, 2003

A. K. Sanyal and B. Modak, "Quantum cosmology with R + R^2 gravity", Class. Quantum Grav. **19**, 515 (2002)

Hisa-aki Shinkai and Sean A. Hayward, "Fate of the first traversible wormhole: Black-hole collapse or inflationary expansion", Phys. Rev. D **66**, 044005 (2002)

Lee Smolin, "How far are we from the theory of quantum gravity", hep-th/0303185 (2003)

S.V. Sushkov, "Wormholes supported by phantom energy", Phys. Rev. D **71**, 043520 (2005)

Leonard Susskind, "The black hole information paradox", Sci. Am. **276**, 52 (1997)

Edward Teo, "Rotating traversable wormholes", Phys. Rev. D 58, 024014 (1998)

Kip Thorne, *Black Holes & Time Warps: Einstein's Outrageous Legacy*, Norton, New York, 1994

Frank J. Tipler, "Singularities and Causality Violation", Ann. Phys. **108**, 1 (1977)

F. J. Tipler, "Rotating cylinders and the possibility of global causality violation", Phys. Rev. D **9**, 2203 (1974)

Douglas Urban and Ken D. Olum, "Averaged null energy condition violation in a formally flat spacetime", Phys. Rev. D **81**, 024039 (2010)

Douglas Urban and Ken D. Olum, "Spacetime averaged null energy condition", Phys. Rev. D **81**, 124004 (2010)

W. J. van Stockum, "Gravitational field of a distribution of particles rotating about an axis of symmetry", Proc. Royal Society Edinburgh **57**, 135 (1937)

Matt Visser, *Lorentzian Wormholes -- From Einstein to Hawking*, AIP Press, Woodbury, New York,1995

Matt Visser, "Traversable wormholes: The Roman ring", Phys. Rev. D. **55**, 5212 (1997)

Matt Visser, Sayan Kar, and Naresh Dadhich, "Traversable wormholes with arbitrarily small energy condition violations", Phys. Rev. Lett. **90**, 201102 (2003)

Dan N. Vollick, "Maintaining a wormhole with a scalar field", Phys. Rev. D **56**, 4724 (1997)

Dan N. Vollick, "Wormholes in string theory", Class. Quantum Grav. **16**, 1599 (1999)

Hermann Weyl, *Was ist Materie?*, p. 57ff, Springer, Berlin (1924)

John Archibald Wheeler with Kenneth Ford, *Geon, Black Holes and Quantum Foam – A Life in Physics,* Norton, New York, 1998

O. B. Zaslavskii, "Exactly solvable model of a wormhole supported by phantom energy", Phys. Rev. D **72**, 061303 (2005)

A1. The Einstein Equations

In case you are curious about what the most beautiful equations in the whole of physics look like, here they are. The Greek indices range in value from 0 to 3 and denote particular spacetime dimensions. Repeated indices denote a sum over the four index values.

$$R_{\mu\nu} - \frac{1}{2} R g_{\mu\nu} = \frac{8\pi G}{c^2} T_{\mu\nu} \qquad \text{(A1)}$$

where,

$$R_{\mu\nu} \equiv$$

$$R^{\alpha}{}_{\mu\beta\nu} = \partial_{\alpha} \Gamma^{\alpha}{}_{\mu\nu} - \partial_{\nu} \Gamma^{\alpha}{}_{\mu\alpha} + \Gamma^{\alpha}{}_{\sigma\alpha} \Gamma^{\sigma}{}_{\mu\nu} - \Gamma^{\alpha}{}_{\sigma\nu} \Gamma^{\sigma}{}_{\mu\alpha}$$

$$R \equiv R_{\mu\nu} g^{\mu\nu}$$

$g_{\mu\nu} \equiv$ the metric tensor

$c \equiv$ the speed of light

$G \equiv$ the gravitational constant

$T_{\mu\nu} \equiv$ the stress energy tensor

and

$$\Gamma^{\alpha}{}_{\mu\nu} \equiv \frac{1}{2} g^{\alpha\beta} \left(\partial_{\nu} g_{\beta\mu} + \partial_{\mu} g_{\beta\nu} - \partial_{\beta} g_{\mu\nu} \right).$$

Translation: The shape of spacetime ($g_{\mu\nu}$) determines the acceleration of falling particles ($\Gamma^{\alpha}{}_{\mu\nu}$), which determines spacetime's curvature ($R^{\alpha}{}_{\mu\beta\nu}$), which determines aggregate measures of this curvature ($R_{\mu\nu}$ and R), which, taken in a particular combination with spacetime's shape($g_{\mu\nu}$), is equal to the stress-energy ($T_{\mu\nu}$) of matter.

What I called "essential curvature" and "stress-energy" in Chapter 3 are respectively the left and right sides of equation (A1).

A2. The Quantum Inequalities

Suppose that an observer passes through a particular region containing a quantum field. He measures the energy density there using a certain device. The device samples density by opening an intake valve. The intake valve remains open for a time T, as measured by the observer. The device measures an energy density ρ. Then

$$<\rho> = \text{average value of } \rho \text{ measured during sampling of duration } T$$

$$\geq -\frac{Kh}{c^3 T^4}$$

where

K is a small ($<<1$) constant that depends on the type of quantum field,

h is Planck's constant, and

c is the speed of light.

Translation: For quantum fields, the degree to which its energy density can be negative is inversely proportion to the fourth power of the time over which the density is experienced.

$<\rho>$ is sometimes expressed in terms of a sampling function $f(t)$. One way to think of $f(t)$ is in terms of a measurement device whose closed intake valve does not suddenly open fully and suddenly slam shut. Rather, it opens gradually and closes gradually in a manner characterized by $f(t)$.

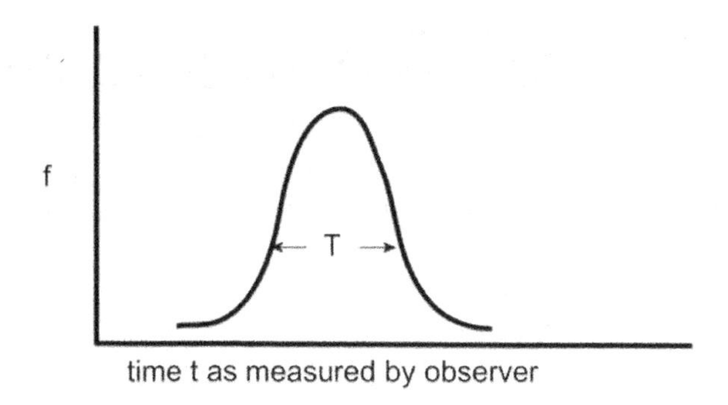

Figure A2.1. Sampling function *f*. The area under the curve of width T is 1.

In terms of the expectation value $<\Psi|\hat{\rho}|\Psi>$ of the density operator $\hat{\rho}$ of a quantum field in state $|\Psi>$, *f(t)* defines the average density as follows

$$<\rho>=\int_{-\infty}^{\infty} f(t) <\Psi|\hat{\rho}|\Psi> dt\,.$$

There is a different *K* and hence a different inequality for each type of quantum field.

A3. Classical Wormhole Metrics

Here are the metrics of the classical wormholes discussed in the text. All are expressed as line elements in spherical coordinates. In each the speed of light c is set to 1.

Schwarzschild

$$ds^2 = -\left(1-\frac{2M}{r}\right)dt^2 + \left(1-\frac{2M}{r}\right)^{-1} dr^2 + r^2\left(d\theta^2 + \sin^2\theta d\phi^2\right)$$

M – mass

Morris-Thorne

$$ds^2 = -e^{2\Phi(r)}dt^2 + \left(1-\frac{b(r)}{r}\right)^{-1} dr^2 + r^2\left(d\theta^2 + \sin^2\theta d\phi^2\right)$$

$\Phi(r)$ – redshift function
$b(r)$ – shape function.

Reissner-Nordstrøm

$$ds^2 = -\left(1-\frac{2M}{r}+\frac{Q^2}{r^2}\right)dt^2 + \left(1-\frac{2M}{r}+\frac{Q^2}{r^2}\right)^{-1} dr^2 + r^2\left(d\theta^2 + \sin^2\theta d\phi^2\right)$$

Q – charge

Kim-Lee

$$ds^2 = -\left(1+\frac{Q^2}{r^2}\right)dt^2 + \left(1-\frac{b_0^2}{r^2}+\frac{Q^2}{r^2}\right)^{-1} dr^2 + r^2(d\theta^2 + \sin^2\theta d\phi^2)$$

b_0 – minimum throat radius.

Kerr

$$ds^2 = -\rho^2\left(\frac{dr^2}{\Delta} + d\theta^2\right) + \left(r^2 + a^2\right)\sin^2\theta d\phi^2 - dt^2 + \frac{2Mr}{\rho^2}\left(a\sin^2\theta d\phi - dt\right)^2,$$

$$\rho^2 \equiv r^2 + a^2\cos^2\theta$$

$$\Delta \equiv r^2 - 2Mr + a^2$$

a – angular momentum per mass

Teo

$$ds^2 = -N^2(r,\theta)dt^2 + \left(1 - \frac{b(r,\theta)}{r}\right)^{-1} dr^2 + r^2K^2(r,\theta)\left[d\theta^2 + \sin^2\theta\left(d\phi - \omega dt\right)^2\right]$$

N – exponentiation of the redshift function
b – shape function with angular dependence
K – determines degree to which wormhole is oblate
ω – function of r and θ that determines wormhole's angular velocity

F

G

H

I

J

K

L

M

N

O

P

Q

R

S

Z

www.eridanuspress.com/PhysicsOfStargatesErrata.html

Made in the USA
Columbia, SC
31 December 2024